Annual Reports in Organic Synthesis—1976

Annual Reports in Organic Synthesis

ANNUAL REPORTS IN ORGANIC SYNTHESIS—1970
John McMurry and R. Bryan Miller, Eds.

ANNUAL REPORTS IN ORGANIC SYNTHESIS—1971
John McMurry and R. Bryan Miller, Eds.

ANNUAL REPORTS IN ORGANIC SYNTHESIS—1972
John McMurry and R. Bryan Miller, Eds.

ANNUAL REPORTS IN ORGANIC SYNTHESIS—1973
R. Bryan Miller and Louis S. Hegedus, Eds.
John McMurry, Series Editor

ANNUAL REPORTS IN ORGANIC SYNTHESIS—1974
Louis S. Hegedus and Stephen R. Wilson, Eds.
R. Bryan Miller, Series Editor

ANNUAL REPORTS IN ORGANIC SYNTHESIS—1975
R. Bryan Miller and L. G. Wade, Jr., Eds.

ANNUAL REPORTS IN ORGANIC SYNTHESIS—1976
R. Bryan Miller and L. G. Wade, Jr., Eds.

Annual Reports in Organic Synthesis—1976

edited by

R. Bryan Miller
University of California, Davis, California

L. G. Wade, Jr.
Colorado State University, Fort Collins, Colorado

ACADEMIC PRESS   New York   San Francisco   London   1977
A Subsidiary of Harcourt Brace Jovanovich, Publishers

Academic Press Rapid Manuscript Reproduction

COPYRIGHT © 1977, BY ACADEMIC PRESS, INC.
ALL RIGHTS RESERVED.
NO PART OF THIS PUBLICATION MAY BE REPRODUCED OR
TRANSMITTED IN ANY FORM OR BY ANY MEANS, ELECTRONIC
OR MECHANICAL, INCLUDING PHOTOCOPY, RECORDING, OR ANY
INFORMATION STORAGE AND RETRIEVAL SYSTEM, WITHOUT
PERMISSION IN WRITING FROM THE PUBLISHER.

ACADEMIC PRESS, INC.
111 Fifth Avenue, New York, New York 10003

United Kingdom Edition published by
ACADEMIC PRESS, INC. (LONDON) LTD.
24/28 Oval Road, London NW1

LIBRARY OF CONGRESS CATALOG CARD NUMBER:

ISBN 0-12-040807-4

PRINTED IN THE UNITED STATES OF AMERICA

CONTENTS

PREFACE xi
JOURNALS ABSTRACTED xiii

I. **CARBON–CARBON BOND FORMING REACTIONS** 1
  A. Carbon–Carbon Single Bonds *(see also:* I.E,
    I.F, I.G) 1
    1. Alkylation of Aldehydes, Ketones, and Their
       Derivatives *(see also:* I.A.7) 1
    2. Alkylation of Nitriles, Acids, and Acid
       Derivatives 5
    3. Alkylation of β-Dicarbonyl and β-Cyanocarbonyl
       Systems 10
    4. Alkylation of N, S, or Se Stabilized Carbanions 13
    5. Alkylation of Organometallic Reagents
       (see also: I.F.2) 22
    6. Other Alkylation Procedures and Reviews 32
    7. Nucleophilic Addition to Electron Deficient
       Carbon *(see also:* I.G) 34
       a. 1,2-Additions 34
          1) Aldol-Type Condensations 34
             a) Intermolecular 34
             b) Intramolecular ....(1975, 55)
          2) Addition of N, S, or Se Stabilized
             Carbanions 46
          3) Grignard-Type Additions 53

CONTENTS

		b.	Conjugate Additions	65
			1) Enolate-Type Additions	65
			2) Organometallic Reagents	69
			3) Other Conjugate Additions	73
	8.	Other Carbon–Carbon Single Bond Forming Reactions		76

B. Carbon–Carbon Double Bonds (*see also:* III.G) 86
    1. Wittig-Type Olefination Reactions 86
    2. Eliminations 94
        a. Alcohols and Derivatives 94
        b. Halides 98
        c. Other Eliminations 100
    3. Other Carbon–Carbon Double Bond Forming Reactions 105
    4. Allene Forming Reactions 110

C. Carbon–Carbon Triple Bonds 114

D. Cyclopropanations 118
    1. Carbene or Carbenoic Addition to a Multiple Bond 118
    2. Other Cyclopropanations 125

E. Thermal Reactions (*see also:* VI.B) 130
    1. Cycloadditions 130
    2. Other Thermal Reactions...................... 136

F. Aromatic Substitutions Forming a New Carbon–Carbon Bond 142
    1. Friedel–Crafts-Type Reactions 142
    2. Coupling Reactions (*see also:* I.G) 146
    3. Other Aromatic Substitutions 148

G. Synthesis *via* Organometallics 151
    1. Organoboranes 151
    2. Carbonylation Reactions 159
    3. Other Syntheses *via* Organometallics 162
    4. Reviews 185

CONTENTS

II. OXIDATIONS 188

    A. C–O Oxidations 188
        1. Alcohol → Ketone, Aldehyde 188
        2. Alcohol, Aldehyde → Acid, Acid Derivative 194

    B. C–H Oxidations 195
        1. C–H → C–O 195
        2. C–H → C–Hal 199
        3. Other C–H Oxidations 202

    C. C–N Oxidations 202

    D. Amine Oxidations 203

    E. Sulfur Oxidations 203

    F. Oxidative Additions to C–C Multiple Bonds 205
        1. Epoxidations 205
        2. Hydroxylation 206
        3. Other 207

    G. Phenol → Quinone Oxidation 209

    H. Oxidative Cleavages 210

    I. Photosensitized Oxygenations 211

    J. Dehydrogenation 212

    K. Other Oxidations and Reviews 212

CONTENTS

III. REDUCTIONS 213

    A. C=O Reductions 213

    B. Nitrile Reductions 222

    C. Reduction of Sulfur Compounds 222

    D. N–O Reductions 223

    E. C–C Multiple Bond Reductions 224
        1. C=C Reductions 224
        2. C≡C Reductions 228
        3. Reduction of Aromatic Rings .... (1975, 285)

    F. Hydrogenolysis of Hetero Bonds 230
        1. C–O → C–H 230
        2. C–Hal → C–H 232
        3. C–S → C–H .... (1975, 293)
        4. C–N → C–H *(see also:* III.H*)* 233

    G. Reductive Eliminations 234

    H. Reductive Cleavages 237

    I. Hydroboration (reduction only) 238

    J. Other Reductions and Reviews 239

IV. SYNTHESIS OF HETEROCYCLES 241

    A. Aziridines 241

    B. Furans, *etc.* 243

    C. Indoles 245

CONTENTS

D.	Lactams	249
E.	Lactones	256
F.	Pyridines, Quinolines, *etc.*	266
G.	Pyrroles, *etc.*	271
H.	Other Heterocycles with One Heteroatom	273
I.	Heterocycles with Two or More Heteroatoms	280
	1. Heterocycles with 2 N's	280
	a. 5-Membered	280
	b. 6-Membered (*see also:* VI.A.15)	284
	c. Other	291
	2. Heterocycles with 1 N and 1 O	291
	3. Heterocycles with 1 N and 1 S	295
	4. Heterocycles with 1 S and 1 O	300
	5. Heterocycles with 3 N's	300
	6. Other Heterocycles	302
J.	General Reviews	304

V.	**PROTECTING GROUPS**	305
A.	Hydroxyl	305
B.	Amine	309
C.	Sulfhydryl	314
D.	Carboxyl	314
E.	Ketone, Aldehyde	317
F.	Phosphate	320

CONTENTS

- **G.** Pi-Bond 321
- **H.** Miscellaneous Protecting Groups 322

VI. USEFUL SYNTHETIC PREPARATIONS 323

- **A.** Functional Group Preparations 323
  1. Acids, Acid Halides, *etc.* (*see also:* II.A.2) 323
  2. Alcohols, Phenols (*see also:* III.A) 324
  3. Alkyl Halides (*see also:* II.B.2) 327
  4. Amides (*see also:* IV.D, VI.A.17) 335
  5. Amines (*see also:* III.B, III.D) 338
  6. Amino Acids and Derivatives 342
  7. Carbenes (*see also:* I.D) 345
  8. Enamines 347
  9. Epoxides (*see also:* II.F.1) 348
  10. Esters (*see also:* IV.E) 349
  11. Ethers 353
  12. Ketones and Aldehydes (*see also:* II.A.1, III.F.1, III.F.4) 354
  13. Nitriles (*see also:* II.D, II.J) 366
  14. Nitro (*see also:* II.D) 371
  15. Nucleotides, *etc.* (*see also:* IV.I.1b) 372
  16. Olefins (*see also:* I.B, II.J, III.G) 378
  17. Peptides (*see also:* V.B, V.D) 380
  18. Vinyl Halides, Vinyl Ethers, Vinyl Esters 384
  19. Sulfur Compounds (*see also:* II.E, III.C) 387

- **B.** Ring Enlargement and Contraction 395
  1. Enlargement 395
  2. Contraction 398

- **C.** Multi-Step Transformations 399
  1. Masked-Carbonyl Systems 399
  2. Other 404

VII. OTHER COMPLETELY MISCELLANEOUS REACTIONS 407

VIII. MISCELLANEOUS REVIEWS 410

AUTHOR INDEX 421

PREFACE

One of the most difficult problems facing chemists today is that of "keeping up with the literature." For several reasons, the problem is particularly severe for the synthetic organic chemist. Bits of information of potential use to the synthetic chemist are scattered throughout common chemistry journals and can be found in any paper, not just those dealing strictly with synthesis. Thus a synthetic chemist must read a large number of journals. He must organize and index what he reads to make the information available for future reference. All snythetic chemists do this; but the task is becoming more difficult each year as the flow of information increases.

The problem however is shared to some extent by all. Most organic chemists are at some time faced with the problem of synthesizing a desired material and, for many, the problems are formidable. Nonspecialists faced with a synthetic problem are most likely not to have kept pace with the developments in synthetic chemistry that may well solve their problems and will not have the necessary information in their files.

Thus, we felt that an organized annual review of snythetically useful information would prove beneficial to nearly all organic chemists, both specialist and nonspecialist in synthesis. It should help relieve some of the information-storage burden of the specialist and should aid the nonspecialist who is seeking help with a specific problem to become rapidly aware of recent synthetic advances. Ideally also, such a review should be minimally priced to be within the means of potential users including graduate students, and it should appear as promptly as possible after the close of the abstracting period.

In producing *Annual Reports in Organic Synthesis—1976* we have abstracted 47 primary chemistry journals, selecting useful synthetic advances. We have tried to present the information in an organized manner, emphasizing rapid visual retrieval. Only the common journals received by our libraries have been extracted. Any journal received after March 1, 1977 will be covered in the next volume. We have

PREFACE

also exercised selectivity in choosing which papers to abstract. Our general guidelines have been to include all reactions and methods that are new, synthetically useful, and reasonably general. Each entry comprises primarily of structures, accompanied by very few comments. The purpose of this is to aid the reader in scanning the book. The mind is capable of absorbing a whole picture in an instant, but is considerably slowed by having to read sentences. If the pictures presented catch the reader's interest, he should then seek details from the original paper.

For the first time we have included an author index to aid the user, but we still have no subject index. To include one would have greatly increased both the cost of the book and the delay time before publication. Instead, we have chosen to use an extensive table of contents. Chapters I–III are organized by reaction type and constitute the major part of the book. The organization of these sections is self-explanatory, and there should be no difficulty in locating a new method of oxidation or a new cyclopropanation procedure. Chapter IV deals with methods of synthesizing heterocyclic systems. Chapter V covers the use of new protecting groups and is also self-explanatory. Chapter VI is divided into three main parts and covers those synthetically useful transformations that do not fit easily into the first three chapters. The first part deals only with functional group synthesis. The second and third parts of Chapter VI are self-explanatory. The third part involves useful multistep sequences, the individual steps of which may be well known. Future volumes of this series will maintain the present table of contents as much as possible. If no entry is found for a particular section, the last volume in which one appears will be cited in the table of contents—see I.A.7.a.1b.

Any undertaking of this type involves a series of compromises. We have chosen to emphasize reasonable cost, rapid publication, and rapid visual retrieval of information at the admitted expense of detail and beauty. This volume is the seventh in an annual series. We welcome suggestions for improvement of future volumes and would like to thank Professor Samuel Danishefsky for his useful comments concerning the author index.

The arduous task of drawing the multitude of structures appearing in this review was carried out by Ms. Linda Benedict and Ms. Sandi Hanson. We thank them very much for their efforts.

R. Bryan Miller
L. G. Wade, Jr.

JOURNALS ABSTRACTED

Accounts of Chemical Research
Acta Chemica Scandinavica
Angewandte Chemie International Edition in English
Annales de Chimie
Australian Journal of Chemistry
Bulletin of the Chemical Society of Japan
Bulletin de Societes Chimiques Belges
Bulletin de la Societe Chimique de France
Canadian Journal of Chemistry
Chemical Communications
Chemical and Pharmaceutical Bulletin
Chemical Reviews
Chemical Society Reviews
Chemische Berichte
Chemistry and Industry
Chemistry Letters
Collection of Czechoslovakian Chemical Communications
Comptes Rendus Hebdomadaires de Seances de l'Academie des Sciences (C)
Doklady Chemistry
Experientia
Fortschritte der Chemischen Forschung
Gazzetta Chimica Italiana
Helvetica Chimica Acta
Indian Journal of Chemistry
Israel Journal of Chemistry
Journal of the American Chemical Society
Journal of the Chemical Society (Perkin I)
Journal of the Chemical Society (Perkin II)
Journal of General Chemistry (USSR)
Journal of Heterocyclic Chemistry
Journal of the Indian Chemical Society

JOURNALS ABSTRACTED

Journal of Organic Chemistry
Journal of Organic Chemistry (USSR)
Journal of Organometallic Chemistry
Journal fur Praktische Chemie
Liebig's Annalen der Chemie
Monatschefte fur Chemie
Pure and Applied Chemistry
Recueil des Travaux Chimiques des Pays-bas
Russian Chemical Reviews
Steroids
Synthesis
Synthetic Communications
Tetrahedron
Tetrahedron Letters
Zeitschrift fur Chemie
Zeitschrift fuer Naturforschung, Teil B

I.A.1-1 J. d'Angelo, Tetrahedron, 32, 2979 (1976).

Review: Ketone Enolates: Regiospecific Preparation and Synthetic Uses.

I.A.1-2 T. Tanaka and T. Mukaiyama, Chem. Lett., 1976, 1259.

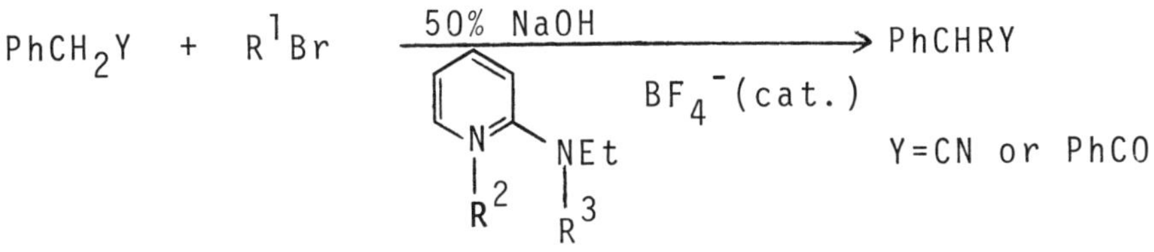

$$PhCH_2Y + R^1Br \xrightarrow[BF_4^- (cat.)]{50\% \text{ NaOH}} PhCHRY$$

(pyridinium catalyst with R^2, NEtR^3)

Y = CN or PhCO

A study of the two-phase catalyst, 2-dialkylaminopyridinium salts, in alkylation reactions was carried out.

I.A.1-3 W. P. Reeves and R. G. Hilbrich, Tetrahedron, 32, 2235 (1976).

$$PhCH_2Y \xrightarrow[RX, 50\% \text{ KOH}]{R_1R_2R_3N,} PhCHRY$$

RX = EtBr, Y = COMe
RX = BuBr, Y = CN

A study of the phase transfer catalysis of various 1°, 2°, and 3° amines was carried out.

I.A.1-4 J. F. Bunnett, X. Creary, and J. E. Sundberg, *J. Org. Chem.*, **41**, 1707 (1976).

$$\text{PhCH=CHBr} \xrightarrow[h\nu]{\underset{\text{CH}_2=\text{CMe}}{\overset{\text{OK}}{|}}} \text{PhCH=CHCH}_2\text{COMe}$$

48%

+

PhCH$_2$CH=CHCOMe

34%

[cyclopentenyl-I] $\xrightarrow[h\nu]{\underset{\text{CH}_2=\text{CMe}}{\overset{\text{OK}}{|}}}$ [cyclopentenyl-CH$_2$COMe]

I.A.1-5 J. M. Fortunato and B. Ganem, *J. Org. Chem.*, **41**, 2194 (1976).

$$\underset{\substack{\text{1) Selectride} \\ \text{THF, }-70° \\ \text{2) R}^2\text{X}}}{\text{R}^1\text{-cyclohexenone}} \longrightarrow \text{R}^1,\text{R}^2\text{-cyclohexanone}$$

Selectride	R^1	R^2X	% Yield
K	H	CH$_2$=CHCH$_2$Br	55
L	H	CH$_2$=CHCH$_2$Br	85
L	Me	MeO$_2$CCH=CHCH$_2$I	60

α,β-Unsaturated esters can also be reductively alkylated.

I.A.1-6 P. A. Chaloner and A. B. Holmes, J. Chem. Soc., Perkin Trans. I, 1976, 1838.

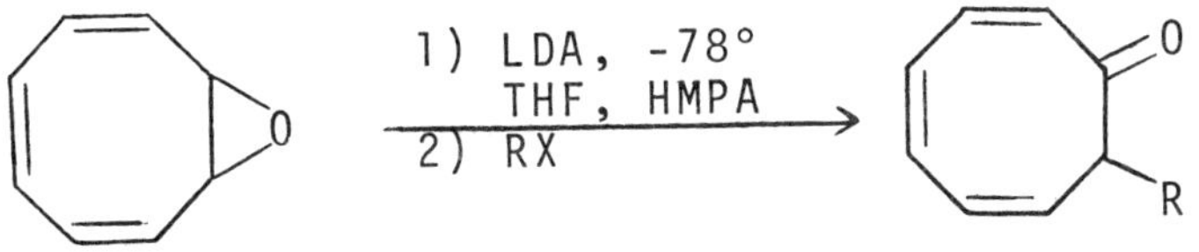
O
1) LDA, -78°
THF, HMPA
2) RX
O
R

RX	% Yield
MeI	78
CH_2=$CHCH_2Br$	74
EtI	81
$PhCH_2I$	37 (51% of R=H)

I.A.1-7 M. Hajek and J. Malek, Coll. Czech. Chem. Commun., 41, 746 (1976).

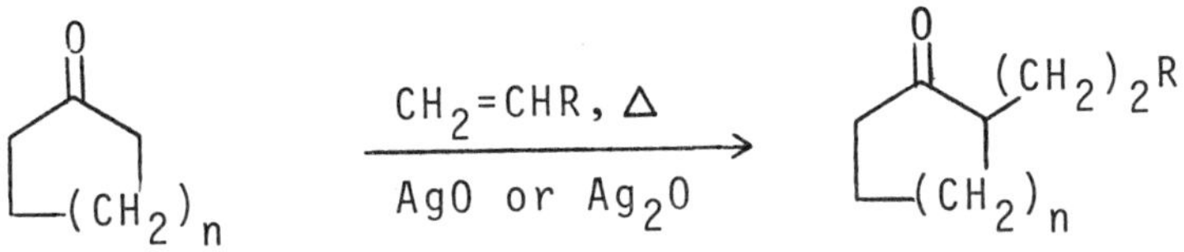
O
(CH2)n
CH2=CHR, Δ
AgO or Ag2O
O
(CH2)2R
(CH2)n

The procedure uses a ten-fold excess of ketone.

I.A.1-8 H.-J. Bestmann and R. W. Saalfrank, Chem. Ber., 109, 403 (1976).

$$R^1CH_2COR^2 + Ph_3P=C=C(OEt)_2 \longrightarrow Ph_3P=CHC(OEt)=CR^1COR^2$$

R^1	R^2	% Yield
H	Ph	65
H	Me	55
H	OEt	74
$-(CH_2)_4-$		69

I.A.1-9 Y. Leroux and C. Jacquelin, Synth. Commun., 6, 597 (1976).

$$PhC \equiv CCH(OEt)_2 \quad \xrightarrow[\substack{2) \ RX \\ 3) \ H_2O}]{\substack{1) \ 2 \text{ eq. n-BuLi-TMEDA} \\ THF, \ -78°}} \quad PhCR=CHCO_2Et$$

RX	% Yield
Me_2SO_4	79
EtI	85
i-PrBr	86
$EtOCH_2Cl$	57

CARBON—CARBON BONDS FORMING REACTIONS

I.A.2-1 Y. Masuyama, Y. Ueno, and M. Okawara, Tetrahedron Lett., 1976, 2967.

$$NCCHR^1SC(S)NMe_2 \xrightarrow[\text{aq. NaOH}]{R^2X} NCCR^1R^2SC(S)NMe_2$$

R^1	R^2X	% Yield
H	MeI	82
H	$CH_2=CHCH_2Cl$	86
Me	$n-C_5H_{11}Br$	94
Et	$n-C_6H_{13}Br$	88

The products are readily converted to the corresponding ketones or aldehydes.

I.A.2-2 A. J. Birch and J. Slobbe, Aust. J. Chem., 29, 2737 (1976).

$$MeCH=CHCH=CHCO_2H \xrightarrow[\substack{2)\ RX \\ 3)\ \text{esterification}}]{1)\ Li,\ NH_3(\ell)} \underset{H}{\overset{H}{EtC=CCHRCO_2Me}}$$

RX	% Yield
MeI	∼ 60
i-PrBr	∼ 70
$CH_2=CHCH_2Br$	∼ 60

Protonated material (R=H) was a significant by-product (25-40%).

I.A.2-3 P. Hullot, T. Cuvigny, M. Larcheveque, and H. Normant, Can. J. Chem., 54, 1098 (1976).

$$R^1R^2CHCONMe_2 \xrightarrow[\text{2) } R_3X, Et_2O \text{ or THF}]{\text{1) } Et_2NLi, PhH, HMPA} R^1R^2R^3CCONMe_2$$

R^1	R^2	R^3X	% Yield
H	H	$C_6H_{13}Br$	82
H	i-Pr	MeBr	70
Me	Me	n-BuBr	86
Me	Me	$CH_2=CHCH_2Br$	80
H	i-Pr	$CH_2=C(Br)CH_2Br$	69

I.A.2-4 R. A. Gorski, G. J. Wolber, and J. Wemple, Tetrahedron Lett., 1976, 2577.

$$R^1CH_2\overset{O}{\underset{\|}{C}}SR^2 \xrightarrow[\text{2) } R^3X, -78°]{\text{1) LDA, THF, } -78°} R^1R^3CH\overset{O}{\underset{\|}{C}}SR^2$$

R^1	R^2	R^3X	% Yield
H	$PhCH_2$	MeI	75
H	t-Bu	$CH_2=CHCH_2Br$	71
$-(CH_2)_2-$		MeI	74
$-(CH_2)_2-$		n-BuI	35

I.A.2-5 B. M. Trost and L. S. Melvin, Jr. *J. Am. Chem. Soc.*, **98**, 1204 (1976).

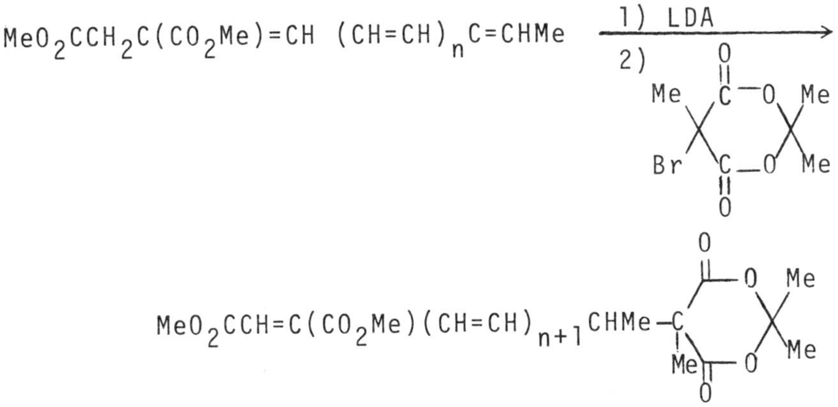

$$MeO_2CCH_2C(CO_2Me)=CH\ (CH=CH)_nC=CHMe \xrightarrow[2)]{1)\ LDA}$$

[reagent 2: cyclic structure with Me, Br, C=O, O-Me groups]

$$MeO_2CCH=C(CO_2Me)(CH=CH)_{n+1}CHMe-[\text{cyclic product}]$$

n = 0, 66%
n = 1, 38%

I.A.2-6 K. Saigo, M. Osaki, and T. Mukaiyama, *Chem. Lett.*, **1976**, 769.

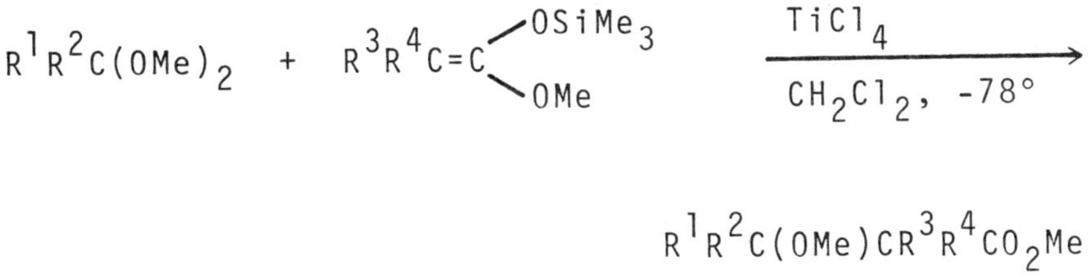

$$R^1R^2C(OMe)_2\ +\ R^3R^4C=C\begin{matrix}OSiMe_3\\OMe\end{matrix} \xrightarrow[CH_2Cl_2,\ -78°]{TiCl_4}$$

$$R^1R^2C(OMe)CR^3R^4CO_2Me$$

R^1	R^2	R^3	R^4	% Yield
$PhCH_2$	H	Ph	H	100
$PhCH_2$	H	$-(CH_2)_5-$		92
Me	Me	$PhCH_2$	Me	94
H	MeO	$PhCH_2$	H	77

I.A.2-7 E. Nakamura, M. Shimizu, and I. Kuwajima, Tetrahedron Lett., 1976, 1699.

$$Me_3SiCH_2CO_2Et \xrightarrow[Bu_4NF]{R^1R^2CO} R^1R^2C(OSiMe_3)CH_2CO_2Et$$

R^1	R^2	% Yield
PhCH=CH	H	81
n-PrCH=CH	H	82
Ph	Ph	88
Ph	Me	0

If R^1R^2CO has acidic α-hydrogens the reaction fails.

I.A.2-8 L. M. Lerner, J. Org. Chem., 41, 2228 (1976).

$$(EtS)_2CHCO_2Me \xrightarrow[2)\ RX]{1)\ NaH,\ DMF} (EtS)_2CRCO_2Me$$

RX = MeI, 100%
RX = i-PrI, 91%

I.A.2-9 W. Beck and M. Girnth, Chem. Ber., 109, 965 (1976).

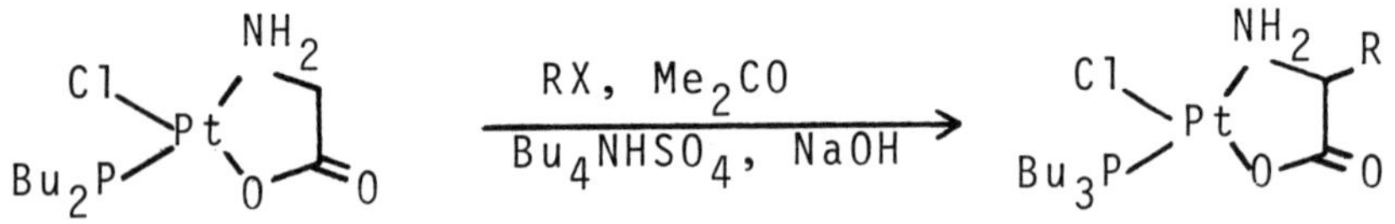

RX = MeOSO$_3$Me or PhCH$_2$Cl

70-80%

CARBON-CARBON BONDS FORMING REACTIONS

I.A.2-10 A. P. Krapcho and E. A. Dundulis, Tetrahedron Lett., 1976, 2205.

$$PhCONHCH_2CO_2H \xrightarrow[\text{2) } RCH_2X]{\text{1) 3 eq. LDA, TMEDA, THF}} PhCONHCH(CH_2R)CO_2H$$
$$23\%\text{-}60\%$$

$$PhCONHCH_2CO_2Et \xrightarrow[\text{2) } RCH_2X]{\text{1) 2 eq. LDA, TMEDA, THF}} PhCONH\overset{\overset{\displaystyle CH_2R}{|}}{C}HCO_2Et$$
$$48\%\text{-}71\%$$

The products can be hydrolyzed to α-amino acids.

I.A.2-11 G. Stork, A.Y.W. Leong, and A. M. Touzin, J. Org. Chem., 41, 3491 (1976).

$$PhCH=NCH_2CO_2Et \xrightarrow[\text{2) } RX]{\text{1) LDA or t-BuOK, THF}} PhCH=NCHRCO_2Et$$

$$R = n\text{-}C_8H_{17}, \ 90\%$$
$$R = i\text{-}Pr, \ 75\%$$
$$R = CH_2CO_2Et, \ 82\%$$

A one-flask sequential dialkylation procedure is also described.

I.A.3-1 H. Oediger and F. Möller, <u>Justus Liebigs Ann. Chem.</u>, <u>1976</u>, 348.

$$R^1CH_2R^2 + 2\ R^3Br \xrightarrow{DBU}{DMF} R^1R^2CR^3_2$$

R^1	R^2	R^3	% Yield
CO_2Me	CN	$PhCH_2$	77
$COMe$	$COMe$	$PhCH_2$	84
CN	CN	$PhCH_2$	91
CN	SO_2Ph	n-Bu	75

$$R^1CH_2R^2 + Br(CH_2)_nBr \xrightarrow{DBU}{DMF} (CH_2)_n\!\!-\!\!CR^1R^2$$

n = 4 or 5   ~60%

I.A.3-2 T. Kurihara, Y. Nakajima, and O. Mitsunobu, <u>Tetrahedron Lett.</u>, <u>1976</u>, 2455.

$$NCCH_2CO_2Et + HO(CH_2)_nOH \xrightarrow{EtO_2CN=NCO_2Et}{Ph_3P,\ THF}$$

$$(CH_2)_n\!\!-\!\!C\!\!\begin{array}{l}CN\\CO_2Et\end{array}$$

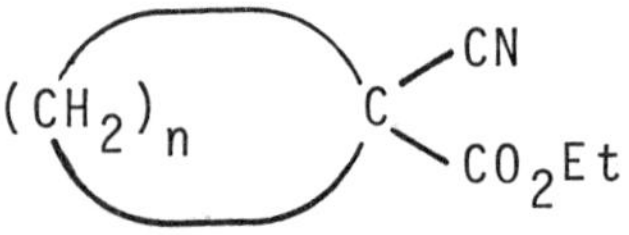

n = 4, 46%
n = 5, 40%
n = 6, 21%

CARBON–CARBON BONDS FORMING REACTIONS

I.A.3-3 J. Nakayama, J. Chem. Soc., Perkin Trans.
I, 1976, 540.

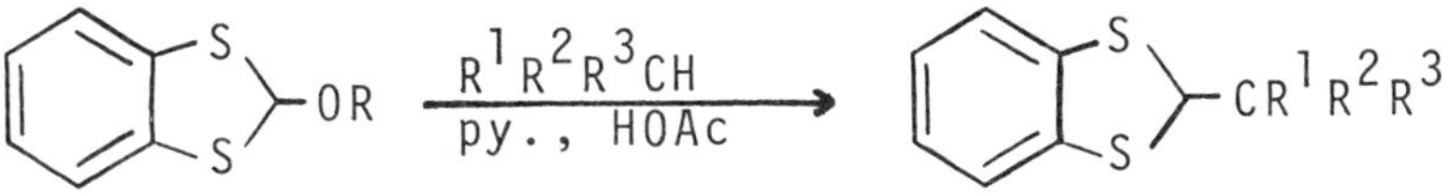

benzodithiole-OR + $R^1R^2R^3CH$ $\xrightarrow{\text{py., HOAc}}$ benzodithiole-$CR^1R^2R^3$

R^1	R^2	R^3	% Yield
H	MeCO	MeCO	90
Me	MeCO	MeCO	73
CHO	$-CO(CH_2)_3-$		55

I.A.3-4 B. Eistert, A. Schmitt, and T. J. Arackal,
Chem. Ber., 109, 1549 (1976).

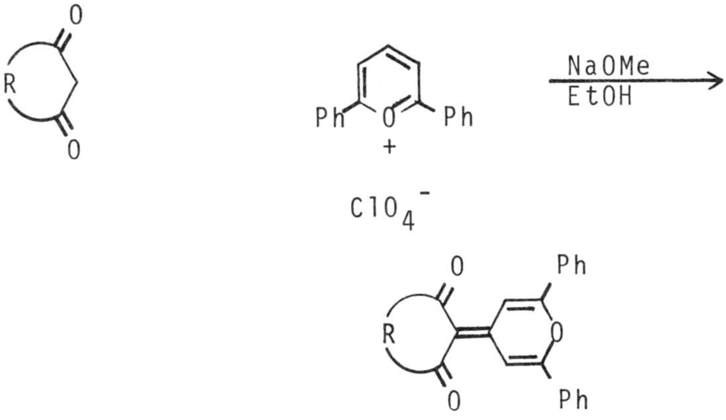

cyclic β-diketone + pyrylium (Ph, Ph) ClO_4^- $\xrightarrow{\text{NaOMe, EtOH}}$ product

69-97%

Ring size of β-diketone is 6-membered.

I.A.3-5 J. Schnekenburger, D. Heber, and E. Heber-Brunschweiger, Justus Liebigs Ann. Chem., 1976, 1799.

[Pyridinium salt with substituents R^1, R^2, R^3 on pyridine ring, N-OMe, ClO_4^- counterion]

$$\xrightarrow[\text{DMF or HMPA}]{CH_2R^4R^5,\ Et_3N}$$

$$R^4R^5CH=CHCR^3=CR^2CR^1=NCOMe$$

R^1	R^2	R^3	R^4	R^5	% Yield
CN	H	H	$COCO_2Et$	CO_2Et	48
H	CN	H	COMe	CN	31
H	H	CN	COMe	COMe	37
CN	H	H	$COCH_2CMe_2CH_2CO$		38

I.A.3-6 N. Ono, H. Eto, R. Tamura, J. Hyami, and A. Kaji, Chem. Lett., 1976, 757.

$$NaCR^1(CN)(CO_2Et)\ +\ R^2R^3C(Br)NO_2\ \xrightarrow[120°]{HMPA}$$

$$R^2R^3C=CR^1CN$$

R^1	R^2	R^3	% Yield
i-Pr	Me	Me	70
i-Pr	$+(CH_2)_5+$		63
$PhCH_2$	Me	Et	75

CARBON—CARBON BONDS FORMING REACTIONS

I.A.3-7 F. Cooke and P. Magnus, *J. Chem. Soc., Chem. Commun.*, **1976**, 519.

$$PhCH_2COCH_2SO_2Ph \xrightarrow[\substack{2) \text{ n-BuLi} \\ 3) \text{ Br}(CH_2)_3Br}]{1) \text{ NaH or LDA}} \underset{60-75\%}{Ph\overset{(CH_2)_3Br}{\underset{|}{C}}HCOCH_2SO_2Ph}$$

$\longrightarrow$ [cyclohexanone with Ph and SO_2Ph substituents]

or

[dihydropyran with Ph and SO_2Ph substituents]

A study of conditions for C- vs. O-alkylation was carried out.

I.A.4-1 H. H. Wasserman, M. J. Hearn, B. Haveaux, and M. Thyes, *J. Org. Chem.*, **41**, 153 (1976).

[bicyclic cyclopropane with X,X substituents, $(CH_2)_n$] $\xrightarrow{\text{MeI} \atop RCH_2NO_2}$ [bicyclic cyclopropane with X, CHRNO$_2$ substituents, $(CH_2)_n$]

n	X	R	% Yield
1	N-pyrrolidino-	H	78
1	N-pyrrolidino-	Me	45
0	N-piperidino-	H	50
0	N-piperidino-	Me	26

I.A.4-2 R. Henning, F. Lehr, and D. Seebach, *Helv. Chim. Acta*, $\underline{59}$, 2213 (1976).

$$ArCH_2CH_2NO_2 \xrightarrow[\text{2) RX}]{\text{1) 2 eq. n-BuLi, THF, HMPA, } -78°} ArCHRCH_2NO_2$$

$R = n\text{-}C_6H_{13}$, $Ar = p\text{-}ClC_6H_4$, 76%

$R = PhCH_2$, $Ar = Ph$, 75%

$$PhCH_2CH_2NO_2 \xrightarrow[\text{2) PhCH=CHNO}_2]{\text{1) 2 eq. n-BuLi, THF, HMPA, } -78°}$$

$$PhCH(CH_2NO_2)CH(CH_2NO_2)Ph$$
80%

I.A.4-3 M. E. Jung, P. A. Blair, and J. A. Lowe, *Tetrahedron Lett.*, $\underline{1976}$, 1439.

$$R^1\underset{\underset{CH_2R^2}{\|}}{C}\!\!=\!\!N\text{-}OH \xrightarrow[\text{2) } R^3I, -78°]{\text{1) 2 eq. BuLi, THF}} R^1\underset{\underset{CHR^2R^3}{\|}}{C}\!\!=\!\!N\text{-}OH$$

R^1	R^2	R^3	% Yield
Me	H	Me	73
Me	Me	Me	81
Me	Me	i-Pr	46
$\text{-}(CH_2)_4\text{-}$		Me	74

If the *syn* alkyl group is disubstituted the reaction does not proceed cleanly.

I.A.4-4 W. G. Kofron and M.-K. Yeh, J. Org. Chem., 41, 439 (1976).

$$R^1\overset{NOH}{\underset{\parallel}{C}}CH_2R^2 \xrightarrow[\substack{2)\ R^3X\\3)\ hydrolysis}]{1)\ 2\ eq.\ n\text{-BuLi, THF}} R^1COCHR^2R^3$$

R^1	R^2	R^3X	% Yield
Me	H	$PhCH_2Cl$	72
H	Me	$PhCH_2Cl$	37
$\text{-}(CH_2)_4\text{-}$		$CH_2=CHCH_2Br$	65
$\text{-}(CH_2)_3\text{-}$		$n\text{-BuI}$	59

I.A.4-5 R. R. Fraser and K. L. Dhawau, J. Chem. Soc., Chem. Commun., 1976, 674.

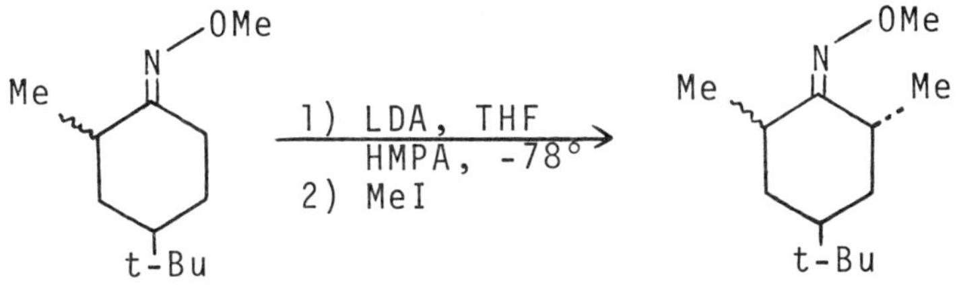

86-94%

Only *syn*-axial alkylation was observed independent of the relative stereochemistry of the starting material.

I.A.4-6 E. J. Corey and D. Enders, *Tetrahedron Lett.*, 1976, 3.

$$R^1CH_2\overset{NNMe_2}{\overset{\|}{C}}R^2 \xrightarrow[\text{2) }R^3X, -78°]{\text{1) LDA, THF, 0°}} R^1R^3CH\overset{NNMe_2}{\overset{\|}{C}}R^2$$

R^1	R^2	R^3X	% Yield
H	n-C_5H_{11}	MeI	95
H	n-C_5H_{11}	i-PrI	83
Ph	H	MeI	95
-(CH$_2$)$_3$CHMe-		MeI	95

A mild oxidative hydrolysis procedure of the products to carbonyl compounds is also described.

I.A.4-7 T. Cuvigny, J. F. LeBorgne, M. Larcheveque, and H. Normant, *Synthesis*, 1976, 237.

$$R^1CH_2CH=NNMe_2 \xrightarrow[\text{2) }R^2Br]{\text{1) Et}_2\text{NLi, PhH, HMPT}}$$

$$R^1R^2CHCH=NNMe_2$$

R^1	R^2	% Yield
Et	Et	75
Et	CH$_2$=CHCH$_2$	76
Bu	Bu	79

I.A.4-8 J. F. LeBorgne, T. Cuviguy, M. Larcheveque, and H. Normant, Synthesis, 1976, 238.

$$R^1R^2CHCH=NNMe_2 \xrightarrow{\begin{array}{l}1)\ Et_2NLi,\ PhH,\ HMPT\\ 2)\ R^3Br\end{array}} R^1R^2R^3CCN$$

R^1	R^2	R^3	% Yield
Me	Me	n-Bu	88
Et	Et	Et	93
Et	Et	$CH_2=CHCH_2$	87

I.A.4-9 G. R. Kieczykowski, R. H. Schlessinger, and R. B. Sulsky, Tetrahedron Lett., 1976, 597.

$$MeCH=CHCH=NC_6H_{11} \xrightarrow{\begin{array}{l}1)\ LDA,\ THF,\ HMPA,\ 0°\\ 2)\ RX,\ -78°\end{array}} CH_2=CHCHRCH=NC_6H_{11}$$

RX = MeI, 98%
RX = $CH_2=CHCH_2Br$, 100%

$$MeCH=CHCH=NC_6H_{11} \xrightarrow{\begin{array}{l}1)\ LDA,\ THF,\ HMPA,\ 0°\\ 2)\ BrCH_2CO_2Me\end{array}} \underset{100\%}{MeCH=CCH=NC_6H_{11} \atop |\ \ \ \ \ \ \ \ \ \ CH_2CO_2Me}$$

Alkylation of the initial products leads to quaternary α-carbon atoms.

I.A.4-10 S. Brenner and H. G. Viehe, Tetrahedron Lett., 1976, 1617.

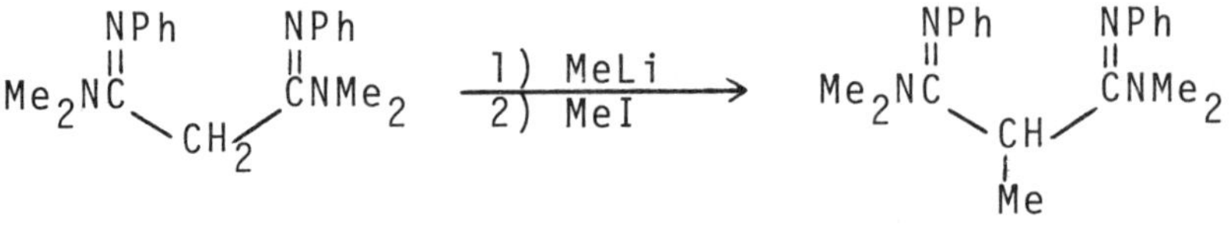

$$Me_2NC(=NPh)-CH_2-C(=NPh)NMe_2 \xrightarrow{\text{1) MeLi} \atop \text{2) MeI}} Me_2NC(=NPh)-CH(Me)-C(=NPh)NMe_2$$

80%

Use of $NaNH_2$ as the base gave exclusive N-alkylation.

I.A.4-11 H. Hoberg and U. Griebsch, Synthesis, 1976, 830.

$$PhCH=NAlEt_2 \xrightarrow{\text{1) 2 K, naphthalene (cat.), THF} \atop \text{2) LiBr, Et}_2\text{O} \atop \text{3) RX}} PhCHRNH_2$$

R = Me, 41%
R = $CH_2=CHCH_2$, 61%

I.A.4-12 R. R. Schmidt and J. Talbiersky, Angew. Chem. Int. Ed. Engl., 15, 171 (1976).

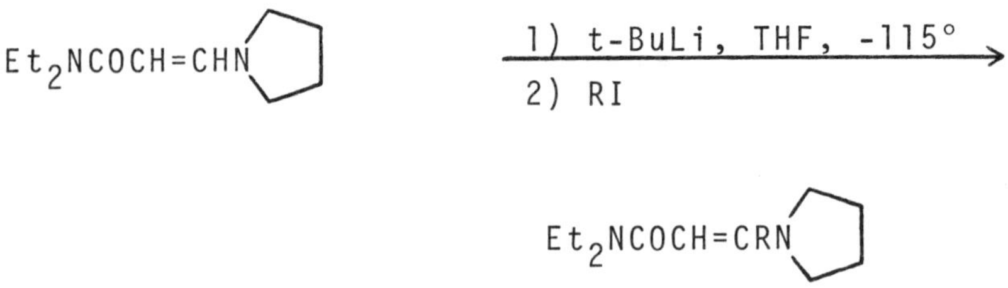

$$Et_2NCOCH=CHN\text{(pyrrolidine)} \xrightarrow{\text{1) t-BuLi, THF, } -115° \atop \text{2) RI}}$$

$$Et_2NCOCH=CRN\text{(pyrrolidine)}$$

R = Me, 95%
R = Et, 60%

CARBON-CARBON BONDS FORMING REACTIONS

I.A.4-13   D. Seebach and W. Lubosch, Angew. Chem. Int. Ed. Engl., 15, 313 (1976).

$$\text{t-BuC(=S)NMe}_2 \xrightarrow[\text{2) RX}]{\text{1) sec-BuLi, -78°, THF, TMEDA}} \text{t-BuC(=S)NMeCH}_2\text{R}$$

RX	% Yield
MeI	80
$n\text{-C}_5\text{H}_{11}\text{I}$	82
PhCH_2Br	44

I.A.4-14   C. F. Beam, C. W. Thomas, R. M. Sandifer, R. S. Foote, and C. H. Hauser, Chem. Ind. (London), 1976, 487.

$$\underset{\text{RCH CPh}}{\overset{\text{Li }\; \text{N}_2\text{Li}_2}{|\quad\;\;\|}} \xrightarrow[\text{2) HCl, H}_2\text{O}]{\text{1) 3 eq. PhCH}_2\text{Cl}} \underset{\text{RCHCOPh}}{\overset{\text{CH}_2\text{Ph}}{|}}$$

R = H, 23%

R = Ph, 69%

I.A.4-15   T. Cohen, D. A. Bennett, and A. J. Mura, Jr., J. Org. Chem., 41, 2506 (1976).

$$\text{PhSCHR}^2\text{CR}^1\text{=C(SPh)R}^3 \xrightarrow[\text{2) R}^4\text{X}]{\text{1) sec-BuLi, THF, -20°, (HMPA)}}$$

$$\text{PhSCR}^2\text{R}^4\text{CR}^1\text{=C(SPh)R}^3$$

90-98%

I.A.4-16 M. Julia and D. Uguen, Bull. Soc. Chim. Fr., 1976, 513.

$$PhSO_2CH_2R^1 \xrightarrow[2)]{1)\ n\text{-BuLi},\ -78°} \overset{O}{\triangle}\!\!\!\!\!\!<^{R^2}_{R^3}$$

$$PhSO_2CHR^1CH_2CH(OH)R^2R^3$$

R^1	R^2	R^3	% Yield
H	Et	H	100
Ph	Me	H	98
$PhCH_2CH_2$	Et	H	94
$Me_2C=CH$	Me	$CH=CH_2$	97

I.A.4-17 Y. Gaoni, Tetrahedron Lett., 1976, 503.

$$ArSO_2(CH_2)_2\!\!-\!\!\overset{O}{\underset{R^1\ R^3}{\triangle}}\!\!-\!\!R^2 \xrightarrow[\text{hexane}]{n\text{-BuLi},\ -15°}$$

$$ArSO_2\!-\!\!\triangle\!\!<^{R^1}_{CR^2R^3OH}$$
$$80\text{-}95\%$$

$$PhSO_2\!-\!\triangle\!\triangle\!\!\overset{O}{\underset{Me}{}} \xrightarrow[\text{hexane}]{n\text{-BuLi},\ -15°} PhSO_2\!-\!\!\diamond\!\!<^{Me}_{CH_2OH}$$

$$25\%$$

I.A.4-18 B. Corbel and T. Durst, J. Org. Chem., **41**, 3648 (1976).

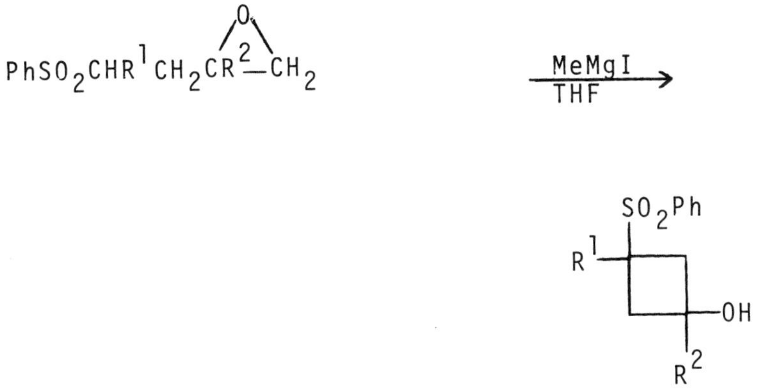

$$PhSO_2CHR^1CH_2CR^2\overset{O}{-}CH_2 \quad \xrightarrow{MeMgI}{THF}$$

<pre>
        SO_2Ph
    R^1─┬─
        │
        └─OH
        R^2
</pre>

$R^1 = R^2 = H$, 96%
$R^1 = H$, $R^2 = Me$, 51%
$R^1 = Ph$, $R^2 = H$, 98%

I.A.4-19 K. Sachdev and H. S. Sachdev, Tetrahedron Lett., **1976**, 4223.

$$Me_3SiCH_2SePh \quad \xrightarrow{\substack{1)\ LDA,\ THF,\ -78° \\ 2)\ RCH_2Br}}$$

$$Me_3SiCH(SePh)CH_2R$$

R = n-Bu, 94%
R = Ph(CH_2)_2, 88%

The products can be oxidized (30% H_2O_2) to aldehydes.

I.A.5-1 E. V. Ermilova, L. A. Remizova, I. A. Favorskaya, and N. L. Tregubova, J. Org. Chem. USSR, 11, 517 (1975).

$$HC\equiv CCH_2OH \xrightarrow[RCH_2Br]{LiNH_2,\ NH_3(\ell)} RCH_2C\equiv CCH_2OH$$
$$40\text{-}60\%$$

Use of the lithium salt allows direct C-alkylation without protection of the hydroxyl group.

I.A.5-2 E. M. Kaiser and J. D. Petty, J. Organomet. Chem., 107, 219 (1976).

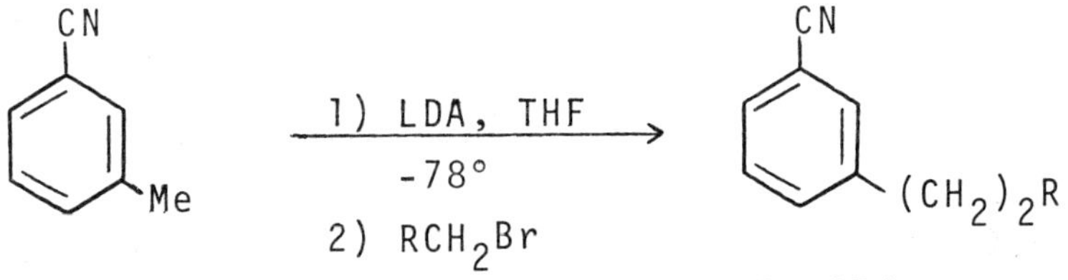

[m-tolunitrile] → 1) LDA, THF, -78°   2) RCH_2Br → [3-cyanophenyl-$(CH_2)_2R$]

30-37%

I.A.5-3 J. E. Stemke, A. R. Chamberlin and F. T. Bond, Tetrahedron Lett., 1976, 2947.

$$n\text{-}C_6H_{13}\underset{\|}{C}Me \quad \xrightarrow[2)\ RX]{1)\ 3\ eq.\ BuLi,\ TMEDA}$$
(NNHTs)

$$n\text{-}C_6H_{13}CR=CH_2$$

R = Me, 71%
R = n-Bu, 65%

CARBON–CARBON BONDS FORMING REACTIONS

I.A.5-4 H. W. Thompson and B. S. Huegi, *J. Chem. Soc., Perkin Trans. I*, <u>1976</u>, 1603.

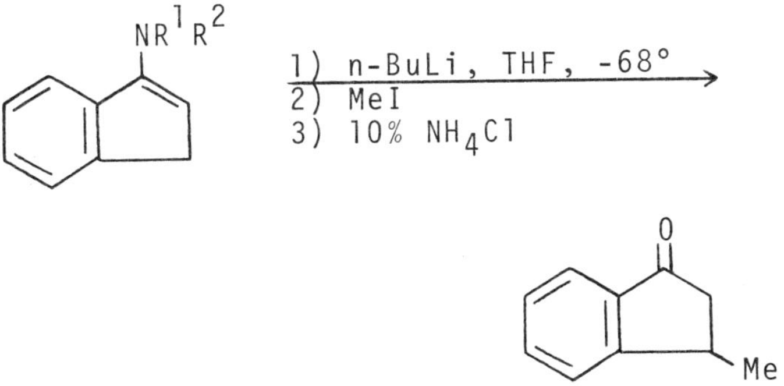

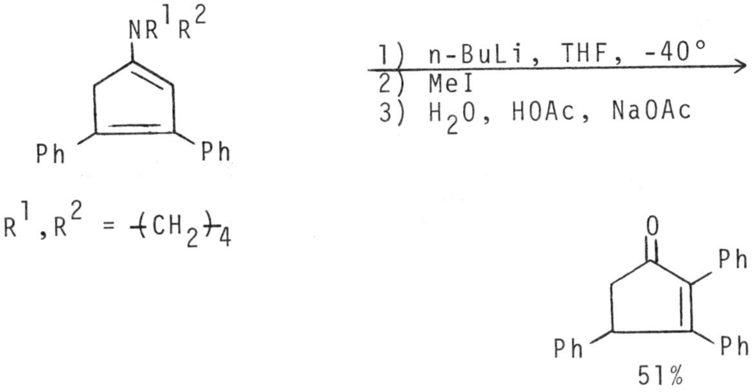

Indene-NR^1R^2

1) n-BuLi, THF, -68°
2) MeI
3) 10% NH_4Cl

→ 3-methyl-1-indanone (Me)

2,5-diphenyl-cyclopentadien-NR^1R^2, Ph, Ph

$R^1, R^2 = +(CH_2)_4$

1) n-BuLi, THF, -40°
2) MeI
3) H_2O, HOAc, NaOAc

→ 2,3,4-triphenylcyclopent-2-enone, 51%

Also works well for 2-indanone.

I.A.5-5 D. Seebach and N. Meyer, *Angew. Chem. Int. Ed. Engl.*, <u>15</u>, 438 (1976).

$n\text{-}Bu_3SnCH_2OH$ $\xrightarrow{\text{1) 2 eq. n-BuLi, n-}C_6H_{14}}_{\text{2) RBr}}$ RCH_2OH

45%

$R = n\text{-}C_8H_{17}$ or $PhCH_2$

I.A.5-6 I. Vlattas, L. D. Vecchia, and A. O. Lee, J. Am. Chem. Soc., 98, 2008 (1976).

PhSCH=CHOEt $\xrightarrow{\text{1) t-BuLi}}_{\text{THF, }-70°}$

$\xrightarrow{\text{n-BuI}}$ $\underset{\text{EtO}}{\overset{\text{H}}{>}}$C=C$\underset{\text{n-Bu}}{\overset{\text{SPh}}{<}}$ 55%

PhSC(Br)=CHOEt $\xrightarrow{\text{1) n-BuLi}}_{\text{Et}_2\text{O, }-70°}$

I.A.5-7 J. J. Eisch and J. E. Galle, J. Organomet. Chem., 121, C10 (1976); see also: ibid, J. Am. Chem. Soc., 98, 4646 (1976).

$\underset{R^3}{\overset{R^2}{>}}\!\!\overset{O}{\triangle}\!\!\underset{H}{\overset{R^1}{<}}$ $\xrightarrow[\text{2) MeI}]{\text{1) base}}$ $\underset{R^3}{\overset{R^2}{>}}\!\!\overset{O}{\triangle}\!\!\underset{Me}{\overset{R^1}{<}}$

R^1	R^2	R^3	base	% Yield
Ph_3Si	H	H	n-BuLi	73
Me_3Si	$n\text{-}C_6H_{13}$	H	t-BuLi, TMEDA	80
$PhSO_2$	H	Ph	n-BuLi	46
$(EtO)_2P(O)$	H	H	LDA	37

I.A.5-8   T. H. Chan and B. S. Ong, Tetrahedron Lett., 1976, 319.

$$R^1R^2C\genfrac{}{}{0pt}{}{OSiMe_3}{SPh} \xrightarrow[HMPA]{R^3Li} R^1R^2R^3COSiMe_3$$

R^1	R^2	R^3	% Yield
Ph	H	Me	86
Ph	H	n-Bu	83
Me	Me	n-Bu	38
$(CH_2)_5$		Me	32

I.A.5-9   W. Oppolzer and R. L. Snowden, Tetrahedron Lett., 1976, 4187.

$$(CH_2=CH)_2CHOSiEt_3 \xrightarrow[2) RX]{1) sec-BuLi, THF, -78°} \underset{A}{CH_2=CH\overset{OSiEt_3}{\underset{|}{C}}=CHCH_2R}$$

$$+$$

$$\underset{B}{(CH_2=CH)_2CROSiEt_3}$$

RX	% Yield	A:B
MeI	86	100:0
$CH_2=CH(CH_2)_2I$	77	66:33
$CH_2=CH(CH_2)_2OTs$	68	10:90
$PhCH_2Br$	84	80:20

I.A.5-10 J. A. Katzenellenbogen and A. I. Crumrine, J. Am. Chem. Soc., $\underline{98}$, 4925 (1976).

$R^1CMe=CHCO_2H$ $\xrightarrow{\begin{array}{l}1)\ LDA,\ THF\\2)\ CuI,\ THF\\3)\ R^1R^3C=CR^4CH_2Br\end{array}}$

$R^1C=CHCO_2H$
  $|$
  $CH_2CH_2CR^4=CR^2R^3$

Study of both regioselectivity and mode of substitution of allylic alkylating agent was carried out. The Cu enolates give high proportions of γ-alkylation.

I.A.5-11 W. Priester, R. West, and T. L. Chwang, J. Am. Chem. Soc., $\underline{98}$, 8413 (1976); see also: W. Priester and R. West, *ibid*, 8421 and 8426.

$Me_nCH_{3-n}C\equiv CH$ $\xrightarrow{n-BuLi,\ hexane}$ $Me_nC_3Li_{4-n}$

$\xrightarrow{Et_2SO_4}$ $Et_{3-n}Me_nCC\equiv CEt$

Use of EtI as the alkylating agent caused oxidative coupling.

CARBON—CARBON BONDS FORMING REACTIONS

I.A.5-12   S. Bhanu, E. A. Khan, and F. Scheinmann, J. Chem. Soc., Perkin Trans. I, 1976, 1609.

$$BuCH_2C\equiv CI \xrightarrow[\substack{\text{hexane, Et}_2O \\ 2)\ n\text{-}C_5H_{11}Br}]{1)\ 3\ eq.\ n\text{-}BuLi,\ -60°} BuCHC\equiv CH$$
$$\underset{C_5H_{11}}{|}$$
52%

A comparison of the reactions of 1-iodo and 1-boroalkynes was carried out.

I.A.5-13   A.J.G. Sagar and F. Scheinmann, Synthesis, 1976, 321.

$$n\text{-}PrCH_2C\equiv CH \xrightarrow[\substack{2)\ n\text{-}C_6H_{13}Br \\ 3)\ Me_2SO_4}]{1)\ 2\ eq.\ n\text{-}BuLi} n\text{-}PrCH(C_6H_{13})C\equiv CMe$$

50%

I.A.5-14   J. Millon and G. Linstrumelle, Tetrahedron Lett., 1976, 1095.

$$\underset{O}{\bigcirc}\!\!-\!R^1 \xrightarrow[\substack{\text{or } R_2^2CuLi \\ 0°,\ 26\ hr}]{R^2Li,\ 10\%\ CuI} R^2(CH_2)_3CHR^1OH$$

R^1	R^2	% Yield
H	n-Bu	66-68
H	n-Pr	60
Me	n-Bu	41

Similar reactions with tetrahydropyran gave an 8% yield of analogous product.

I.A.5-15 K. Kitatani, T. Hiyama, and H. Nozaki, $\underline{J.}$
$\underline{Am. Chem. Soc.}$, $\underline{98}$, 2362 (1976).

$$\underset{R^1 \diagup\!\!\!\diagdown R^2}{Br \quad Br} \xrightarrow{\substack{1) R^3_2CuLi, Et_2O \\ 2) MeI}} \underset{R^1 \diagup\!\!\!\diagdown R^2}{R^3 \quad Me}$$

R^1	R^2	R^3	% Yield
Ph	H	n-Bu	100
Ph	H	sec-Bu	43
n-C_6H_{13}	H	n-Bu	50
$-(CH_2)_4-$		n-Bu	82

A stereoselective one-pot procedure is described.

I.A.5-16 P. A. Grieco, C.-L.J. Wang, and G. Majetich, $\underline{J. Org. Chem.}$, $\underline{41}$, 726 (1976).

$$CH_2=C \underset{CH(OEt)}{\overset{M}{\diagup}} \xrightarrow{R^1R^2C=CR^3CH_2Br}$$

$$R^1R^2C=CR^3CH_2\overset{CH(OEt)_2}{\underset{|}{C}}=CH_2$$

$M = PhSCu^-, Li^+$, 74-85%

$M = Cu$, 50-90%

CARBON-CARBON BONDS FORMING REACTIONS

I.A.5-17 L. Blaszczak, J. Winkler, and S. O'Kuhn, Tetrahedron Lett., 1976, 4405.

[cyclohexene with OP(O)(OPh)$_2$ and t-Bu substituents] $\xrightarrow{\text{n-Bu}_2\text{CuLi}}{\text{THF, }-30°}$ [cyclohexene with n-Bu and t-Bu substituents]

I.A.5-18 T. A. Baer and R. L. Carney, Tetrahedron Lett., 1976, 4697.

$$Br(CH_2)_nCO_2MgCl \xrightarrow[\text{Li}_2\text{CuCl}_4 \text{ (cat.)}]{\text{RMgBr}} R(CH_2)_nCO_2H$$

$R = C_5H_{11}$, $n = 10$, 94%
$R = $ sec-Bu, $n = 4$, 88%

I.A.5-19 G. M. Whitesides and F. D. Gutowski, J. Org. Chem., 41, 2882 (1976).

[cyclohexane with two (CH$_2$)$_n$Cl chains] $\xrightarrow[\text{2) AgTf}]{\text{1) Mg, THF}}$ [bicyclic product with (CH$_2$)$_{2n}$]

$n = 1$, 61%
$n = 2$, 57%

This procedure does not work for formation of 7-membered or higher ring systems.

29

I.A.5-20 G. N. Dorofeenko, L. V. Mezheritskaya, and E. S. Matskovskaya, J. Org. Chem. USSR, 11, 2476 (1975).

$$\text{Ar}\underset{ClO_4^-}{\overset{O^+}{\diamondsuit}} \xrightarrow[2)\ H_3O^+]{1)\ RMgX,\ Et_2O} \text{ArCOR} \quad 65\text{-}89\%$$

I.A.5-21 I. Degani and R. Fochi, J. Chem. Soc., Perkin Trans I, 1976, 1886.

$$\underset{S}{\overset{S}{\bigodot}}\!\!-R^1 \xrightarrow[Et_2O]{R^2MgX} \underset{S}{\overset{S}{\bigodot}}\!\!\underset{R^2}{\overset{R^1}{\diagup}}$$

R^1	R^2	% Yield
H	Ph	87
H	n-Pr	88
Ph	Me	66
Ph	Ph	30

I.A.5-22 G. Zweifel and R. A. Lynd, Synthesis, 1976, 816.

$$RC\equiv CH \xrightarrow[2)\ ClCH_2OEt]{1)\ i\text{-}Bu_2AlH,\ n\text{-}C_6H_{14}} \underset{H}{\overset{R}{\diagdown}}C=C\underset{CH_2OEt}{\overset{H}{\diagup}}$$

R = n-Bu, 80%
R = t-Bu, 75%

CARBON—CARBON BONDS FORMING REACTIONS

I.A.5-23 S. Baba, D. E. Van Horn, and E. Negishi, Tetrahedron Lett., 1976, 1927.

$$R^1C\equiv CH \xrightarrow[\substack{2)\ n\text{-BuLi} \\ 3)\ R^2X,\ THF}]{1)\ i\text{-Bu}_2AlH} \quad \substack{R^1 \\ H} C=C \substack{H \\ R^2}$$

R^1	R^2X	% Yield
$n\text{-}C_5H_{11}$	MeI	65
$n\text{-}Bu$	$n\text{-}C_8H_{17}I$	49
$n\text{-}Bu$	$CH_2=CHCH_2Br$	73
$n\text{-}Bu$	$PhCH_2Br$	46

I.A.5-24 J. J. Eisch and G. A. Damasevitz, J. Org. Chem., 41, 2214 (1976); see also: K. Uchida, K. Utimoto, and H. Nozaki, *ibid*, 2215.

$$R^1C\equiv CR^2 \xrightarrow[\substack{2)\ MeLi \\ 3)\ R^3X}]{1)\ i\text{-Bu}_2AlH} \quad \substack{R^1 \\ H} C=C \substack{R^2 \\ R^3}$$

R^1	R^2	R^3	% Yield
Ph	Ph	$CH_2=CHCH_2$	50-70
$n\text{-}C_6H_{13}$	H	Me	60
Ph	$SiMe_3$	$CH_2=CHCH_2$	40*

*Depending on reaction conditions either the E or Z isomer can be synthesized stereoselectively.

I.A.5-25 K. Uchida, K. Utimoto, and H. Nozaki, J. Org. Chem., **41**, 2941 (1976).

$$R^1C\equiv CSiMe_3 \xrightarrow[\substack{\text{2) MeLi}\\\text{3) CuI}\\\text{4) }R^2X}]{\text{1) (cyclohexyl)}_2BH} \underset{H}{\overset{R^1}{>}}C=C\underset{R^2}{\overset{SiMe_3}{<}}$$

R^1	R^2X	% Yield
$n\text{-}C_6H_{13}$	MeI	94
$n\text{-}C_6H_{13}$	$CH_2=CHCH_2Cl$	93
$THPO(CH_2)_4$	EtI	85

I.A.6-1 M. Tokuda, T. Taguchi, O. Nishio, and M. Itoh, J. Chem. Soc., Chem. Commun., **1976**, 606.

$$PhC\equiv CH \xrightarrow[\substack{n\text{-}Bu_4NI, \text{ HMPA}\\RX}]{\text{electrolysis}} PhC\equiv CR$$

R = Me, Et, n-Bu

79-100% (39-81% conversion)

I.A.6-2 R. Alkabets and I. Granoth, J. Chem. Soc., Perkin Trans. I, **1976**, 2380.

$$Ar_2CH_2 \xrightarrow[\text{(MeOCH}_2CH_2)_2O]{LiAlH_4, \ 150°} Ar_2CMe_2$$

50-80%

I.A.6-3 N. M. Przhiyalgovskaya, E. N. Sidorenko, and A. T. Prudchenko, J. Org. Chem. USSR, 11, 1671 (1975).

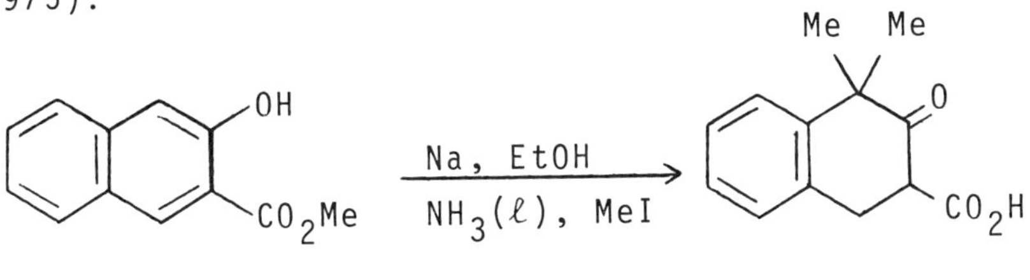

$$\text{2-HO-3-MeO}_2\text{C-naphthalene} \xrightarrow[\text{NH}_3(\ell), \text{MeI}]{\text{Na, EtOH}} \text{1,1-Me}_2\text{-2-oxo-3-CO}_2\text{H-tetralin}$$

25%

I.A.6-4 H. Kise, Y. Arase, S. Shiraishi, M. Senō, and T. Asahara, J. Chem. Soc., Chem. Commun., 1976, 299.

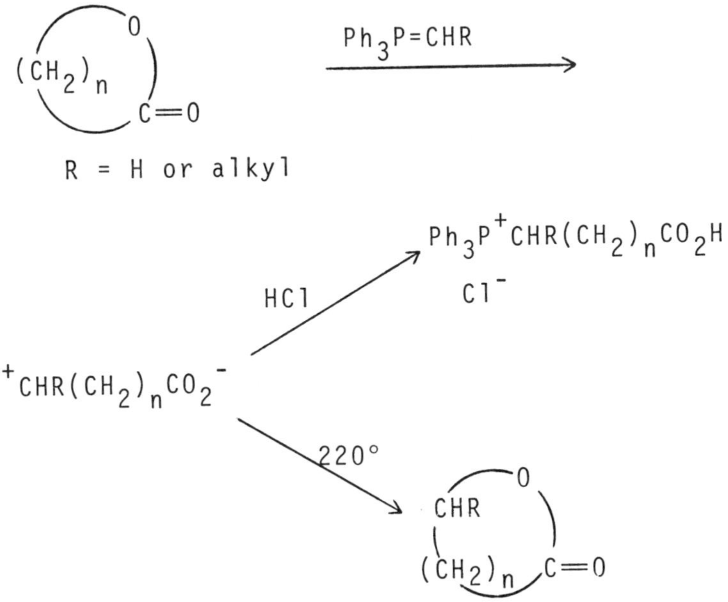

$$\underset{R = H \text{ or alkyl}}{(CH_2)_n\!\!\!\bigg\rangle\!\!\!\underset{C=O}{\overset{O}{\big\rangle}}} \xrightarrow{Ph_3P=CHR}$$

$$Ph_3P^+CHR(CH_2)_nCO_2^- \xrightarrow{HCl} Ph_3P^+CHR(CH_2)_nCO_2H \cdot Cl^-$$

$$\xrightarrow{220°} \underset{(CH_2)_n}{\overset{CHR}{\big\langle}}\!\!\!\bigg\rangle\!\!\!\underset{C=O}{\overset{O}{\big\rangle}}$$

I.A.6-5 J. M. Midgley, J. S. Millership, and W. B. Whalley, *J. Chem. Soc., Perkin Trans. I*, <u>1976</u>, 1384.

$$\text{ArCH(OSiMe}_3)\text{C}\equiv\text{CH} \xrightarrow[\text{2) CO}_2]{\text{1) EtMgBr, THF, reflux}} \text{ArCHEtC}\equiv\text{CCO}_2\text{H}$$

$$\text{ArCH}_2\text{OSiMe}_3 \xrightarrow{\text{NaCH(CO}_2\text{Et)}_2} \text{ArCH}_2\text{CH(CO}_2\text{Et)}_2$$

$$\text{Ar} = 2,4,6\text{-(MeO)}_3\text{C}_6\text{H}_2$$

I.A.7.a.1a-1 G. Wittig, *Fortschr. Chem. Forsch.*, <u>67</u>, 1 (1976).

Review: Old and New in the Field of Directed Aldol Condensations.

I.A.7.a.1a-2 G. W. Gokel, S. A. Di Biase, and B. A. Lipisko, *Tetrahedron Lett.*, <u>1976</u>, 3495.

$$R^1\text{COR}^2 \xrightarrow[\text{MeCN}]{\text{powdered KOH}} R^1R^2\text{C=CHCN}$$

$$R^1 = \text{Ph}, R^2 = \text{H}, 82\%$$
$$R^1, R^2 = -(\text{CH}_2)_5-, 50\%$$

CARBON–CARBON BONDS FORMING REACTIONS

I.A.7.a.1a-3 I. Kuwajima, T. Sato, M. Arai, and N. Minami, Tetrahedron Lett., 1976, 1817; see also: I. Kuwajima, N. Minami, and T. Sato, *ibid*, 2253.

$$R^1COMe + R^2CHO \xrightarrow{Me_2CHCH_2C(SiMe_3)_2OLi}_{THF,\ -40°}$$

$$R^1COCH_2CH(OH)R^2$$
71-86%

Use of the hindered base allows the directed aldol to proceed without prior formation of the ketone enolate.

I.A.7.a.1a-4 T. Mukaiyama and T. Inoue, Chem. Lett., 1976, 559.

$$R^1COCH_2R^2 \xrightarrow[\substack{R_3N,\ Et_2O \\ 2)\ R^3CHO}]{1)\ Bu_2BOTf} R^1COCHR^2CH(OH)R^3$$

R^1	R^2	R^3	% Yield
Ph	Et	Ph	80
Ph	H	$PhCH_2CH_2$	75
i-Bu	H	C_5H_{11}	70
$+CH_2+_4$		$PhCH_2CH_2$	72

I.A.7.a.1a-5 Ae. de Groot and B.J.M. Jansen, Recl. Trav. Chim. Pays-Bas, 95, 81 (1976).

$$i\text{-BuO} \underset{}{\overset{R^1}{\diagup\hspace{-0.2em}\diagdown}} O \xrightarrow[\text{base}]{R^2COR^3} R^2CO \underset{}{\overset{O \quad R^1}{\diagup\hspace{-0.2em}\diagdown}} O\text{-}i\text{-Bu}$$

R^1	R^2	R^3	Base	% Yield
H	Me	OEt	$(Me_3Si)_2NLi$	78
H	H	O-i-Bu	NaH	95
Me	H	O-i-Bu	NaH	68

I.A.7.a.1a-6 G. Stork and G. A. Kraus, J. Am. Chem. Soc., 98, 2351 (1976).

$$MeCOCH=C(OMe)Me \xrightarrow[\text{2) } R^1COR^2, \text{ THF, } -78°]{\text{1) LDA, THF, } -78°}$$

$$R^1R^2C(OH)CH_2COCH=C(OMe)Me$$

The product can be converted by reduction and acid hydrolysis to vinylogous aldols ($R^1R^2C(OH)CH_2CH=CHCOMe$).

CARBON—CARBON BONDS FORMING REACTIONS

I.A.7.a.1a-7 T. M. Harris, G. P. Murphy, and A. J. Poje, J. Am. Chem. Soc., **98**, 7733 (1976).

$$RCOCH_2COCH_2COMe \xrightarrow[\text{2) } CO_2(g)]{\text{1) excess LDA, THF, 0°}} $$

$$RCOCH_2COCH_2COCH_2CO_2H$$

R = Ph, 82%
R = Me, 72%
R = Et, 42%

I.A.7.a.1a-8 G. P. Chiusoli and F. Gasparoni, Gazz. Chim. Ital., **106**, 201 (1977).

$$RCH_2CO_2Li \xrightarrow[\text{2) } CH_2O]{\text{1) } Et_2NLi, \text{ hexane}} HO_2CCHRCH_2CHRCO_2H$$

63%

I.A.7.a.1a-9 A. G. Schultz and M. H. Berger, J. Org. Chem., **41**, 585 (1976).

$$Me_2CHCO_2CH_2OMe \xrightarrow{LDA, THF} Me_2C(CH_2OH)CO_2Me$$

69%

$$MeCH=CMeCO_2CH_2OMe \xrightarrow{LDA, THF} CH_2=CHCMe(CH_2OH)CO_2Me$$

61%

I.A.7.a.1a-10   D. A. Konen, P. E. Pfeffer, and L. S. Silbert, Tetrahedron, 32, 2507 (1976).

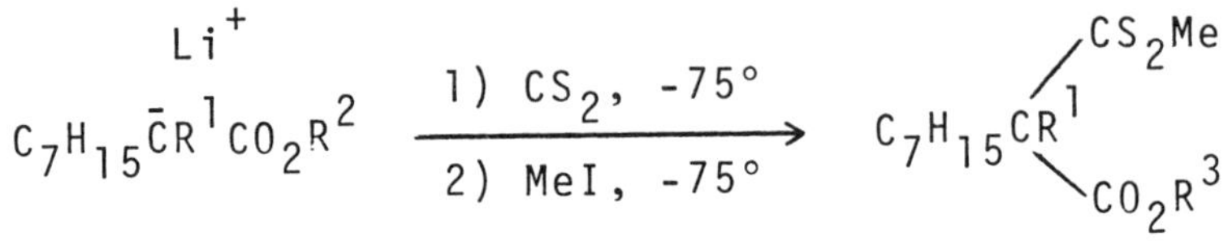

$$C_7H_{15}\bar{C}R^1CO_2R^2 \; Li^+ \xrightarrow[\text{2) MeI, -75°}]{\text{1) } CS_2, \text{-75°}} C_7H_{15}CR^1(CS_2Me)(CO_2R^3)$$

R^1	R^2	R^3	% Yield
H	Li	H	88
H	Me	Me	85
Me	Li	H	77
Me	Me	Me	75

I.A.7.a.1a-11   S. L. Hartzell and M. W. Rathke, Tetrahedron Lett., 1976, 2757.

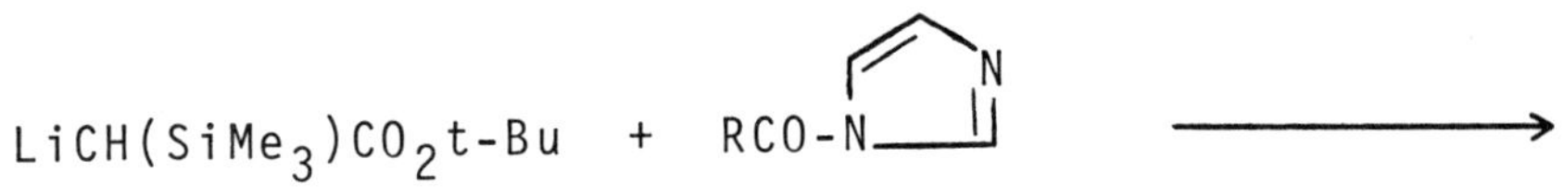

$$\text{LiCH(SiMe}_3\text{)CO}_2\text{t-Bu} + \text{RCO-N}\diagdown\!\!\diagup\!\!\text{N} \longrightarrow$$

$$\text{RCOCH}_2\text{COt-Bu}$$

R = Me, 94%
R = PhCH=CH, 50%

I.A.7.a.1a-12 S. L. Hartzell and M. W. Rathke, Tetrahedron Lett., 1976, 2737.

$(Me_3Si)_2CHCO_2t\text{-}Bu \quad \xrightarrow{\text{1) LDA, THF, -78°}}_{\text{2) RCHO}}$

$RCH=C(SiMe_3)CO_2t\text{-}Bu$

70-90%

I.A.7.a.1a-13 W. Beck and M. Girnth, Chem. Ber., 109, 965 (1976).

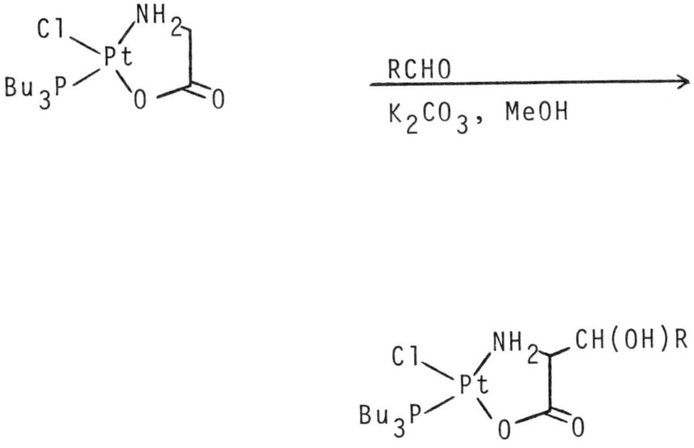

[Pt complex with Cl, Bu$_3$P, NH$_2$, O, O ligands] $\xrightarrow{RCHO}_{K_2CO_3,\ MeOH}$ [Pt complex with CH(OH)R substituent]

R = Me or Ph, 95%

I.A.7.a.1a-14 D. A. Evans, L. K. Truesdale, and K. G. Grimm, J. Org. Chem., **41**, 3335 (1976).

$$RCHO \xrightarrow[KCN \cdot 18\text{-crown-6}]{M_3SiC(N_2)CO_2Et, \; CHCl_3}$$

$$RCH(OSiMe_3)C(N_2)CO_2Et$$

R = Ph, 86%
R = $n\text{-}C_5H_{11}$, 86%
R = MeCH=CMe, 44%

I.A.7.a.1a-15 K. Smith and K. Swaminathan, J. Chem. Soc., Chem. Commun., **1976**, 387.

$$HCON(i\text{-}Pr)_2 \xrightarrow[\substack{\text{Trapp mixture} \\ 2)\; R^1COR^2}]{1)\; t\text{-BuLi},\; -95°} R^1R^2C(OH)CON(i\text{-}Pr)_2$$

R^1	R^2	% Yield
Ph	Ph	85
Ph	H	80
Et	H	62
PhCH=CH	H	68

CARBON-CARBON BONDS FORMING REACTIONS

I.A.7.a.1a-16 D. Seebach, W. Lubosch, and D. Enders, Chem. Ber., 109, 1309 (1976).

$$\underset{HCNMe_2}{\overset{S}{\|}} \xrightarrow[2) R^1COR^2]{1) LDA, THF, -100°} R^1R^2C(OH)\overset{S}{\underset{CNMe_2}{\|}}$$

R^1	R^2	% Yield
Et	H	65
Ph	H	65
Me	Me	75
Ph	Ph	80
$-(CH_2)_5-$		50

I.A.7.a.1a-17 G. Kinast and L.-F. Tietze, Angew. Chem. Int. Ed. Engl., 15, 239 (1976).

$$R^1COCHR^2R^3 \xrightarrow[MeCN, 82°]{CH_2=\overset{+}{N}Me_2 Cl^-} R^1COCR^2R^3CH_2\overset{+}{N}HMe_2 Cl^-$$

R^1	R^2	R^3	% Yield
t-Bu	H	Me	58
H	Me	Me	81
$-(CH_2)_3-$		H	79
$-(CH_2)_3-$		Me	79

I.A.7.a.1a-18 S. Danishefsky, T. Kitahara, R. McKee, and P. F. Schuda, J. Am. Chem. Soc., **98**, 6715 (1976).

$$R^1C(OSiMe_3)=CHR^2 \xrightarrow[\text{2) aq. NaOH}]{\text{1) } Me_2\overset{+}{N}=CH_2 I^-, CH_2Cl_2}$$

$$R^1COCHR^2CH_2NMe_2$$

$$R^1, R^2 = \{CH_2\}_4, \ 87\%$$

$$R^1 = CH=CHOMe, \ R^2 = H, \ 95\%$$

The dimethyl(methylene)ammonium iodide also reacts with enolates of lactones.

I.A.7.a.1a-19 T. Sato, J. Hanna, H. Nakamura, and T. Mukaiyama, Bull. Chem. Soc. Japan, **49**, 1055 (1976).

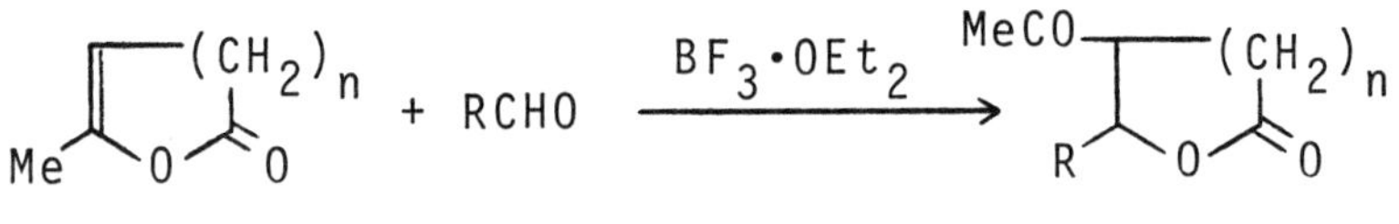

Me-C(=CH-(CH_2)_n-O-C(=O))-O + RCHO →[BF_3·OEt_2] MeCO-CHR-(CH_2)_n-O-C(=O)

n	R	% Yield
1	Ph	62
2	Ph	73
1	H	93
2	Et	78
1	ClCH$_2$	90

I.A.7.a.1a-20   C. Bischoff and H. Herma, J. Prakt. Chem., 318, 773 (1976).

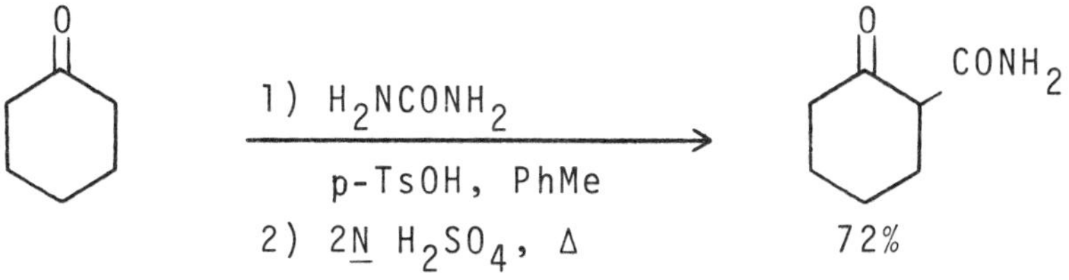

[cyclohexanone] → 1) H_2NCONH_2, p-TsOH, PhMe; 2) $2\underline{N}\ H_2SO_4$, Δ → [2-carbamoylcyclohexanone]   72%

I.A.7.a.1a-21   T. Mukaiyama, T. Sato, S. Suzuki, T. Inoue, and H. Nakamura, Chem. Lett., 1976, 95.

$R^1COCHR^2CO_2CH_2CCl_3$ $\xrightarrow[\text{Zn, DMSO}]{R^3CHO}$ $R^1COCHR^2CH(OH)R^3$

R^1	R^2	R^3	% Yield
Me	n-Bu	Ph	86
Me	n-Bu	Me	90
Et	Me	Ph	92
Et	Me	i-Pr	61

I.A.7.a.1a-22 H. Singh and A. S. Cheema, *Chem. Ind.*
(London), **1976**, 413.

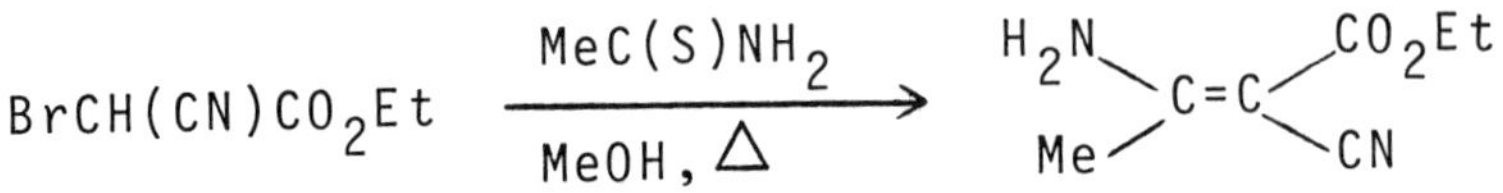

$$BrCH(CN)CO_2Et \xrightarrow[MeOH, \Delta]{MeC(S)NH_2} \underset{Me}{\overset{H_2N}{>}}C=C\underset{CN}{\overset{CO_2Et}{<}}$$

40-45%

I.A.7.a.1a-23 J. Liebscher and H. Hartmann, *J.*
Prakt. Chem., **318**, 705 (1976).

$$ArC(Cl)=CHCH=\overset{+}{N}Me_2 \; ClO_4^- \xrightarrow[HOAc, \, Et_3N]{YCH_2CN}$$

$$ArC(CL)=CHCH=C(Y)CN$$

$$Y = CO_2H, \; CO_2Et, \; CONH_2,$$

$$CN, \; CSNH_2$$

50-80%

I.A.7.a.2-1 D. Seebach and F. Lehr, Angew. Chem. Int. Ed. Eng., 15, 505 (1976).

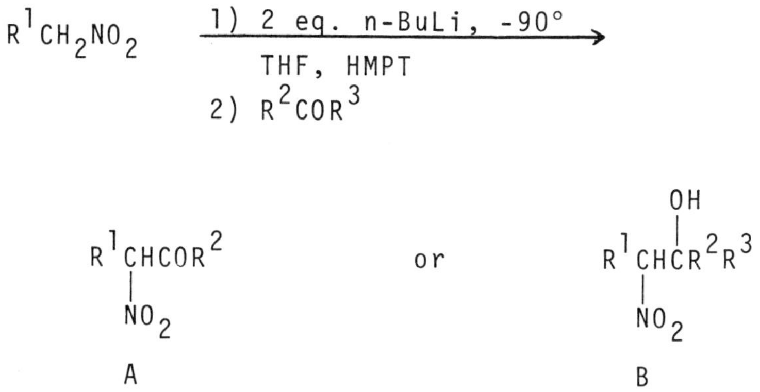

$R^1CH_2NO_2$ $\xrightarrow[\text{2) } R^2COR^3]{\text{1) 2 eq. n-BuLi, -90°, THF, HMPT}}$

$R^1\underset{NO_2}{CHCOR^2}$ or $R^1\underset{NO_2}{CH\overset{OH}{\underset{}{C}}R^2R^3}$

A   B

R^1	R^2	R^3	% Yield
Et	MeO	MeO	55(A)
i-Pr	MeO	Cl	65(A)
i-Pr	Me	OAc	30(A)
i-Pr	n-Pr	OMe	60(A)
Et	Ph	H	80(B)
Et	Me	Me	40(B)
Et	$-(CH_2)_5-$		65(B)

I.A.7.a.2-2 R. Henning, F. Lehr, and D. Seebach, Helv. Chim. Acta, $\underline{59}$, 2213 (1976).

$$R^1CH_2CHR^2NO_2 \xrightarrow[\text{2) } R^3COR^4]{\text{1) 2 eq. n-BuLi, THF, HMPA, } -78°}$$

$$R^2CH(NO_2)CHR^1C(OH)R^3R^4$$

R^1	R^2	R^3	R^4	% Yield
p-ClC$_6$H$_4$	H	Et	H	75
Ph	H	Ph	H	80
Ph	H	$-(CH_2)_4-$		69
H	Me	Ph	H	50

I.A.7.a.2-3 W. G. Kofron and M.-K. Yeh, J. Org. Chem., $\underline{41}$, 439 (1976).

$$R^1\overset{NOH}{\overset{\|}{C}}CH_2R^2 \xrightarrow[\substack{\text{2) } R^3COR^4 \\ \text{3) hydrolysis}}]{\text{1) 2 eq. n-BuLi, THF}} R^1COCR^2{=}CR^3R^4$$

R^1	R^2	R^3	R^4	% Yield
Me	H	Ph	Ph	55
Me	H	Me	CH=CH$_2$	17
$-(CH_2)_4-$		Me	Me	48
$-(CH_2)_4-$		Me	H	45

I.A.7.a.2-4 R. M. Sandifer, S. E. Davis, and C. F. Beam, Synth. Commun., 6, 339 (1976).

$$PhC(Me)=NNHR^1 \xrightarrow{R^2R^3CO} Ph\overset{NNHR^1}{\underset{}{C}}CH_2CR^2R^3OH$$

R^1	R^2	R^3	% Yield
PhCO	Ph	p-ClC$_6$H$_4$	84
PhSO$_2$	Ph	Ph	78

I.A.7.a.2-5 E. J. Corey, D. Enders, and M. G. Bock, Tetrahedron Lett., 1976, 7; see also: E. J. Corey and D. Enders, ibid, 11.

$$Me_3SiCHR^1CH=X \xrightarrow[\substack{2) R^2CHO \\ 3) H_3O^+}]{1) LDA, THF} R^2CH=CR^1CHO$$

R^1	R^2	X	% Yield
H	n-C$_5$H$_{11}$	Nt-Bu	94
Me	i-Pr	Nt-Bu	88
H	n-C$_5$H$_{11}$	NNMe$_2$	90
H	Ph	NNMe$_2$	90

I.A.7.a.2-6 J. F. LeBorgne, T. Cuvigny, M. Larcheveque, and H. Normant, Synthesis, 1976, 238.

$$R^1R^2CHCH=NNMe_2 \xrightarrow[2)\ R^3COR^4]{1)\ Et_2NLi,\ PhH,\ HMPT}$$

$$R^1R^2C(CN)CR^3R^4OH$$

R^1	R^2	R^3	R^4	% Yield
Me	Me	H	n-Pr	87
Et	Et	Me	Me	88
Me	Me	Ph	Ph	90

I.A.7.a.2-7 C. F. Beam, C. W. Thomas, R. M. Sandifer, R. S. Foote, and C. R. Hauser, Chem. Ind. (London), 1976, 487.

$$\underset{PhCCH_2Li}{\overset{N_2Li_2}{\|}} \xrightarrow[2)\ HCl-H_2O]{1)\ Ph_2CO}$$

$$PhCOCH_2C(OH)Ph_2$$

24%

CARBON–CARBON BONDS FORMING REACTIONS

I.A.7.a.2-8 U. Schöllkopf and H.-U. Scholz, Synthesis, 1976, 271.

$$H-\underset{\underset{SiMe_3}{\|}}{C}{=}N_2 \xrightarrow[\text{2) } R^1COR^2]{\text{1) n-BuLi, THF, pentane, }-100°} R^1R^2C(OH)\underset{\underset{SiMe_3}{\|}}{C}{=}N_2$$

R^1	R^2	% Yield
Me	Me	65
H	Ph	80
$\text{-}(CH_2)_5\text{-}$		65

The product can be converted into epoxides and aldehydes by known procedures.

I.A.7.a.2-9 D. Seebach and W. Lubosch, Angew. Chem. Int. Ed. Engl., 15, 313 (1976).

$$t\text{-}Bu\underset{\underset{S}{\|}}{C}NMe_2 \xrightarrow[\text{2) } R^1COR^2]{\text{1) sec-BuLi, }-78°\text{, THF, TMEDA}} t\text{-}Bu\underset{\underset{S}{\|}}{C}NMeCH_2COHR^1R^2$$

R^1	R^2	% Yield
Ph	H	70
Ph	Ph	63
i-Pr	H	23
$\text{-}(CH_2)_5\text{-}$		17

I.A.7.a.2-10 R. R. Schmidt and J. Talbiersky, Angew. Chem. Int. Ed. Engl., 15, 171 (1976).

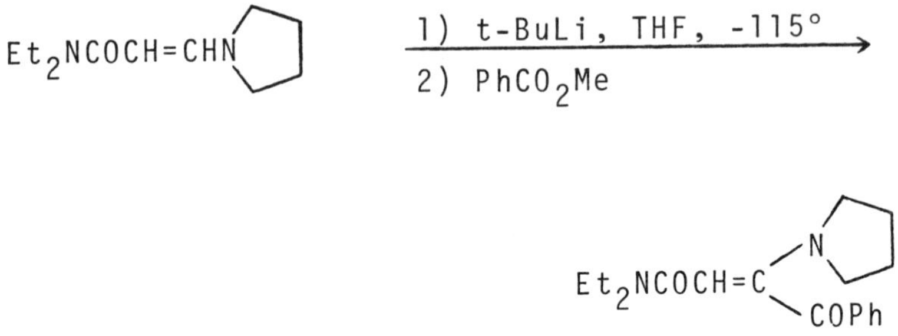

Et$_2$NCOCH=CHN(pyrrolidine) $\xrightarrow{\text{1) t-BuLi, THF, -115°} \atop \text{2) PhCO}_2\text{Me}}$

Et$_2$NCOCH=C(N-pyrrolidine)(COPh)

95%

I.A.7.a.2-11 H. Hoberg and U. Griebsch, Synthesis 1976, 830.

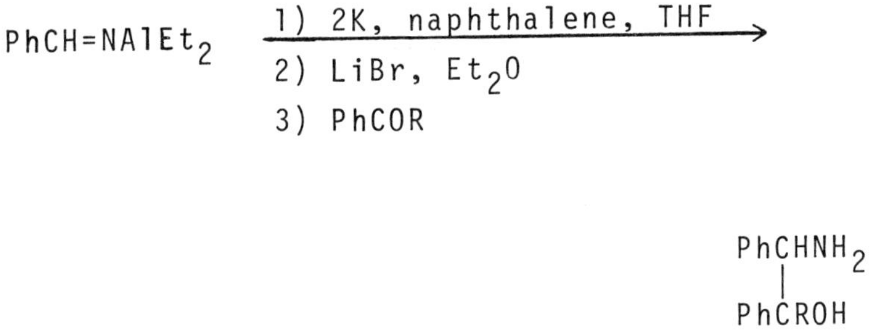

PhCH=NAlEt$_2$ $\xrightarrow{\text{1) 2K, naphthalene, THF} \atop \text{2) LiBr, Et}_2\text{O} \atop \text{3) PhCOR}}$

PhCHNH$_2$
|
PhCROH

R = H, 58%

R = Ph, 75%

I.A.7.a.2-12 M. Schmidt and E. Weissflog, Z. Naturforsch., Teil B, 31b, 136 (1976).

[cyclic (CH$_2$S)$_n$ dithioacetal]–Li $\xrightarrow{\text{1) CO}_2 \text{ or CS}_2 \quad \text{2) HCl}}$ [cyclic (CH$_2$S)$_n$ dithioacetal]–CO$_2$H (or CS$_2$H)

n = 1, 2, or 3 (70–95%)

(MeS)$_2$CHLi $\xrightarrow{\text{1) CO}_2 \quad \text{2) HCl}}$ (MeS)$_2$CHCO$_2$H

75%

I.A.7.a.2-13 A. Bongini, D. Savoia, and A. Umani-Ronchi, J. Organomet. Chem., 112, 1 (1976).

RSO$_2$Me $\xrightarrow{\text{1) 2 eq. n-BuLi, THF,} \quad \text{2) PhCHO}}$ RO$_2$S–CH(Ph)–CH(OH)–Ph (threo/erythro diol)

R = Ph, 60%
R = Me$_2$N, 50%

PhSO$_2$Me $\xrightarrow{\text{1) 2 eq. n-BuLi, THF,} \quad \text{2) MgI}_2 \text{ or BF}_3\text{·Et}_2\text{O} \quad \text{3) R}^1\text{COR}^2}$ PhO$_2$SCH=CR^{1}R^2 + PhSO$_2$CH$_2$C(OH)R^{1}R^2

∼35% each

I.A.7.a.2-14 K. Tanaka, R. Tanikaga, and A. Kaji, Chem. Lett., 1976, 917.

$$EtO\overset{S}{C}SCHR^1CO_2Et \xrightarrow[2) R^2COR^3]{1) LDA, THF, -78°} R^2R^3C=CR^1CO_2Et$$

R^1	R^2	R^3	% Yield
H	Et	H	52
H	Me	Me	85
H	$-(CH_2)_5-$		78
Me	Ph	H	82
Et	i-Pr	H	82

I.A.7.a.2-15 J. N. Denis, W. Dumont, and A. Krief, Tetrahedron Lett., 1976, 453.

$$(R^1Se)_2CR^2R^3 \xrightarrow[2) R^4COR^5]{1) n-BuLi} R^1SeCR^2R^3COR^4$$

R^1	R^2	R^3	R^4	R^5	% Yield
Ph	$-(CH_2)_2Ct$-$Bu(CH_2)_2-$		H	NMe_2	46
Me	H	C_6H_{13}	Ph	Cl	65
Ph	Me	Me	OMe	Cl	80
Me	H	C_6H_{13}	OMe	Cl	40

CARBON-CARBON BONDS FORMING REACTIONS

I.A.7.a.2-16 W. Dumont and A. Krief, *Angew. Chem. Int. Ed. Engl.*, 15, 161 (1976).

$$R_3^1SiCHR^2SeMe \xrightarrow[\substack{2) R^3COR^4 \\ 3) H_3O^+}]{1)\ n\text{-BuLi, THF, }0°} \underset{R^2}{\overset{H}{\diagdown}}C=C\underset{R^4}{\overset{R^3}{\diagup}}$$

R^1	R^2	R^3	R^4	% Yield
Me	C_6H_{13}	H	C_6H_{13}	45
Me	C_6H_5	$-(CH_2)_5-$		40
Et	Me	$-(CH_2)_2CH(t\text{-Bu})(CH_2)_2-$		40

I.A.7.a.3-1 E. Nakamura and I. Kuwajima, *Angew. Chem. Int. Ed. Engl.*, 15, 498 (1976).

$$PhC\equiv CSiMe_3 \xrightarrow[Bu_4N^+F^-\ (cat.)]{R^1COR^2,\ THF} PhC\equiv C\overset{OSiMe_3}{\underset{|}{C}}R^1R^2$$

R^1	R^2	% Yield
Ph	H	76
n-C_7H_{15}	H	70
Ph	Ph	79
$-(CH_2)_5-$		87

I.A.7.a.3-2 D. Abenhaim, G. Boireau, C. Bernardon, A. Deberly, and C. Germain, *Tetrahedron Lett.*, **1976**, 993.

$$RCOMe \xrightarrow{\text{NaAlEt}_4,\ \text{NiCl}_2\ (\text{cat.}),\ C_5H_{12}} RC(OH)EtMe$$

R = Ph, 84%
R = PhCH$_2$, 81%

I.A.7.a.3-3 J. F. Ruppert and J. D. White, *J. Org. Chem.*, **41**, 550 (1976).

$$R^1CH=CHCH_2Br\ +\ R^2COR^3 \xrightarrow[\text{flow system}]{Zn}$$

$$R^2R^3C(OH)CHR^1CH=CH_2$$

R^1	R^2	R^3	% Yield
H	Ph	H	96
H	Ph	Me	97
Me	Et	Et	96
H	$-(CH_2)_5-$		97

I.A.7.a.3-4 A. A. Ponaras, Tetrahedron Lett., 1976, 3105.

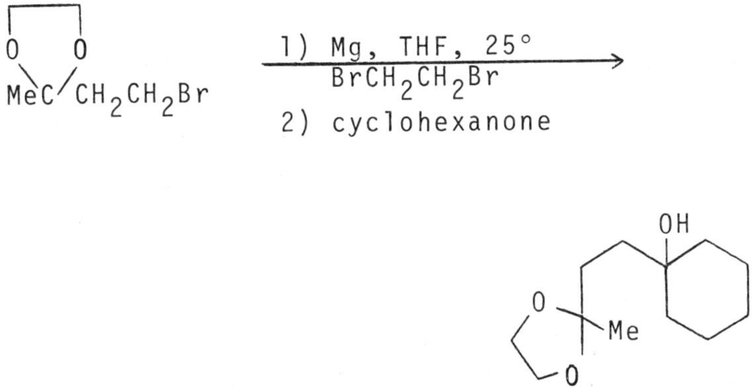

MeC(OCH₂CH₂O)CH₂CH₂Br

$$\xrightarrow[\text{2) cyclohexanone}]{\text{1) Mg, THF, 25°, BrCH}_2\text{CH}_2\text{Br}}$$

[dioxolane-CH₂CH₂-C(OH)(cyclohexyl), Me substituent]

97%

I.A.7.a.3-5 D. Seebach, K.-H. Geiss, and M. Pohmakotr, Angew. Chem. Int. Ed. Engl., 15, 437 (1976).

$$[CH_2=CH-CHS]^= \; 2Li^+ \quad \xrightarrow[\text{3) MeI}]{\text{1) MgBr}_2 \;\; \text{2) R}^1COR^2}$$

$$R^1R^2C(OH)CH(SMe)CH=CH_2$$

R^1	R^2	% Yield
Et	H	91
Ph	H	73
Ph	Me	89
${(CH_2)}_4$		94

I.A.7.a.3-6 N. L. Holy, Synth. Commun., 6, 539 (1976).

$$CH_2=\overset{+}{N}Me_2 \ CF_3CO_2^- \xrightarrow[Et_2O]{RM} RCH_2NMe_2$$

RM = PhMgBr, 85%
RM = n-BuLi, 72%

I.A.7.a.3-7 H. G. Richey, Jr., R. C. McLane, and C. J. Phillips, Tetrahedron Lett., 1976, 233.

$$R^1R^2C=NOH \xrightarrow{R^3Li} R^1R^2R^3CNOH$$

R^1	R^2	R^3	% Yield
Ph	Ph	n-Bu	63
Ph	H	Ph	63
Me	Me	n-Bu	17
Pr	H	Me	46

I.A.7.a.3-8   D. S. Matteson and P. K. Jesthi, <u>J.</u>
Organomet. Chem., <u>110</u>, 25 (1976).

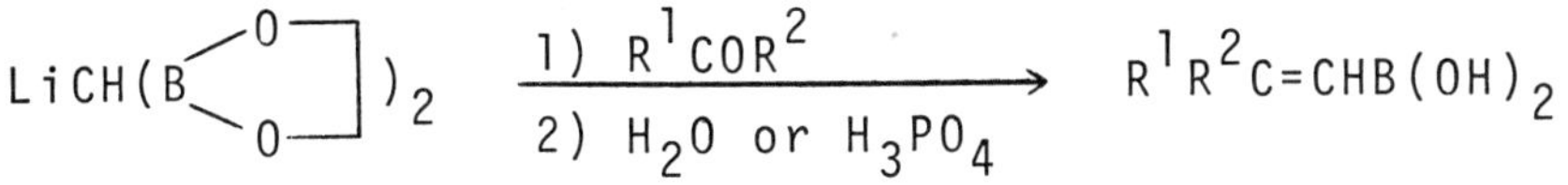

$$\text{LiCH}(B{\overset{O}{\underset{O}{\diagup\!\!\diagdown}}})_2 \xrightarrow[\text{2) } H_2O \text{ or } H_3PO_4]{\text{1) } R^1COR^2} R^1R^2C=CHB(OH)_2$$

R^1	R^2	% Yield
Me	H	50-67
Ph	H	87
Me	Me	65
$\text{-(CH}_2\text{)}_6\text{-}$		84

I.A.7.a.3-9   D. Seebach and N. Meyer, <u>Angew. Chem.
Int. Ed. Engl.</u>, <u>15</u>, 438 (1976).

$$n\text{-Bu}_3\text{SnCH}_2\text{OH} \xrightarrow[\text{2) } R^1COR^2]{\text{1) 2 eq. } n\text{-BuLi, } n\text{-}C_6H_{14}}$$

$$R^1R^2C(OH)CH_2OH$$

$R^1 = Ph, R^2 = H, 62\%$
$R^1, R^2 = \text{-(CH}_2\text{)}_5\text{-}, 46\%$

I.A.7.a.3-10 W. C. Still and T. L. Macdonald, J. Org. Chem., 41, 3620 (1976).

$$CH_2=CHCH_2OSiMe_3 \xrightarrow[\text{2) } R^1COR^2]{\text{1) sec-BuLi, THF, HMPA, }-40°}$$

$$R^1R^2C(OH)CH(OSiMe_3)CH=CH_2$$

$R^1, R^2 = -(CH_2)_5-$, 76%

$R^1 = Pr$, $R^2 = H$, 80%

I.A.7.a.3-11 W. Oppolzer and R. L. Snowden, Tetrahedron Lett., 1976, 4187.

$$(CH_2=CH)_2CHOSiEt_3 \xrightarrow[\text{2) } R^1COR_2]{\text{1) sec-BuLi, THF, }-78°}$$

$$CH_2=CHC(OSiEt_3)=CHCH_2CR^1R^2OH$$

$R^1 = Me$, $R^2 = H$, 92%

$R^1, R^2 = -(CH_2)_5-$, 73%

I.A.7.a.3-12   H. Neumann and D. Seebach, Tetrahedron Lett., 1976, 4839.

$$\underset{H}{\overset{R}{>}}C=C\underset{Br}{\overset{H}{<}} \quad \xrightarrow[\text{2) PhCHO}]{\text{1) 2 eq. t-BuLi, -120°, Trepp mixture}} \quad \underset{H}{\overset{R}{>}}C=C\underset{CH(OH)Ph}{\overset{H}{<}}$$

R = H, 76%
R = Me, 72%
R = Ph, 73%

I.A.7.a.3-13   J. E. Stemke, A. R. Chamberlin, and F. T. Bond, Tetrahedron Lett., 1976, 2947.

$$R^1\overset{NNHT_s}{\underset{\|}{C}}CH_2R^2 \quad \xrightarrow[\text{2) }R^3COR^4]{\text{1) 3 eq. n-BuLi, TMEDA}} \quad R^3R^4C(OH)CR^1=CHR^2$$

R^1	R^2	R^3	R^4	% Yield
C_6H_{13}	H	Ph	H	78
C_6H_{13}	H	$-(CH_2)_3CH=CH-$		67
$-CHMe(CH_2)_3$		$-(CH_2)_5$		48

α,β-Unsaturated carboxylic acids can be synthesized by reaction with CO_2.

I.A.7.a.3-14   C. N. Skold, Synth. Commun., **6**, 119 (1976).

$$\text{MeOCH=C(Br)OMe} \xrightarrow[\substack{2)\ R^1COR^2 \\ 3)\ H_3O^+}]{1)\ BuLi,\ Et_2O,\ -70°}$$

$$R^1R^2C=C(OMe)CHO$$

R^1	R^2	% Yield
Ph	H	76
Ph	Me	51
$-(CH_2)_5-$		68

I.A.7.a.3-15   I. Vlattas, L. D. Vecchia, and A. O. Lee, J. Am. Chem. Soc., **98**, 2008 (1976).

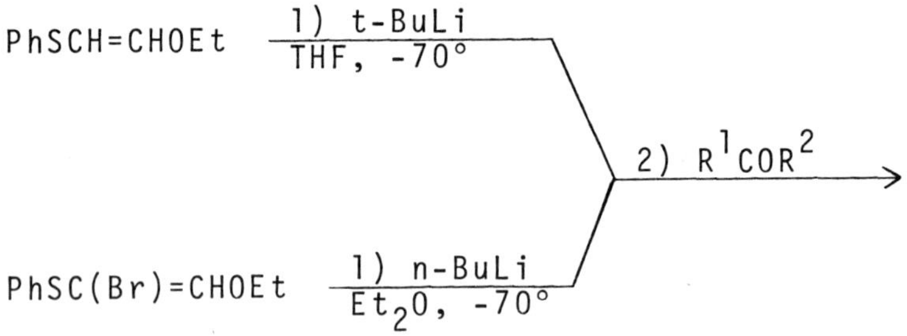

$$\text{PhSCH=CHOEt} \xrightarrow{\substack{1)\ t\text{-BuLi} \\ THF,\ -70°}}$$

$$\xrightarrow{2)\ R^1COR^2}$$

$$\text{PhSC(Br)=CHOEt} \xrightarrow{\substack{1)\ n\text{-BuLi} \\ Et_2O,\ -70°}}$$

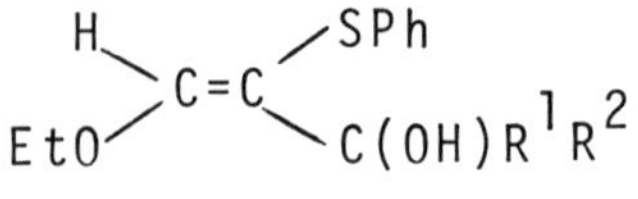

$$\underset{EtO}{\overset{H}{>}}C=C\underset{C(OH)R^1R^2}{\overset{SPh}{<}}$$

R^1 = Ph, R^2 = H, 75%

$R^1, R^2 = -(CH_2)_4-$, 78%

I.A.7.a.3-16 W. E. Parham and L. D. Jones, *J. Org. Chem.*, **41**, 2704 (1976).

[structure: 4-bromo-benzoate t-Bu ester] CO_2t-Bu / Br

$$\xrightarrow{\begin{array}{c}1) \text{ n-BuLi, THF, } -100°\\ 2) R^1COR^2\end{array}}$$

$R^1R^2C(OH)$—[C$_6$H$_4$]—CO_2t-Bu

$R^1 = R^2 = Ph$, 76%

$R^1, R^2 = +(CH_2)_5$, 75%

I.A.7.a.3-17 M. J. Manning, P. W. Raynolds, and J. S. Swenton, *J. Am. Chem. Soc.*, **98**, 5008 (1976).

[structure: 1-bromo-3,3,6,6-tetramethoxy-1,4-cyclohexadiene]

$$\xrightarrow{\begin{array}{c}1) \text{ n-BuLi, THF, } -70°\\ 2) R^1COR^2\end{array}}$$

[structure: 1-X-3,3,6,6-tetramethoxy-1,4-cyclohexadiene]

$R^1, R^2 = +(CH_2)_5$, $X = C(OH)R^1R^2$, 81%

$R^1 = Ph$, $R^2 = OMe$, $X = PhCO$, 78%

$R^1 = Ph$, $R^2 = Cl$, $X = PhCO$, 67%

I.A.7.a.3-18 K. Okuhara, J. Org. Chem., **41**, 1487 (1976).

$$\text{ArLi} \xrightarrow[\substack{2) \text{ n-BuLi, } -50°\\ 3) \text{ CO}_2}]{1) \text{ CF}_2=\text{CCl}_2} \text{ArC}\equiv\text{CCO}_2\text{H}$$

72-84%

I.A.7.a.3-19 P. C. Traas, H. Boelens, and H. J. Takken, Tetrahedron Lett., **1976**, 2287.

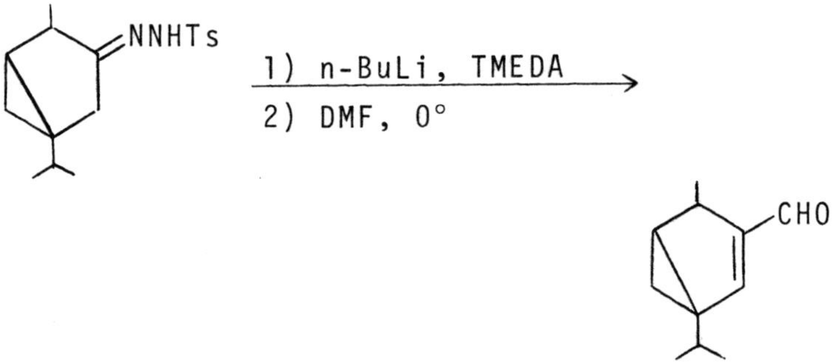

1) n-BuLi, TMEDA
2) DMF, 0°

60%

I.A.7.a.3-20 J. J. Eisch and J. E. Galle, J. Am. Chem. Soc., **98**, 4646 (1976).

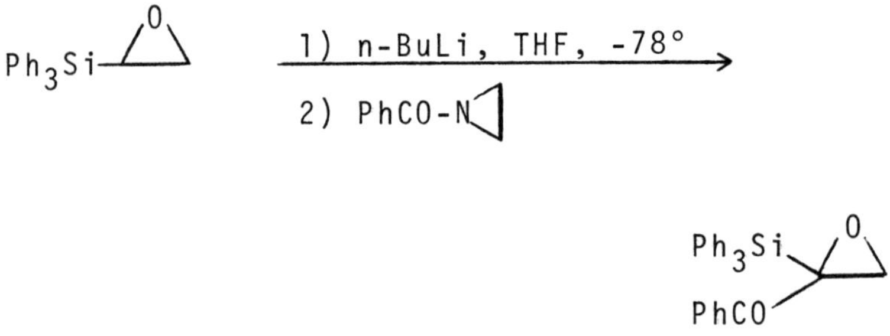

$$\text{Ph}_3\text{Si}\text{-}\triangle^O \xrightarrow[\substack{2) \text{ PhCO-N}\triangleleft}]{1) \text{ n-BuLi, THF, } -78°} \begin{array}{c}\text{Ph}_3\text{Si}\\ \text{PhCO}\end{array}\triangle^O$$

61%

I.A.7.a.3-21 P. Savignac and F. Mathey, Tetrahedron Lett., 1976, 2829.

$$(R^1O)_2P(Y)Me \xrightarrow[\substack{\text{2) CuI, } -35°\\ \text{3) } R^2COCl}]{\text{1) n-BuLi, } -70°} (R^1O)_2P(Y)CH_2COR^2$$

Y	R^1	R^2	% Yield
O	Me	Me	80
O	Et	t-Bu	95
O	Et	CMe=CH_2	100
S	Me	Me	93

I.A.7.a.3-22 D. E. Bergbreiter and J. M. Killough, J. Org. Chem., 41, 2750 (1976).

$$R^1MgX \xrightarrow[\text{2) } R^2COCl]{\text{1) MeCu}} R^1COR^2$$

R^1	R^2	% Yield
n-Bu	Ph	85
PrCHMe	Me	90
t-Bu	Ph	93
Ph	t-Bu	100

I.A.7.a.3-23 C. Wiaux-Zamar, J.-P. Dejonghe, L. Ghosez, J. F. Normant, and J. Villieras, Angew. Chem. Int. Ed. Engl., 15, 371 (1976).

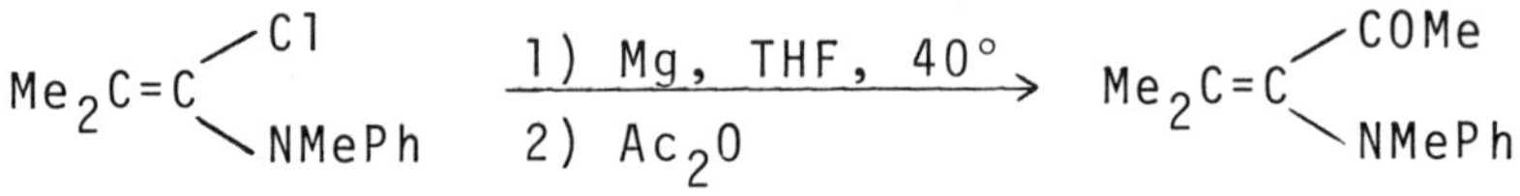

$$Me_2C=C\begin{smallmatrix}Cl\\NMePh\end{smallmatrix} \xrightarrow[2)\ Ac_2O]{1)\ Mg,\ THF,\ 40°} Me_2C=C\begin{smallmatrix}COMe\\NMePh\end{smallmatrix}$$

59%

I.A.7.a.3-24 H. Viola and R. Mayer, Z. Chem., 16, 335 (1976).

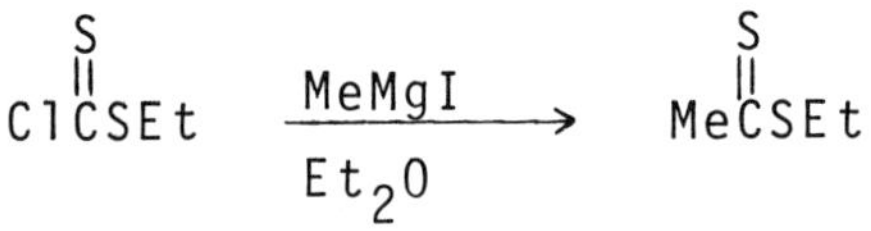

$$ClC(S)SEt \xrightarrow{MeMgI,\ Et_2O} MeC(S)SEt$$

40%

I.A.7.a.3-25 G. Zweifel and R. A. Lynd, Synthesis, 1976, 625.

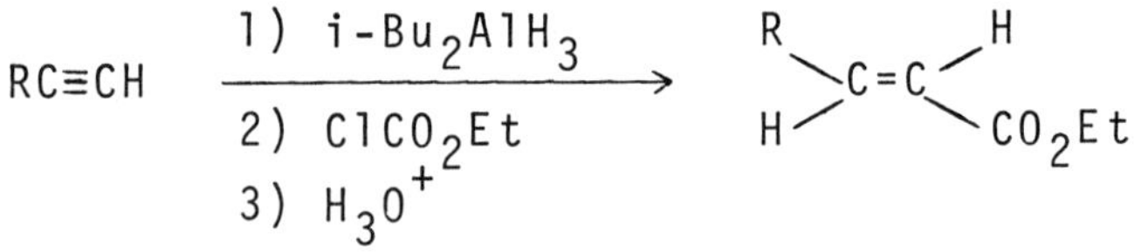

$$RC\equiv CH \xrightarrow[\substack{2)\ ClCO_2Et\\3)\ H_3O^+}]{1)\ i\text{-}Bu_2AlH_3} \begin{smallmatrix}R\\H\end{smallmatrix}C=C\begin{smallmatrix}H\\CO_2Et\end{smallmatrix}$$

R^1	% Yield
n-Bu	64
i-Pr	64
t-Bu	72

I.A.7.a.3-26   G. Cahiez, A. Masuda, D. Bernard, and
J. F. Normant, Tetrahedron Lett., <u>1975</u>, 3155.

$$R^1MnI \xrightarrow[Et_2O]{R^2COCl} R^1COR^2$$

$$R^1, R^2 = alkyl$$

60-91%

I.A.7.a.3-27   K. Saigo, K. Kawata, and T. Mukaiyama,
Chem. Lett., <u>1976</u>, 771.

$$R^1CHO \xrightarrow[2)\ t-BuOLi,\ NCS]{1)\ R^2MgBr} R^1COR^2$$

R^1	R^2	% Yield
Ph	n-Bu	82
C_6H_{13}	Ph	78
PhCH=CH	n-Bu	67

I.A.7.b.1-1   R. E. Gawley, Synthesis, <u>1976</u>, 777.

Review: The Robinson Annelation and Related
Reactions.

I.A.7.b.1-2   H. Stetter, Angew. Chem. Int. Ed. Engl., 15, 639 (1976).

Review: Catalyzed Addition of Aldehydes to Activated Double Bonds — A New Synthetic Approach.

I.A.7.b.1-3   K. Saigo, M. Osaki, and T. Mukaiyama, Chem. Lett., 1976, 163.

$$R^1COCH=CR^2R^3 + R^4R^5C=C(OMe)OSiMe_3 \xrightarrow[CH_2Cl_2,\ -78°]{1)\ TiCl_4}$$

$$2)\ 5\%\ K_2CO_3$$

$$R^1COCH_2CR^2R^3CR^4R^5CO_2Me$$

R^1	R^2	R^3	R^4	R^5	% Yield
Ph	Ph	H	Me	Me	100
Me	Me	Me	PhCH$_2$	H	79

I.A.7.b.1-4   K. Narasaka, K. Soai, Y. Aikawa, and
T. Mukaiyama, Bull. Chem. Soc. Japan, **49**, 779 (1976).

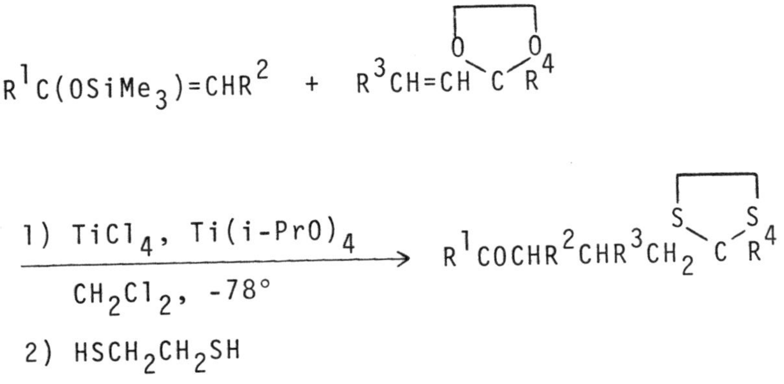

$$R^1C(OSiMe_3)=CHR^2 \;+\; R^3CH=CH\overset{\displaystyle O\frown O}{C}R^4$$

$$\xrightarrow[\substack{CH_2Cl_2,\; -78° \\ 2)\; HSCH_2CH_2SH}]{1)\; TiCl_4,\; Ti(i\text{-}PrO)_4} R^1COCHR^2CHR^3CH_2\overset{\displaystyle S\frown S}{C}R^4$$

R^1	R^2	R^3	R^4	% Yield
Ph	H	H	Me	78
Me	Ph	H	Me	61
Me	Ph	$-(CH_2)_3-$		58

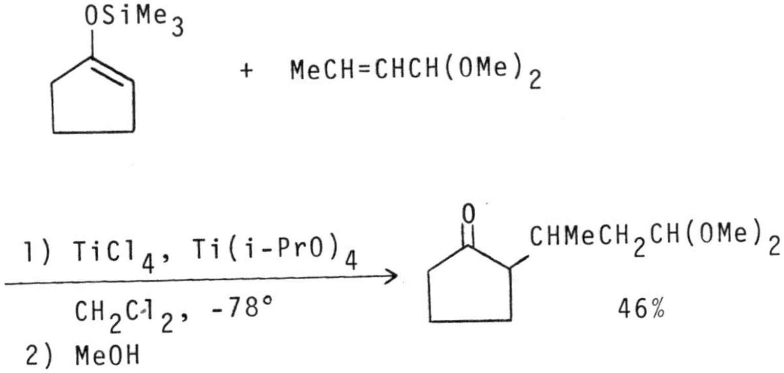

$$\text{(1-trimethylsilyloxycyclopentene)} \;+\; MeCH=CHCH(OMe)_2$$

$$\xrightarrow[\substack{CH_2Cl_2,\; -78° \\ 2)\; MeOH}]{1)\; TiCl_4,\; Ti(i\text{-}PrO)_4} \text{2-(CHMeCH}_2\text{CH(OMe)}_2\text{)cyclopentanone} \quad 46\%$$

I.A.7.b.1-5 M. Miyashita, T. Yanami, and A. Yoshi-koshi, J. Am. Chem. Soc., 98, 4679 (1976).

Reagents/conditions:
1) Lewis acid, CH_2Cl_2
2) H_2O, reflux

Starting material: cyclic enol silyl ether with Me and $OSiMe_3$ groups, $(CH_2)_n$

Plus:

$R^1CH_2C(NO_2)=CHR^2$

Product: bicyclic diketone with R^2 and CH_2R^1 substituents, $(CH_2)_n$

n	R^1	R^2	Lewis Acid	% Yield
2	H	H	$SnCl_4$	85
2	Me	H	$TiCl_4$	76
2	H	Me	$AlCl_3$	63
1	H	H	$SnCl_4$	41

I.A.7.b.1-6 H. Hagiwada, T. Kodama, H. Kosugi, and H. Uda, J. Chem. Soc., Chem. Commun., 1976, 413.

Starting cyclohexenone with R^1, R^2, R^3 substituents.

1) LDA, HMPA, THF
2) $MeCH=CBrCO_2Me$

Product: tricyclic ketone with R^1, R^2, R^3, Me, and CO_2Me substituents, 20–55%

CARBON—CARBON BONDS FORMING REACTIONS

I.A.7.b.2-1 W. G. Kofron and J. Mathew, J. Org. Chem., 41, 114 (1976).

$$R^1R^2C=CR^3CO_2Me \xrightarrow[Et_2O]{NaCHPh_2} Ph_2CHCR^1R^2CHR^3CO_2Me$$

R^1	R^2	R^3	% Yield
H	H	H	40
H	H	Me	40
Me	H	H	87
Me	Me	H	90
Ph	H	Me	76

I.A.7.b.2-2 G. Stork, A.Y.W. Leong, and A. M. Touzin, J. Org. Chem., 41, 3491 (1976).

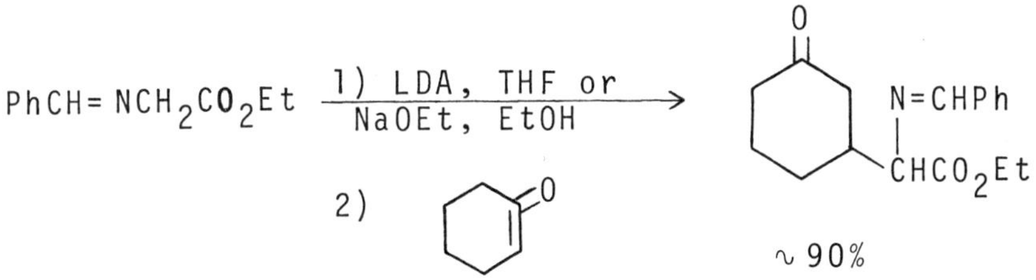

PhCH=NCH$_2$CO$_2$Et $\xrightarrow[\text{NaOEt, EtOH}]{\text{1) LDA, THF or}}$

2) [cyclohexenone]

→ [3-substituted cyclohexanone with N=CHPh and CHCO$_2$Et side chain]

∼ 90%

I.A.7.b.2-3 I.G.C. Coutts and M. Hamblin, J. Chem. Soc., Chem. Commun., 1976, 58.

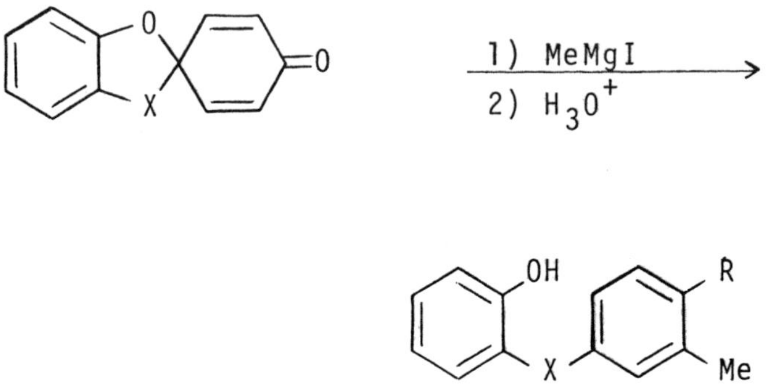

[reactant spiro compound] → 1) MeMgI 2) H_3O^+ →

[product: 2-(substituted)phenol with OH, X, R, Me groups]

R = OH and Me
R = O and NTs

Unusual 1,3-addition of Grignard to the enone system.

I.A.7.b.2-4 E. Elkik and M. Imbeaux-Oudotte, Bull. Soc. Chim. Fr., 1976, 439.

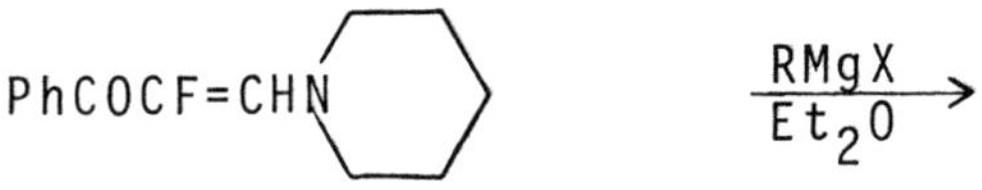

PhCOCF=CHN⟨piperidine⟩ $\xrightarrow{\text{RMgX}}{\text{Et}_2\text{O}}$ PhCOCF=CHR

R = Ph, 82%
R = Et, 72%

I.A.7.b.2-5 Y. Ittah and I. Shahak, Synthesis, 1976, 320.

$(MeS)_2C=C(CO_2Et)_2 \xrightarrow[Et_2O]{RMgX, CuBr (cat.)}$

$R_3CCH(CO_2Et)_2$

A

or

$R_2C=C(CO_2Et)_2$

B

R	% Yield
Me	78-A
Et	72-A
Ph	60-A
i-Pr	60-B
PhCH$_2$	42-B

I.A.7.b.2-6 R. D. Clark and C. H. Heathcock, J. Org. Chem., 41, 636 (1976).

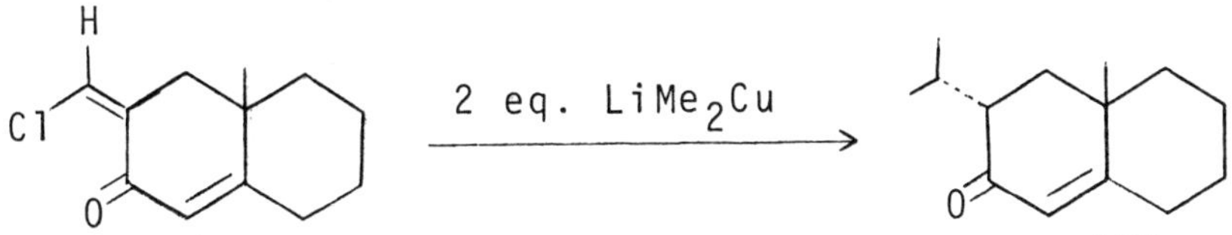

Other reactions of β-chloro-α,β-unsaturated ketones are described.

I.A.7.b.2-7 A. Alexakis, A. Commercon, J. Villiéras and J. F. Normant, Tetrahedron Lett., 1976, 2313.

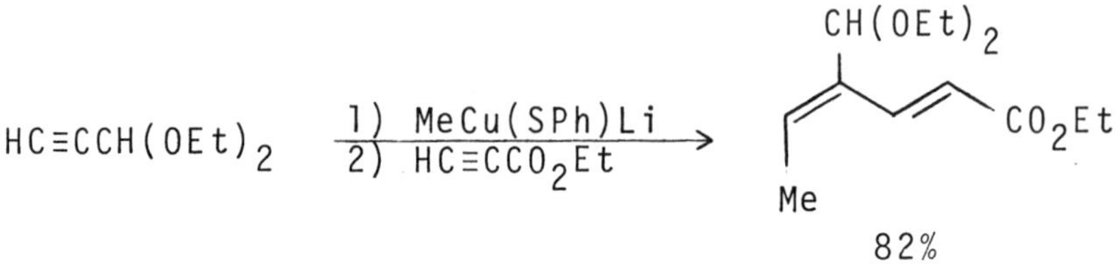

$HC\equiv CCH(OEt)_2 \xrightarrow{\begin{array}{l}1)\ MeCu(SPh)Li\\ 2)\ HC\equiv CCO_2Et\end{array}}$ (diene product with $CH(OEt)_2$, Me, CO_2Et) 82%

I.A.7.b.2-8 A. Alexakis, J. Normant, and J. Villiéras, Tetrahedron Lett., 1976, 3461.

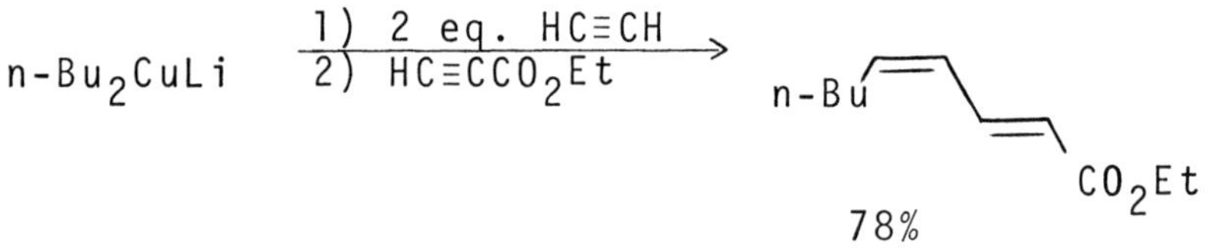

$n\text{-}Bu_2CuLi \xrightarrow{\begin{array}{l}1)\ 2\ eq.\ HC\equiv CH\\ 2)\ HC\equiv CCO_2Et\end{array}}$ n-Bu–diene–CO_2Et 78%

I.A.7.b.2-9 N. Miyaura, M. Itoh, and A. Suzuki, Tetrahedron Lett., 1976, 255.

$CH_2=CHCN \xrightarrow{[R_3BMe]Cu} RCH_2CH_2CN$

R = n-Pr, 84%
R = i-Bu, 88%

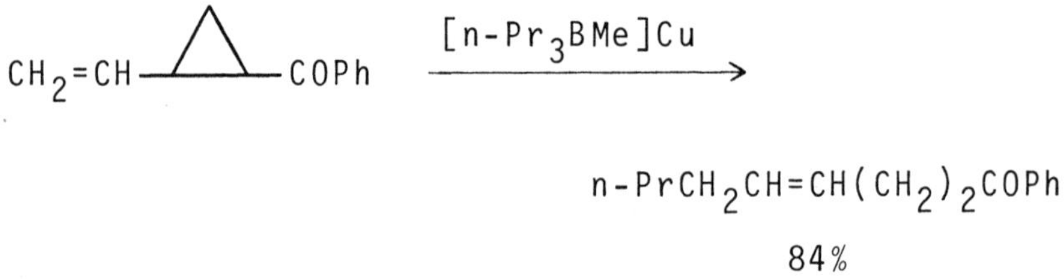

$CH_2=CH\text{-}\triangle\text{-}COPh \xrightarrow{[n\text{-}Pr_3BMe]Cu}$

$n\text{-}PrCH_2CH=CH(CH_2)_2COPh$

84%

CARBON–CARBON BONDS FORMING REACTIONS

I.A.7.b.3-1   D. R. Hicks, R. C. Anderson, and B. Fraser-Reid, Synth. Commun., 6, 417 (1976).

$$R^1R^2C=CHCOR^3 \xrightarrow[h\nu]{R^4CHO} R^4COCR^1R^2CH_2COR^3$$

R^1	R^2	R^3	R^4	% Yield
Me	Me	Me	Me	31
H		$-(CH_2)_3-$	n-Pr	42
H		$CH(OEt)OCH(CH_2OTr)$	Ph	58

I.A.7.b.3-2   H. Stetter and H. Kuhlmann, Chem. Ber., 109, 2890 and 3426 (1976); see also: H. Stetter, M. Schreckenberg, and K. Wiemann, *ibid*, 541.

$$R^1CHO + R^2CH=CHX \xrightarrow[\text{reflux}]{\substack{HOCH_2CH_2\diagdown\quad Me \\ \overset{S}{\underset{}{\bigotimes}}\!\!=\!\!N\text{-}R\;\;X^- \\ Et_3N,\;EtOH}} R^1COCHR^2CH_2X$$

R^1	R^2	X	% Yield
Me	H	CN	30
Pr	Ph	COPh	70
2-Furyl	H	CO_2Et	31
2-Furyl	Ph	COMe	80

I.A.7.b.3-3 P. Knittel and J. Warkentin, Can. J. Chem., 54, 1341 (1976).

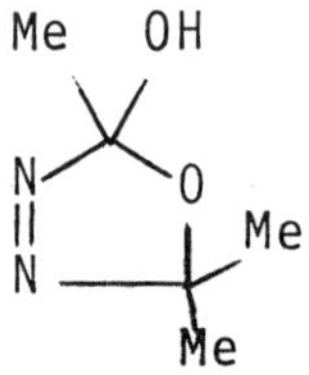

[structure: 2,2-dimethyl-5-methyl-5-hydroxy-Δ³-1,3,4-oxadiazoline] + $RCH=CHX$ $\xrightarrow{\text{PhH}, 60°}$

$$AcOCMe_2CHRCH_2X$$

R = H, Me or CO_2Me

R = CN, CHO, COME or CO_2Me

I.A.7.b.3-4 D.W.K. Yeung and J. Warkentin, Can. J. Chem., 54, 1345 (1976).

$Ph_2C(OH)N=Nt-Bu$ + $RCH=CHX$ $\xrightarrow{\text{PhH}, 50°}$ $t\text{-}BuCHRCH_2X$

R	X	% Yield
H	CN	63
H	COMe	31
Me	CHO	32

CARBON—CARBON BONDS FORMING REACTIONS

I.A.7.b.3-5 K. Yamamura, J. Chem. Soc., Chem. Commun., 1976, 438.

$$PhCH=C(COPh)_2 \xrightarrow[Pd(OAc)_2]{PhH-HOAc} PhCHCH(COPh)_2$$

52%

I.A.7.b.3-6 D. A. White, J. Chem. Soc., Perkin Trans. I, 1976, 1926.

$$R^1CH=CR^2(X) \xrightarrow[2)\ MeY]{1)\ Et_4\overset{+}{N}\ NC\cdot CO_2^-} NCCHR^1CR^2(X)CO_2Me$$

R^1	R^2	X	Y	% Yield
H	H	CO_2Me	I	77
H	Me	CO_2Me	I	82
Me	H	CO_2Me	I	68
H	H	CN	SO_4Me	78

I.A.8-1 W. E. Parham and Y. A. Sayed, <u>Synthesis</u>, <u>1976</u>, 116.

[o-HOCH$_2$-C$_6$H$_4$-CR^1(OH)CH$_2$R^2] $\xrightarrow[\Delta]{48\% \text{ HBr}}$ [indene with R^1 at 1-position, R^2 at 2-position]

R^1	R^2	% Yield
Me	Me	84
Et	Me	79
-(CH$_2$)$_4$-		81
Ph	H	52

Not useful for the preparation of 2- or 3-monoalkyl-indenes.

I.A.8-2 I. Fleming, A. Pearce and R. L. Snowden, <u>J. Chem. Soc., Chem. Commun.</u>, <u>1976</u>, 182.

PhCH(OH)CEtCH$_2$SiMe$_3$ $\xrightarrow[\text{PhH}]{p\text{-TsOH}}$ Ph$_2$P(O)CHPhCEt=CH$_2$
         |
      P(O)Ph$_2$

(MeO)$_2$CH(CH$_2$)$_3$C(CH$_2$SiMe$_3$)=CH$_2$ $\xrightarrow[\text{CCl}_4]{\text{SnCl}_4}$ [cyclohexane with =CH$_2$ and OMe substituents]

Use of Me$_3$Si- group to control products in carbonium ion reactions is demonstrated.

I.A.8-3   I. A. Savost'yanova, M. S. Burmistrova, and
Y. I. Baukov, J. Gen. Chem. USSR, 45, 2308 (1975).

$$Me_3SiC\equiv COEt + RCHO \xrightarrow[Et_2O]{Et_2O \cdot BF_3} RCH=C(SiMe_3)CO_2Et$$

$$R = CCl_3, 76\%$$
$$R = Ph, 72\%$$

I.A.8-4   O. V. Lubinskaya, A. S. Shashkov, V. A.
Chertkov, and W. A. Smit, Synthesis, 1976, 742.

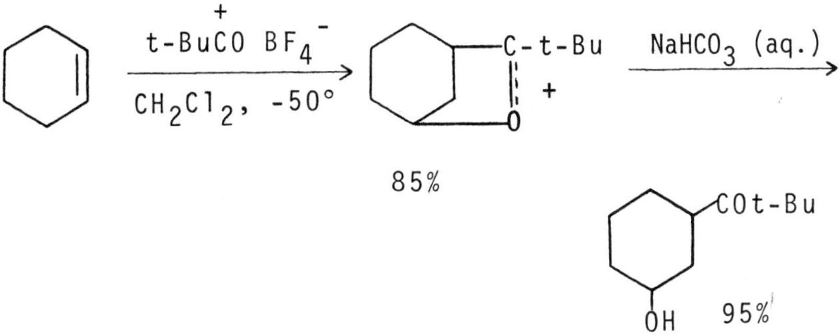

85%

I.A.8-5   M. Grignon-Dubois, J. Dunogues, and R.
Calas, Synthesis, 1976, 737.

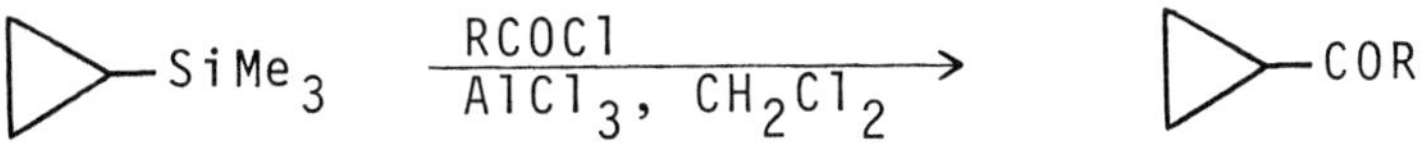

R	% Yield
Me	60
Ph	50
$Me_2C=CH$	75

I.A.8-6   M. Hojo, R. Masuda, Y. Kokuryo, H. Shioda, and S. Matsuo, Chem. Lett., 1976, 499.

$$RYCH=CH_2 \xrightarrow{(CX_3CO)_2O} RYCH=CHCOCX_3$$

RY	X	% Yield
EtO	F	100
EtO	Cl	71
p-MeC$_6$H$_4$SO$_2$NPh	F	92
MeSO$_2$NPh	Cl	94

I.A.8-7   M. Hojo, R. Masuda, and Y. Kamitori, Tetrahedron Lett., 1976, 1009.

$$RSCY=CH_2 \xrightarrow[py., CHCl_3]{(CX_3CO)_2O} RSCY=CHCOCX_3$$

R	Y	X	% Yield
p-MeC$_6$H$_4$	p-MeC$_6$H$_4$S	F	93
p-MeC$_6$H$_4$	p-MeC$_6$H$_4$S	Cl	96
p-ClC$_6$H$_4$	H	F	97
Me	H	F	100
Ph	Ph	F	79

I.A.8-8 V. I. Gorbutenko, and N. V. Mel'nichenko,
J. Org. Chem. USSR, 11, 2261 (1975).

$$ROCH=CH_2 \xrightarrow[\text{2) Et}_3\text{N, Et}_2\text{O, -20°}]{\text{1) ClC(O)NCO, Et}_2\text{O}} ROCH=CHC(O)NCO$$

36-43%

I.A.8-9 L. Capuano, T. Tammer, and R. Zander,
Chem. Ber., 109, 3497 (1976).

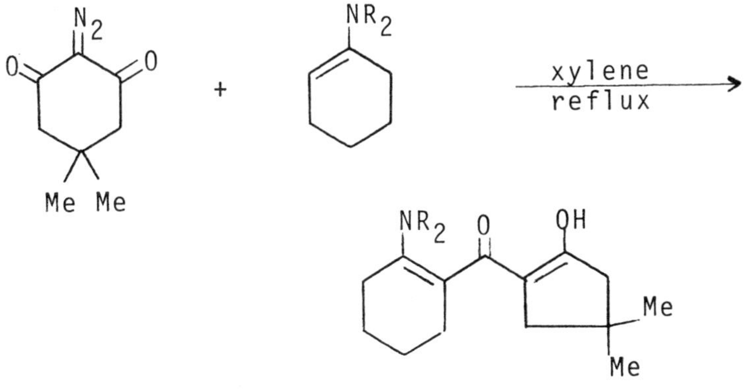

[Reaction: 2-diazo-5,5-dimethyl-1,3-cyclohexanedione + 1-(NR$_2$)cyclohexene, xylene reflux, yields product shown]

$R_2 = -(CH_2)_2O(CH_2)_2-$, 62%

I.A.8-10 B. Sket and M. Zupan, <u>Synth. Commun.</u>, <u>6</u>, 309 (1976).

$$\underset{R^1}{\overset{Ph}{>}}C=C\underset{R^2}{\overset{H}{<}} \quad \xrightarrow[\text{polymer supported iodobenzene}]{h\nu,\ CH_2Cl_2} \quad \underset{R^1}{\overset{Ph}{>}}C=C\underset{R^2}{\overset{R^2}{<}}\underset{R^2}{>}C=C\underset{Ph}{\overset{R^1}{<}}$$

R^1	R^2	% Yield
Ph	H	80
Ph	Me	65
H	Me	50

I.A.8-11 A. Couture, K. Ho, M. Hoshino, P. de Mayo, R. Suau, and W. R. Ware, <u>J. Am. Chem. Soc.</u>, <u>98</u>, 6218 (1976).

$$Ph\overset{S}{\overset{\|}{C}}CMe_2CH_2CHR^2CH_2R^1 \quad \xrightarrow{h\nu} \quad$$

[cyclopentane ring with Ph, SH, R^1, R^2, Me, Me substituents]

70-90%

I.A.8-12 T. Sato, G. Izumi, and T. Imamura, J. Chem. Soc., Perkin Trans. I, 1976, 788.

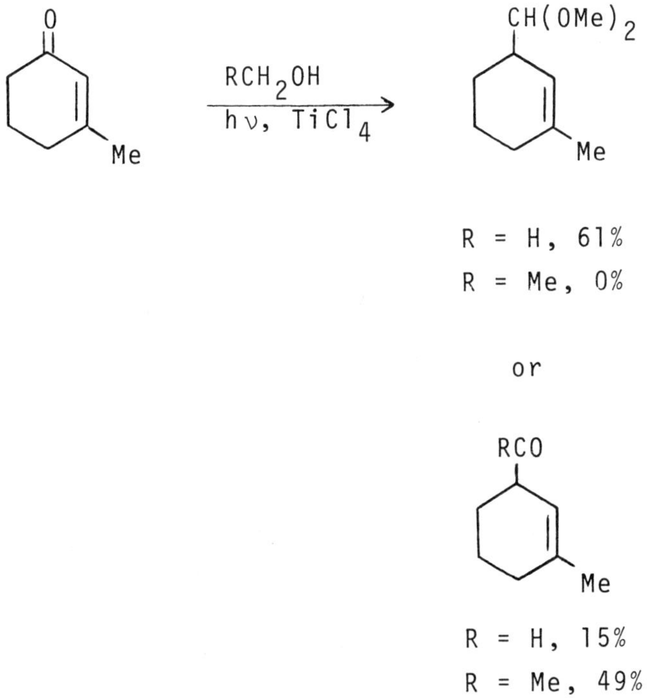

[cyclohexenone with Me] + RCH_2OH →($h\nu$, $TiCl_4$) [cyclohexene with $CH(OMe)_2$ and Me]

R = H, 61%
R = Me, 0%

or

[cyclohexene with RCO and Me]

R = H, 15%
R = Me, 49%

Dihydrofurans are isolated in cases with a cisoid α,β-unsaturated ketone and in some open chain systems.

I.A.8-13 T. Shono, I. Nishiguchi, and H. Omizu, Chem. Lett., 1976, 1233.

$R^2COCH_2CR^1_2CHC\equiv CH$ $\xrightarrow[\text{DMF, Et}_4\text{NOTs}]{\text{electrolysis}}$

[cyclopentane product with CH_2R^2, OH, and two R^1 substituents]

85-95%

I.A.8-14 J. J. Habeeb and D. G. Tuck, J. Chem. Soc., Chem. Commun., 1976, 698.

PhCOBr + EtBr $\xrightarrow[\substack{\text{Cd anode}\\\text{Et}_3\text{NClO}_4\text{, diglyme}}]{\text{electrolysis}}$ PhCOEt

64% (based on Cd consumed.

I.A.8-15 E. Yoshii, T. Koizumi, and T. Kawazoe, Chem. Pharm. Bull., 24, 1957 (1976).

$R^1CH_2CR^2C=C(OMe)OSiEt_3$ $\xrightarrow[\text{PhH or DME}]{\text{PhHgCBrCl}_2}$

$R^1CH_2CR^2=C(Cl)CO_2Me$

R^1	R^2	% Yield
H	H	80
Me	Me	90
Ph	H	52

I.A.8-16 P. Kuhl, M. Mühlstädt, and J. Graefe, Synthesis, 1976, 825.

[cyclohexanone] + CHCl$_3$, 50% NaOH / H$_2$O, TEBAC → [1-chloro-cyclohexane-1-carboxylic acid]

45%

I.A.8-17 G. F. Luteri and W. T. Ford, J. Organomet. Chem., 105, 139 (1976).

$CH_2=CPhCH_2MgOPh$
  +
$PhCH=CHPh$

THF
HMPA or
[2.1.1]cryptate
→ [1,2,4-triphenylcyclopentane]

37-45%

In the absence of HMPA or [2.1.1]cryptate the reaction does not proceed.

I.A.8-18 P. L. Fuchs, J. Org. Chem., 41, 2935 (1976); see also: G. Stork and A. A. Ponaras, ibid, 2937.

[4,4-dimethyl-2-methyl-cyclohexanone tosylhydrazone epoxide]

1) n-BuLi, THF, -78°
2) PhCu, -20° to 0°

→ [product with NNHTs, Ph, OH]

85%

Products can be dehydrated and carbonyl exchanged to give α-aryl-α,β-unsaturated ketones.

I.A.8-19 F. Mathey and P. Savignac, *Synthesis*, **1976**, 766.

$(EtO)_2P(O)Me$ $\xrightarrow{\begin{array}{l}\text{1) n-BuLi, THF, } -65° \\ \text{2) } R^1X, -78° \text{ to } 25° \\ \text{3) n-BuLi, } -60° \\ \text{4) CuI, } -35°, 1 \text{ hr} \\ \text{5) } R^2COCl, Et_2O, -40°\end{array}}$

$$(EtO)_2P(O)CHR^1COR^2$$

R^1X	R^2	% Yield
MeI	Me	87
$CH_2=CHCH_2Br$	Me	75
MeI	i-Bu	70

A "one-pot" synthesis of diethyl 1-alkyl-2-oxoalkanephosphonates.

CARBON–CARBON BONDS FORMING REACTIONS

I.A.8-20 P.-L. Compagnon and B. Grosjean, Synthesis, 1976, 448.

$$PhC\equiv CCu \xrightarrow[Et_2O,\ MeCN]{BrCN} PhC\equiv CCN$$
$$60\%$$

I.A.8-21 N. DeKimpe, R. Verhe, L. DeBuyck, H. Hasma, and N. Schamp, Tetrahedron, 32, 3063 (1976).

$$R^1R^2C(Cl)CH=Nt\text{-}Bu \xrightarrow[reflux]{KCN,\ MeOH}$$

$$R^1R^2C=C(CN)NHt\text{-}Bu$$

R^1	R^2	% Yield
Me	Me	95
Me	Ph	17-36 (depends on conditions)
$-(CH_2)_5-$		94

I.B.1-1 H. J. Bestmann, W. Stransky, and O. Vostrowsky, *Chem. Ber.*, <u>109</u>, 1694 (1976).

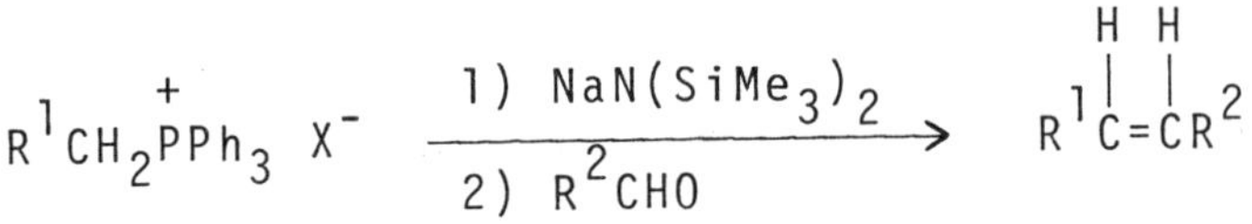

$$R^1CH_2PPh_3^+ \ X^- \xrightarrow[\text{2) } R^2CHO]{\text{1) NaN(SiMe}_3)_2} R^1\overset{H}{\underset{|}{C}}=\overset{H}{\underset{|}{C}}R^2$$

Use of the sodium *bis*(trimethylsilyl)amide base allows preparation of lithium-salt free ylides

---

I.B.1-2 R. Broos and M. Ateunis, *Synth. Commun.*, <u>6</u>, 53 (1976).

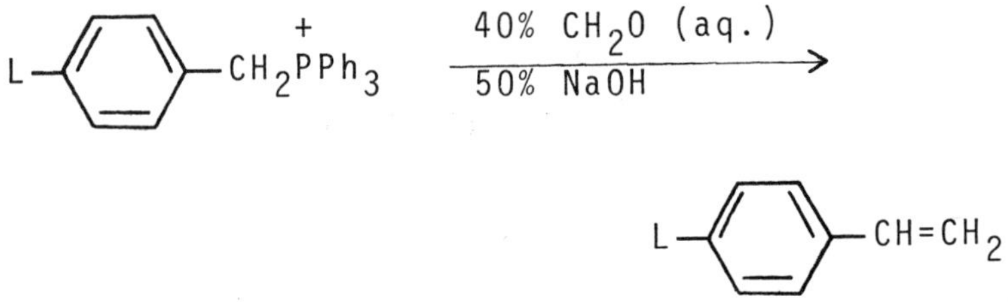

$$L-\text{C}_6\text{H}_4-CH_2PPh_3^+ \xrightarrow[\text{50\% NaOH}]{\text{40\% CH}_2\text{O (aq.)}} L-\text{C}_6\text{H}_4-CH=CH_2$$

L	% Yield
CN	80
Br	82
Me	90
MeO	80

I.B.1-3 V. Subramanyam, E. H. Silver, and A. H. Soloway, *J. Org. Chem.*, **41**, 1272 (1976).

$$Ph_3P=CHR + HCO_2Et \longrightarrow RCH=CHOEt$$

$R = CO_2Et$, 95%
$R = Ph$, 90%
$R = i\text{-}Bu$, 22%

I.B.1-4 L. Kalvoda, *Coll. Czech. Chem. Commun.*, **41**, 2034 (1976).

$$RCOX \xrightarrow[Ph_3P=CHCO_2Et]{HCN,\ CHCl_3} RC(CN)=CHCO_2Et$$

R	X	% Yield
Ph	Cl	81
Me	OAc	58
t-Bu	Cl	83 (only *cis* isomer)

I.B.1-5 P. Bravo, A. Ricca, C. Ticozzi, and O. Vajna de Pava, Gazz. Chim. Ital., 106, 743 (1976).

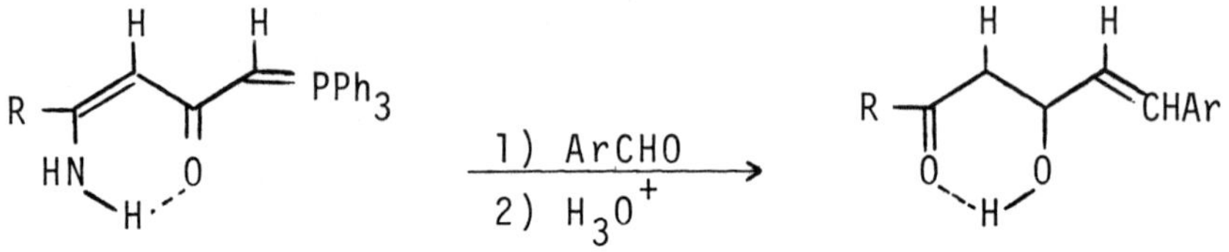

[Reaction scheme: enaminone with PPh$_3$ group reacts with 1) ArCHO, 2) H$_3$O$^+$ to give diketone product with CHAr group]

R	Ar	% Yield
Me	p-NO$_2$C$_6$H$_4$	60
i-Pr	p-MeOC$_6$H$_4$	78
Ph	p-MeOC$_6$H$_4$	75

I.B.1-6 H.-J. Bestmann and R. W. Saalfrank, Chem. Ber., 109, 403 (1976).

$$Ph_3P=CHC(OEt)=CR^1COR^2 \xrightarrow{R^3CHO}$$

$$R^3CH=CHC(OEt)=CR^1COR^2$$

R^1	R^2	R^3	% Yield
H	Ph	Ph	57
H	Ph	Me	71
H	OEt	Ph	66
-(CH$_2$)$_4$-		Ph	51

I.B.1-7 N. Petragnani, R. Rodrigues, and J. V. Comasseto, J. Organomet. Chem., 114, 281 (1976).

$$Ph_3P=CR^1SePh + R^2CHO \xrightarrow{THF} R^2CH=CR^1SePh$$

R^1	R^2	% Yield
H	Ph	98
Me	Ph	99
Me	Et	65

I.B.1-8 G. A. Wheaton and D. J. Burton, Tetrahedron Lett., 1976, 895.

$$Ph_3P=CR^1R^2 \xrightarrow[Et_2O]{HCF_2Cl} F_2C=CR^1R^2$$

R^1, R^2 = H or alkyl

81-100%

The by-product $Ph_3\overset{+}{P}CHR^1R^2 Cl^-$ can be recovered in >90% yield.

I.B.1-9 A. Hercouet and M. LeCorre, <u>Tetrahedron Lett.</u>, <u>1976</u>, 825.

$$Ph_3P=CHCOCH_2P(O)(OEt)_2 \xrightarrow{RCHO}$$

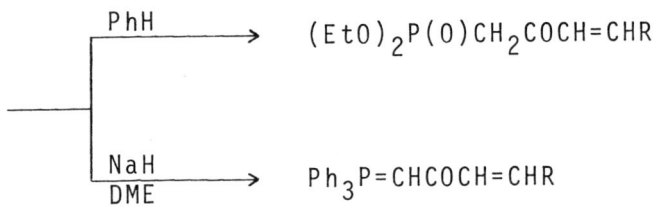

$$\xrightarrow{PhH} (EtO)_2P(O)CH_2COCH=CHR$$

$$\xrightarrow[DME]{NaH} Ph_3P=CHCOCH=CHR$$

I.B.1-10 M. Mikolajczyk, S. Grzejszczak, and A. Zatorski, <u>Tetrahedron Lett.</u>, <u>1976</u>, 2731.

$$(R^1O)_2P(O)CH(SR^2)_2 \xrightarrow{\substack{1)\ n\text{-BuLi, THF} \\ 2)\ R^3COR^4}}$$

$$(R^2S)_2C=CR^3R^4$$

R^3 and R^4 = H, alkyl and aryl

66-96%

I.B.1-11 P. Coutrot, C. Laurenco, J. Petrova, and P. Savignac, Synthesis, 1976, 107.

$$PhSCHClP(O)(OEt)_2 \xrightarrow[\text{2) } R^1COR^2]{\text{1) LiCCl}_3} R^1R^2C=C(Cl)SPh$$

R^1	R^2	% Yield
p-ClC$_6$H$_4$	H	65
PhCH=CH	H	46
-(CH$_2$)$_5$-		60

I.B.1-12 M. Mikolajczy, S. Grzejszczak, W. Midura, and A. Zatorski, Synthesis, 1976, 396.

$$(EtO)_2P(O)CH_2X + RCHO \xrightarrow[CH_2Cl_2]{50\% \text{ NaOH}} XCH=CHR$$

X	R	% Yield
CN	Ph	80
CO$_2$H	Ph	95
MeSO	Ph	51
(EtO)$_2$P(O)	PhCH=CH	70

No phase-transfer catalyst necessary.

I.B.1-13 S. E. Dinizo, R. W. Freerksen, W. E. Pabst, and D. S. Watt, J. Org. Chem., **41**, 2846 (1976).

$$R^1COR^2 \xrightarrow[\text{NaH, THF}]{(EtO)_2P(O)CH(OR^3)CN} R^1R^2C=C(OR^3)CN$$

R^1	R^2	R^3	% Yield
Ph	H	Me	81
Ph	H	t-Bu	94
Ph	Me	t-Bu	93
-(CH$_2$)$_5$-		Me	81

I.B.1-14 B. Lythgoe, T. A. Moran, M.E.N. Nambudiry, and S. Ruston, J. Chem. Soc., Perkin Trans. I, **1976**, 2386; see also: A. H. Davidson and S. Warren, *ibid*, 639.

<chemical reaction>

(Z)-MeCH=CHP(O)Ph$_2$ → 1) n-BuLi, THF, -78°
                            2) cyclohexanone
                            3) rt, 2 hr.

gives cyclohexylidene-CH=CHMe (E)
(∼3% Z-isomer)   48%

(E)-MeCH=CHP(O)Ph$_2$ → same as above →

gives cyclohexylidene-CH=CHMe (Z)
(<2% E-isomer)   72%

I.B.1-15 E. J. Corey, D. Enders and M. G. Bock, Tetrahedron Lett., 1976, 7.

$$Me_3SiCHR^1CH=NNMe_2 \xrightarrow{\begin{array}{l}1)\ LDA,\ -20°\\2)\ R^2COR^3,\ -78°\end{array}}$$

$$R^2R^3C=CR^1CH=NNMe_2$$

85-95%

I.B.1-16 R. R. Schrock, J. Am. Chem. Soc., 98, 5399 (1976).

$$Ta[CH_2t-Bu]_3[CHt-Bu] \xrightarrow{R^1COR^2} R^1R^2C=CHt-Bu$$

R^1	R^2	% Yield
Me	Me	80
Ph	H	90
Me	OEt	60
H	NMe_2	77

Analogously Nb compounds give similar results.

I.B.2.a-1   C. W. Spangler and T. W. Hartford, Synthesis, 1976, 108; see also: ibid, J. Chem. Soc., Perkin Trans. I, 1976, 1792.

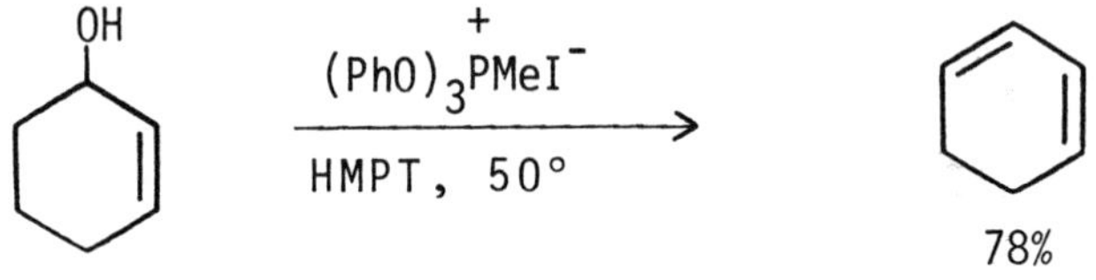

cyclohex-2-en-1-ol $\xrightarrow[\text{HMPT, 50°}]{(PhO)_3PMeI^+\ ^-}$ 1,3-cyclohexadiene

78%

$CH_2=CMeCH(OH)CH_2CH=CH_2 \xrightarrow[\text{HMPT, 50°}]{(PhO)_3PMeI^+\ ^-}$

$CH_2=CMeCH=CHCH=CH_2$

53%

I.B.2.a-2   C. Descoins and D. Samain, Tetrahedron Lett., 1976, 745.

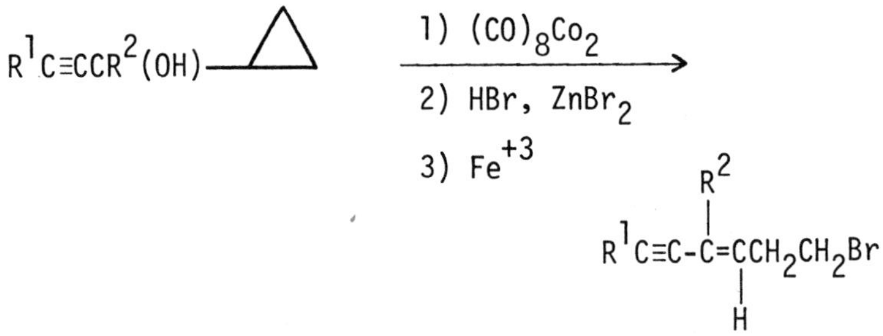

$R^1C\equiv CCR^2(OH)\text{—}\triangle \xrightarrow[\begin{array}{l}\text{2) HBr, ZnBr}_2\\ \text{3) Fe}^{+3}\end{array}]{\text{1) (CO)}_8Co_2}$

$$R^1C\equiv C-\underset{H}{\overset{R^2}{C}}=CCH_2CH_2Br$$

R^1, R^2 = alkyl or H

53-65%

Use of the cobalt complex gives a high stereoselectivity in the formation of the double bond (usually >97% E-isomer).

I.B.2.a-3 K. Utimoto, M. Obayashi, and H. Nozaki, J. Org. Chem., 41, 2940 (1976).

$$R^1COCH(SiMe_3)Am \xrightarrow[\text{or glacial HOAc}]{\begin{array}{c}1)\ R^2Li,\ THF\\ 2)\ t\text{-BuOK, reflux}\end{array}} R^1R^2C=CHAm$$

Stereochemistry of the product is determined by the method of elimination (step 2).

I.B.2.a-4 J. Remion, W. Dumont, and A. Krief, Tetrahedron Lett., 1976, 1385; see also: A. M. Leonard-Coppens and A. Krief, *ibid*, 3227, and J. Remion and A. Krief, *ibid*, 3743.

$$R^1CH(SePh)C(OH)R^2R^3 \xrightarrow[C_5H_{11},\ reflux]{p\text{-TsOH}} R^1CH=CR^2R^3$$

R^1	R^2	R^3	% Yield
H	H	n-$C_{10}H_{21}$	71
Me	H	n-$C_{10}H_{21}$	65
H	\multicolumn{2}{c}{$-(CH_2)_2CH(t\text{-Bu})(CH_2)_2-$}	65

I.B.2.a-5 P. Brownbridge, I. Fleming, A. Pearce, and S. Warren, J. Chem. Soc., Chem. Commun., 1976, 751.

$$\text{PhSCH} \begin{smallmatrix} \diagup CH_2SiMe_3 \\ \diagdown CR^1R^2OH \end{smallmatrix} \xrightarrow[\text{PhH}]{p\text{-TsOH}} \text{PhSCR}^1R^2CH=CH_2$$

R^1 = Me, R^2 = H, 93%

R^1 = R^2 = Me, 96%

I.B.2.a-6 P. Entmayr and G. Köbrich, Chem. Ber., 109, 2175 (1976).

$$R^1COR^2 \xrightarrow[\text{2) Me}_3\text{SiCl}]{\text{1) LiCHX}_2} \underset{A}{R^1R^2C(OSiMe_3)CHX_2} \xrightarrow{R^3Li} \underset{B}{R^1R^2C=CX_2}$$

				% Yield	
X	R^1	R^2	R^3Li	A	B
Cl	Ph	Me	n-BuLi	64	51
Cl	▷	Me	n-BuLi	48	48
Cl	―(CH$_2$)$_5$―		n-BuLi	75	46
Br	Ph	Me	LDA	64	59
Br	―(CH$_2$)$_5$―		LDA	46	61

$$R^1R^2C(OSiMe_3)CHBr_2 \xrightarrow{n\text{-BuLi}} R^1R^2C=CHBr$$

52-93%

I.B.2.a-7   K. Nyberg, Synthesis, 1976, 545.

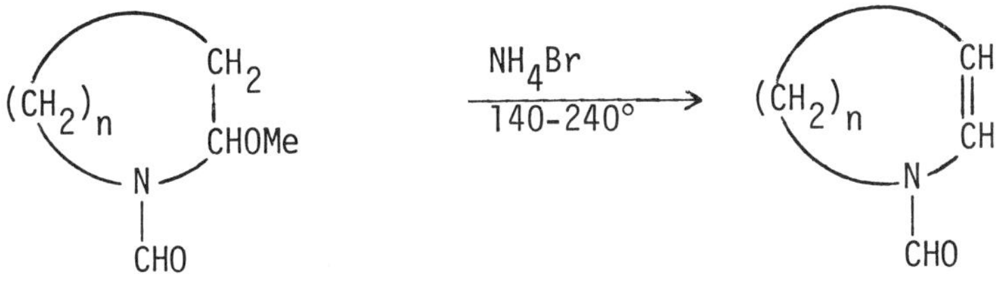

$(CH_2)_n$ ring with CH_2-N(CHO)-CHOMe → $(CH_2)_n$ ring with CH=CH-N(CHO)

Reagents: NH_4Br, 140–240°

n = 2-4, 76-89%

I.B.2.a-8   G. H. Posner and G. M. Gurria, J. Org. Chem., 41, 578 (1976).

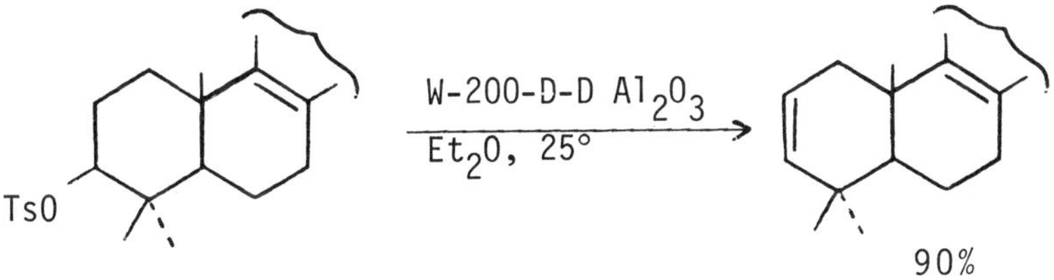

Reagents: W-200-D-D Al_2O_3, Et_2O, 25°

90%

Use of a vacuum-dehydrated alumina enhanced its activity in elimination reactions.

I.B.2.a-9   J. H. Babler and A. E. Moormann, J. Org. Chem., 41, 1477 (1976).

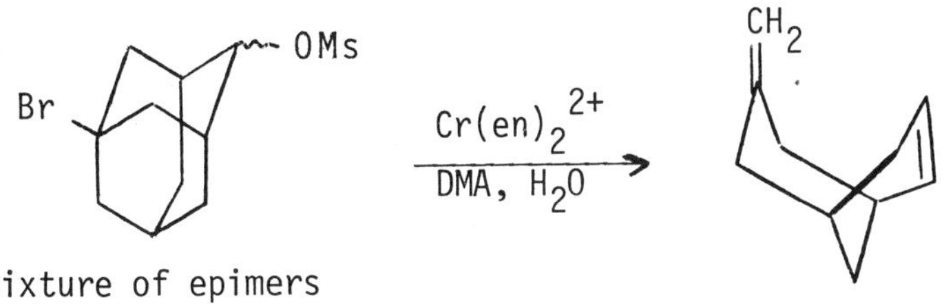

Reagents: $Cr(en)_2^{2+}$, DMA, H_2O

mixture of epimers

40%

I.B.2.b-1 R. N. Warrener, G. J. Collin, G. I. Hutchison, and M. N. Paddon-Row, J. Chem. Soc., Chem. Commun., 1976, 373.

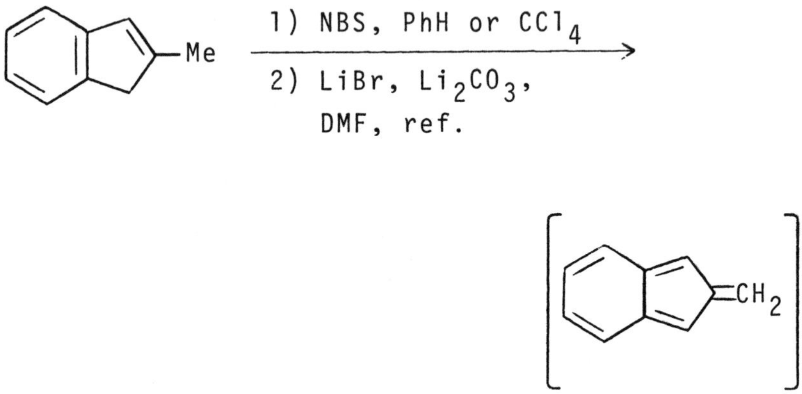

[2-methylindene] $\xrightarrow{\text{1) NBS, PhH or CCl}_4 \quad \text{2) LiBr, Li}_2\text{CO}_3, \text{DMF, ref.}}$ [indene=CH$_2$]

Trapped as N-methyl maleimide adduct.

I.B.2.b-2 P. E. Sonnet and J. E. Oliver, J. Org. Chem., 41, 3284 (1976).

$$R^1CHXCH(Cl)R^2 \xrightarrow{\text{NaI}/\text{DMF}} R^1CH=CHR^2$$

X = Br or Cl

Highly stereoselective *cis*-eliminations were observed.

I.B.2.b-3 T. Kempe, T. Norin, and R. Caputo, Acta Chem. Scand., B30, 366 (1976).

$$RCHBrCHBrR \xrightarrow[\text{DMF, 140°}]{Na_2S_2O_4} R^1CH=CHR^2$$

$R^1 = R^2 = Ph$, 90%

$R^1, R^2 = -(CH_2)_4-$, 60%

I.B.2.b-4 A. Kasal, Coll. Czech. Chem. Commun., 41, 2040 (1976).

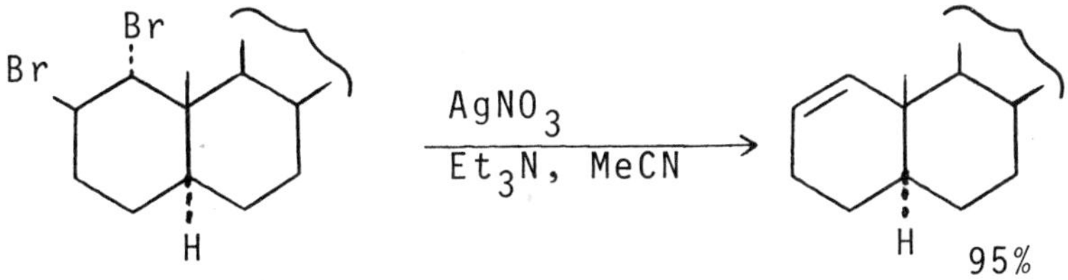

$$\xrightarrow[\text{Et}_3\text{N, MeCN}]{AgNO_3}$$

95%

A series of steroidal vicinal dibromides were studied.

I.B.2.b-5 G. A. Olah and G. K. SuryaPrakash, Synthesis, 1976, 607.

$$R^1CHXCHXR^2 \xrightarrow[\text{THF}]{Ti(II)} R^1CH=CHR^2$$

X	R^1	R^2	% Yield
Br	Ph	Ph	88
Br	$n\text{-}C_6H_{13}$	H	84
Cl	$-(CH_2)_4-$		72
Br	$-(CH_2)_6-$		72

I.B.2.b-6 P. Dowd and L. K. Marwaha, J. Org. Chem., 41, 4035 (1976).

$BrCH_2\overset{\overset{\displaystyle CO_2H}{|}}{C}BrCH_2CHBrCO_2H$ $\xrightarrow{\text{Sn-Cu couple}}{\text{THF}}$

$CH_2=C(CO_2H)CH_2CHBrCO_2H$

81%

$(BrCH_2)_2C=C(CH_2Br)_2$ $\xrightarrow{\text{Sn-Cu couple}}{\text{THF}}$

$CH_2=C(CH_2Br)C(CH_2Br)=CH_2$

Nonactivated dibromides ($BrCH_2CH_2Br$) react very slowly.

I.B.2.c-1 P. B. Dervan and M. A. Shippey, J. Am. Chem. Soc., 98, 1265 (1976).

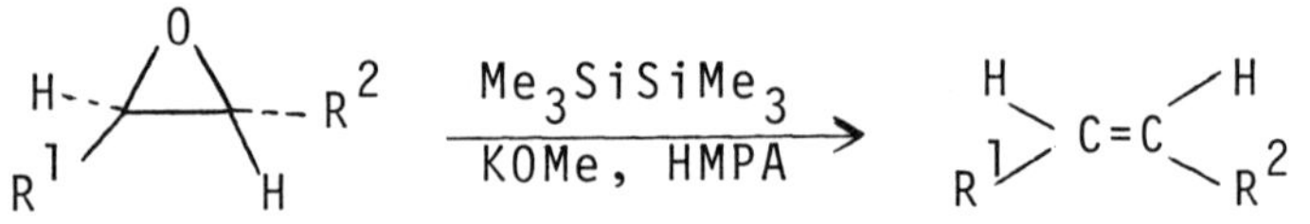

$\xrightarrow{\text{Me}_3\text{SiSiMe}_3}{\text{KOMe, HMPA}}$

75-99%

(>99% stereospecific)

I.B.2.c-2   M. T. Reetz and M. Plachky, Synthesis, 1976, 199.

$$\underset{R^3R^4}{R^1\diagdown\!\!\overset{O}{\diagup\!\!\diagdown}\!\!\diagup R^2} \xrightarrow[\text{THF}]{Me_2PhSiLi} \underset{R^3R^2}{R^1\diagdown\diagup R^4}C=C$$

R^1	R^2	R^3	R^4	% Yield
Ph	H	H	Ph	75
Ph	Ph	H	H	83
Pr	H	H	H	60

I.B.2.c-3   V. Calo, L. Lopez, A. Mincuzzi, and G. Pesce, Synthesis, 1976, 200.

$$\underset{R^3R^4}{R^1\diagdown\!\!\overset{Y}{\diagup\!\!\diagdown}\!\!\diagup R^2} + \text{(N-Me benzothiazol-2-ylidene)}=Se \xrightarrow[\text{CH}_2\text{Cl}_2]{\text{CF}_3\text{CO}_2\text{H}}$$

$$\underset{R^3R^4}{R^1\diagdown\diagup R^2}C=C$$

R^1	R^2	R^3	R^4	Y	% Yield
Cl	H	H	H	O	100
Ph	H	H	Ph	O	97
Ph	H	H	H	S	87
$-(CH_2)_4-$		H	H	O or S	100

I.B.2.c-4 T. Cohen, A. J. Mura, Jr., D. W. Shull, E. R. Fogel, R. J. Ruffner, and J. R. Falck, J. Org. Chem., 41, 3218 (1976).

$$PhSCHR^1CH_2CR^2(SPh_2) \xrightarrow[i-Pr_2NEt, PhH]{Cu_2Ph(CF_3SO_3)_2}$$

$$CH_2=CHCR^3=CHR^4$$

$R^1 = Me, R^2 = R^3 = H, R^4 = PhS, 84\%$

$R^1 = R^4 = H, R^2 = Me, R^3 = PhS, 76\%$

I.B.2.c-5 E. Vedejs and D. A. Engler, Tetrahedron Lett., 1976, 3487.

$$R^1CH_2CR^2(SMe)CO_2R^3 \xrightarrow[2) DBU, DMF, 50°]{1) CF_3SO_3CH_2CO_2Et}$$

$$R^1CH=CR^2CO_2R^3$$

70-94%

$$R^1CH_2CHR^2(SPh) \xrightarrow[2) DBU, DMF, 50°]{1) CF_3SO_3CH_2CO_2Et} R^1CH=CHR^2$$

31-89%

I.B.2.c-6 B. M. Trost and K. Hiroi, *J. Am. Chem. Soc.*, **98**, 4313 (1976); see also: B. M. Trost, T. N. Salzmann, and K. Hiroi, *ibid*, 4887.

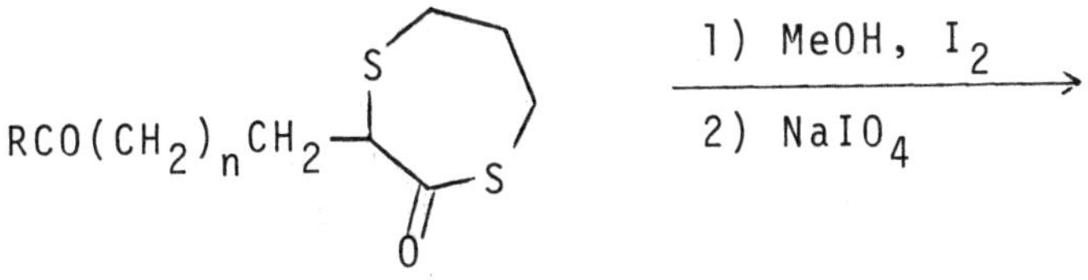

$$RCO(CH_2)_nCH_2-\underset{O}{\text{[cyclic dithiepane-thioester]}} \xrightarrow[\text{2) NaIO}_4]{\text{1) MeOH, I}_2}$$

$$RCO(CH_2)_nCH=CHCO_2Me_3$$

51-55%

I.B.2.c-7 G.M. Ksander and J. E. McMurry, *Tetrahedron Lett.*, **1976**, 4691.

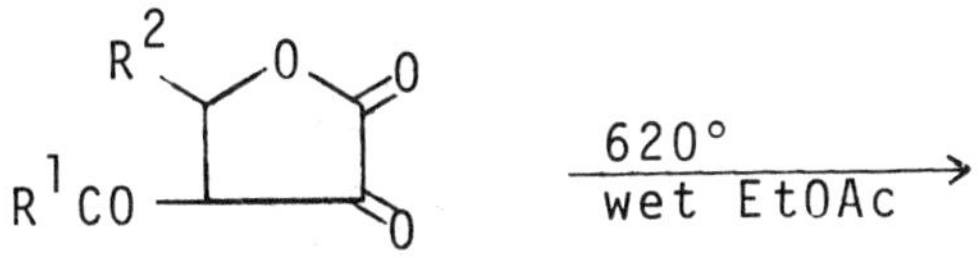

$$R^1CO-\underset{\text{[furanone with R}^2\text{]}}{} \xrightarrow[\text{wet EtOAc}]{620°} R^1COCH=CHR^2$$

R^1 = Ph, R^2 = H, 82%
R^1 = n-C_6H_{13}, R^2 = Me, 84%
R^1 = cyclopropyl, R^2 = H, 65%

I.B.2.c-8 T. Kato, T. Chiba, and T. Chiba, <u>Chem. Pharm. Bull.</u>, <u>24</u>, 3034 (1976).

R^1	R^2	% Yield
Ph	OMe	75
H	OMe	50
H	Me	51

I.B.2.c-9 V. Jäger and H. Grund, <u>Angew. Chem. Int. Ed. Engl.</u>, <u>15</u>, 50 (1976).

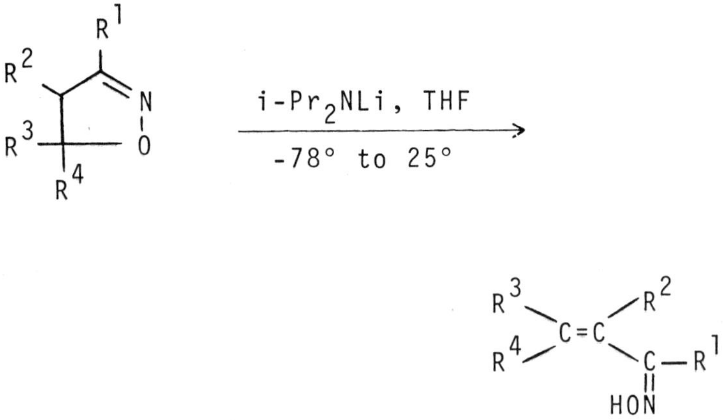

~ 60%

CARBON—CARBON BONDS FORMING REACTIONS

I.B.3-1 J. E. McMurry and M. P. Fleming, *J. Org. Chem.*, **41**, 896 (1976); see also: A. L. Baumstark, E.J.H. Bechara, and M. J. Semigran, *Tetrahedron Lett.*, **1976**, 3265.

$$R^1R^2CO \xrightarrow{Ti°} R^1R^2C=CR^1R^2$$

$R^1, R_2 = \text{-(}CH_2\text{)}_5\text{-}$, 85%
$R^1 = R_2 = i\text{-}Pr$, 40%

$$R^1R^2C(OH)-C(OH)R^1R^2 \xrightarrow{Ti°} R^1R^2C=CR^1R^2$$

$R^1, R^2 = \text{-(}CH_5\text{)}_5\text{-}$, 85%
$R^1 = H, R^2 = Bu$, 80%

The active Ti° is prepared from $TiCl_3$ and K metal.

I.B.3-2 G. A. Olah and G. K. Surya Prakash, *Synthesis*, **1976**, 607.

$$PhCRCl_2 \xrightarrow[THF]{Ti(II)} PhCR=CRPh$$

$R = Ph$, 96%
$R = H$, 79%

I.B.3-3 T. Minami, M. Matsumoto, and T. Agawa, J. Chem. Soc., Chem. Commun., $\underline{1976}$, 1053.

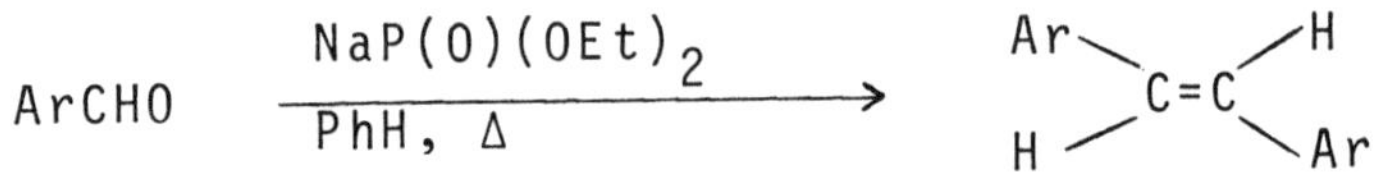

$$ArCHO \xrightarrow[PhH, \Delta]{NaP(O)(OEt)_2} \underset{H}{\overset{Ar}{>}}C=C\underset{Ar}{\overset{H}{<}}$$

22-85%

I.B.3-4 A. J. Fry and L.-L. Chung, Tetrahedron Lett., $\underline{1976}$, 645.

$$(R^1R^2CBr)_2P(O)OMe \xrightarrow[DMSO, TEAB, 20°]{electrolysis} R^1R^2C=CR^1R^2$$

$R^1 = Ph, R^2 = H, 76\%$
$R^1 = R^2 = Me, 40\%$

I.B.3-5 H. J. Bestmann, L. Kisielowski, and W. Distler, Angew. Chem. Int. Ed. Engl., $\underline{15}$, 298 (1976).

$$RCH=PPh_3 + (PhO)_3P-O_3 \xrightarrow{PhMe \text{ or } CH_2Cl_2} RCH=CHR$$

R	% Yield
$n\text{-}C_{11}H_{23}$	61
Ph	89
MeCO	87
MeO_2C	83

I.B.3-6 T. G. Back, D.H.R. Barton, M.R. Britten-Kelly, and F. S. Guziec, Jr., J. Chem. Soc., Perkin Trans. I, 1976, 2079.

$$\underset{t\text{-Bu}}{\overset{t\text{-Bu}}{>}}\!\!C\!\!\underset{N=N}{\overset{Se}{\diagdown\!\!\diagup}}\!\!C\!\!\underset{Ph}{\overset{Ph}{<}} \xrightarrow{CCl_4, \text{ reflux}}$$

$$(t\text{-Bu})_2C=C(Ph)_2$$

67%

Attempts to synthesize tetra-t-butylethylene by this procedure were unsuccessful.

I.B.3-7 A. L. Fridman and F. A. Gabitov, J. Org. Chem. USSR, 11, 1113 (1975).

$$PhCO\overset{I}{\underset{|}{C}}(Ph)C(NO_2)_3 \xrightarrow[\text{reflux}]{PhCl} PhCOCPh=CPhCOPh$$

95%

I.B.3-8 A. G. Hortmann and A. Bhattacharjya, J. Am. Chem. Soc., 98, 7081, (1976).

$$\underset{Me}{\overset{Me}{>}}\!\!\underset{S}{\overset{NNHTs}{\diagup\!\!\!\diagdown}}\!\!\underset{Me}{\overset{Me}{<}} \xrightarrow[2)\ 150°]{1)\ NaH} Me_2\!\!\underset{}{\overset{S}{\triangle}}\!\!=CMe_2$$

40-65%

I.B.3-9 D. A. Konen, P. E. Pfeffer, and L. S. Silbert, Tetrahedron, 32, 2507 (1976).

$$C_7H_{15}CH_2CO_2Me \xrightarrow[\substack{\text{3) n-BuLi (1 eq.)} \\ \text{4) MeI (2 eq.)}}]{\substack{\text{1) LDA, THF, HMPA} \\ \text{2) CS}_2\text{, -75°}}}$$

$$\underset{MeO_2C}{\overset{C_7H_{15}}{\diagdown}}C=C(SMe)_2$$

78%

I.B.3-10 A. C. Brouwer and H.J.T. Bos, Tetrahedron Lett., 1976, 209.

$$R^1R^2C=S \xrightarrow[h\nu]{MeSC\equiv CSMe} R^1R^2C=C(SMe)\overset{S}{\overset{\|}{C}}SMe$$

30-70%

$$R^1, R^2 = aryl$$

I.B.3-11 M. V. Bhatt, C. G. Rao, and S. Rengaraju, J. Chem. Soc., Chem. Commun., 1976, 103.

$$R^1CH_2\overset{NOH}{\overset{\|}{C}}R^2 \xrightarrow[AcCl]{Ac_2O, py.} \underset{R^1}{\overset{AcO}{\diagdown}}C=C\underset{R^2}{\overset{NAc_2}{\diagup}}$$

17-58%

I.B.3-12 J. J. Eisch and J. E. Galle, J. Am. Chem. Soc., 98, 4646 (1976).

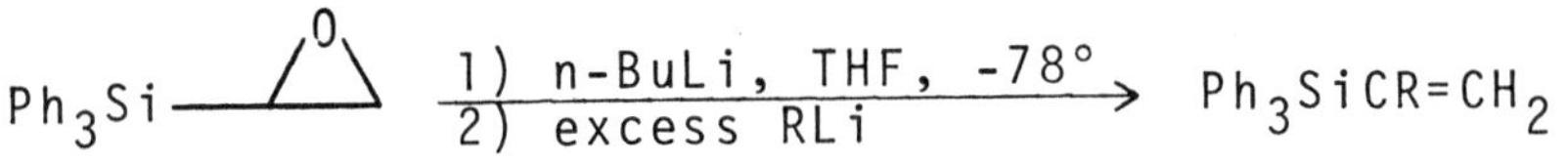

$$Ph_3Si\text{-epoxide} \xrightarrow[\text{2) excess RLi}]{\text{1) n-BuLi, THF, -78°}} Ph_3SiCR=CH_2$$

R = n-Bu, 70%
R = Ph, 66%

I.B.3-13 S. Iwasaki, Helv. Chim. Acta, 59, 2753 (1976).

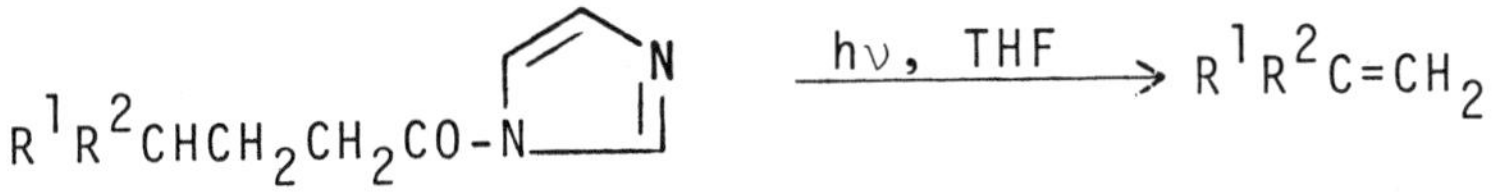

$$R^1R^2CHCH_2CH_2CO\text{-N(imidazole)} \xrightarrow{h\nu, \text{ THF}} R^1R^2C=CH_2$$

$R^1 = C_{14}H_{29}$, $R^2 = H$, 40%
$R^1 = C_{14}H_{29}$, $R^2 = Me$, 62%

Applicable to side-chain degradation of bile acids and lanosterol.

I.B.3-14 G. Ferdinand and K. Schank, Synthesis, 1976, 404.

$$RCH_2O_2SCH(OMe)COPh \xrightarrow[\text{2) } CH_2O \text{ (g)}]{\text{1) NaH or LiH, MeCN}}$$

$$RCHO_2SC(OMe)=CH_2$$

R = H, 70%
R = Ph, 80%

I.B.4-1 H. Yamazaki, J. Chem. Soc., Chem. Commun., 1976, 841.

$$t\text{-BuC}\equiv\text{CH} \xrightarrow[\text{PhH, }100°]{H_2Ru(CO)(Ph_3P)_3} \begin{array}{c}t\text{-Bu}\\ \\ H\end{array}\!\!C\!=\!C\!=\!C\!\!\begin{array}{c}H\\ \\ t\text{-Bu}\end{array}$$

I.B.4-2 F. Delbecq and J. Gore, Angew. Chem. Int. Ed. Engl., 15, 496 (1976); see also: *ibid*, Bull. Soc. Chim. Fr., 1976, 533 and 541.

$$R^1R^2C(OAc)C\equiv CH \xrightarrow[\text{MgBr}_2,\text{ Et}_2O]{R^3\text{MgX}} R^1R^2C=C=CHR^3$$

R^1	R^2	R^3	% Yield
Me	Me	$n\text{-}C_8H_{17}$	64
$\text{-}(CH_2)_5\text{-}$		$n\text{-Pr}$	72
$\text{-}(CH_2)_5\text{-}$		$i\text{-Pr}$	44
$\text{-}(CH_2)_5\text{-}$		$t\text{-Bu}$	80

I.B.4-3 D. J. Pasto, G. F. Hennion, R. H. Shults, A. Waterhouse, and S.-K. Chou, <u>J. Org. Chem.</u>, **41**, 3496 (1976).

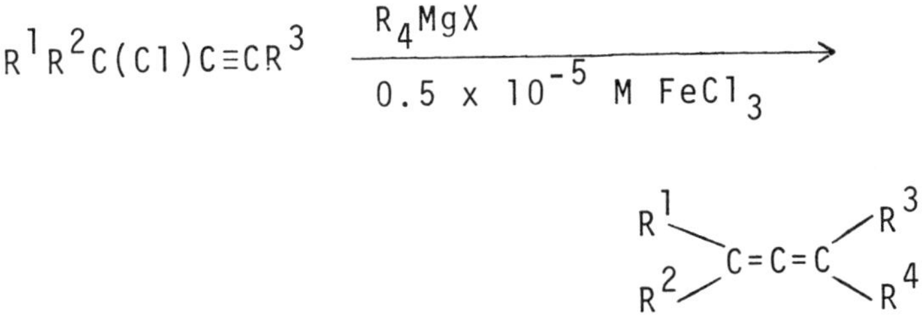

$$R^1R^2C(Cl)C\equiv CR^3 \xrightarrow[0.5 \times 10^{-5} \text{ M FeCl}_3]{R_4MgX}$$

$$\begin{array}{c} R^1 \\ R^2 \end{array} C=C=C \begin{array}{c} R^3 \\ R^4 \end{array}$$

40-90%

I.B.4-4 G. Tadema, P. Vermeer, J. Meijer, and L. Bransdma, <u>Recl. Trav. Chim. Pays-Bas</u>, **95**, 66 (1976).

$$(EtO)_2CHC\equiv CH \xrightarrow[\text{CuBr (cat.), Et}_2O]{RMgBr} EtOCH=C=CHR$$

R	% Yield
n-Bu	78
cyclopentyl	80
t-Bu	79

I.B.4-5   J.-L. Moreau and M. Gaudemar, J. Organomet. Chem., 108, 159 (1976).

$$MeOCR^1R^2C{\equiv}CR^3 \xrightarrow{\displaystyle \frac{R^4MgBr}{Et_2O,\ CuBr\ (10\ mol\%)}}$$

$$\begin{array}{c} R^1 \\ R^2 \end{array}\!\!\!C{=}C{=}C\!\!\!\begin{array}{c} R^3 \\ R^4 \end{array}$$

R^1	R^2	R^3	R^4	% Yield
H	H	H	C_7H_{15}	70
H	H	Me	Ph	82
Me	H	Et	$i\text{-}C_5H_{11}$	45
Me	Me	Me	C_8H_{17}	52

I.B.4-6   P. Crabbé, E. Barreiro, J.-M. Dollat, and J.-L. Luche, J. Chem. Soc., Chem. Commun., 1976, 183.

$$R^1_2CuLi\ +\ R^2C{\equiv}CC(OAc)R^3R^4 \longrightarrow$$

$$R^1R^2C{=}C{=}CR^3R^4\ (A)\quad or\quad R^2CH{=}C{=}CR^3R^4\ (B)$$

Conditions are described for the selective formation of product (A) or (B).

I.B.4-7 J. L. Ripoll, *J. Chem. Soc., Chem. Commun.*, **1976**, 235.

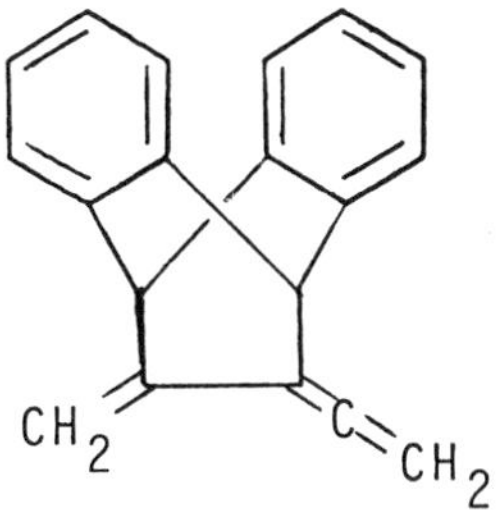

$$CH_2=C=C=C=CH_2 \quad (A)$$

$$\xrightarrow[10^{-3} \text{ Torr}]{700°}$$

$$+$$

$$MeC\equiv C-C\equiv CH \quad (B)$$

$$85\% \ (A:B = 7:3)$$

I.B.4-8 T. L. Gilchrist and D.P.J. Pearson, *J. Chem. Soc., Perkin Trans. I*, **1976**, 989.

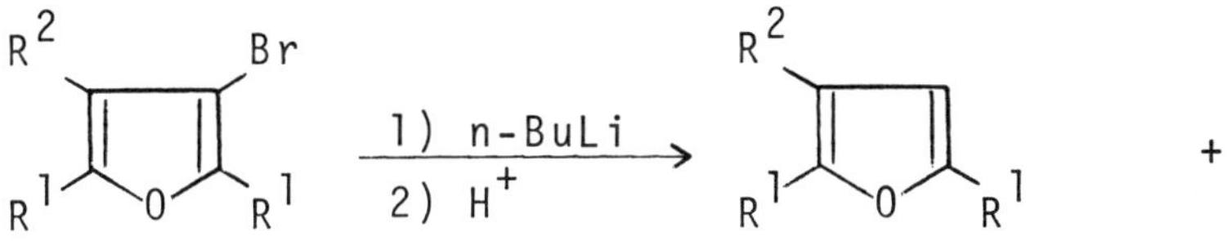

1) n-BuLi
2) H^+

$+$

$R^1COCR^2=C=CHR^1$

$+$

$R^1COCHR^2C\equiv CR^1$

Product distribution is dependent on solvent and reaction conditions.

I.C-1 C. B. Reese and A. Shaw, J. Chem. Soc., Perkin Trans. I, 1976, 890.

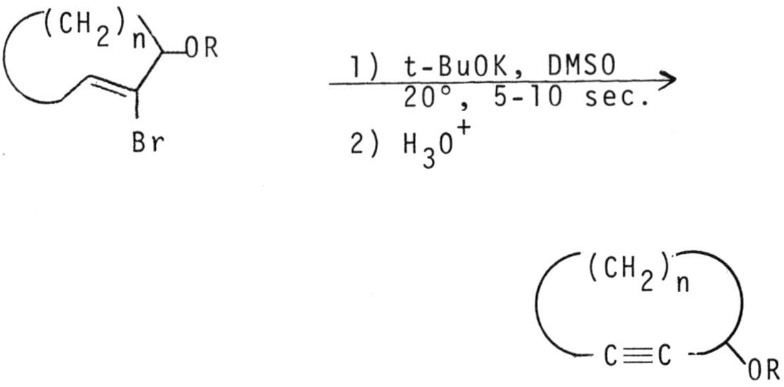

[structure: cycloalkene with (CH$_2$)$_n$ ring, OR group, and Br on alkene]   1) t-BuOK, DMSO, 20°, 5-10 sec.
2) H$_3$O$^+$
→ [cycloalkyne with (CH$_2$)$_n$, C≡C, OR]

n = 5, 74%
n = 6, 70%

I.C-2 D. P. Bauer and R. S. Macomber, J. Org. Chem., 41, 2640 (1976).

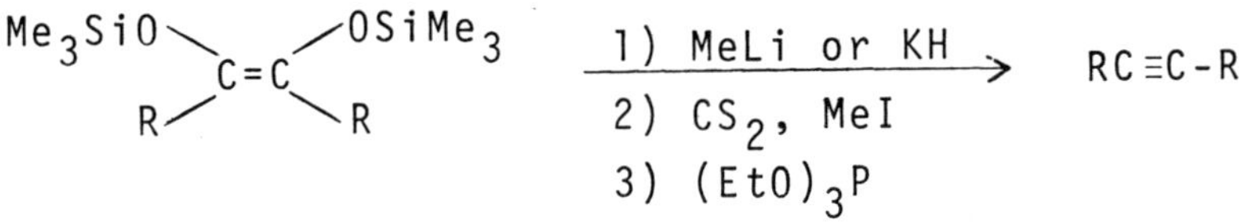

Me$_3$SiO\\C=C/OSiMe$_3$ 1) MeLi or KH   RC≡C-R
R/   \\R   2) CS$_2$, MeI
                            3) (EtO)$_3$P

R = Ph, 35%
R = n-Pr, 28

I.C-3 A. Gorgues and A. LeCoq, <u>Tetrahedron Lett.</u>, <u>1976</u>, 4723.

$$R^1CHBrCHBrR^2 \xrightarrow[\text{50\% NaOH}]{Bu_4NHSO_4,\ n\text{-}C_5H_{12}} R^1C\equiv CR^2$$

$R^1 = H$, $R^2 = CH(OEt)_2$, 80%
$R^1 = R^2 = Ph$, 75%

$$(EtO)_2CHCH=CCl_2 \xrightarrow[\text{50\% NaOH}]{BuNHSO_4,\ n\text{-}C_5H_{12}} (EtO)_2CHC\equiv CCl$$

70%

I.C-4 A. Gorgues and A. Le Coq, <u>Bull. Soc. Chim. Fr.</u>, <u>1976</u>, 125.

$$R^1CX=CHR^2 \xrightarrow[\text{PhH or PhH, } CH_2Cl_2]{Triton\ B} R^1C\equiv CR^2$$

R^1	R^2	X	% Yield
p-MeC$_6$H$_4$	H	Cl	70
Br	CCl$_2$=CCl	Br	70
Br	CH(OEt)$_2$	Br	85
COMe	PhC≡C	I	64
COMe	CBr$_2$=CCl	Br	51

I.C-5 T. Suzuki, T. Sonoda, S. Kobayashi, and H. Taniguchi, J. Chem. Soc., Chem. Commun., 1976, 180.

$$Ar_2C=CHBr \xrightarrow{h\nu} ArC \equiv CAr$$
$$50-86\%$$

Ar must be oxygen substituted (Ar = $MeOC_6H_4$)

I.C-6 P. J. Kocienski, J. M. Ansell, and B. E. Norcross, J. Org. Chem., 41, 3650 (1976).

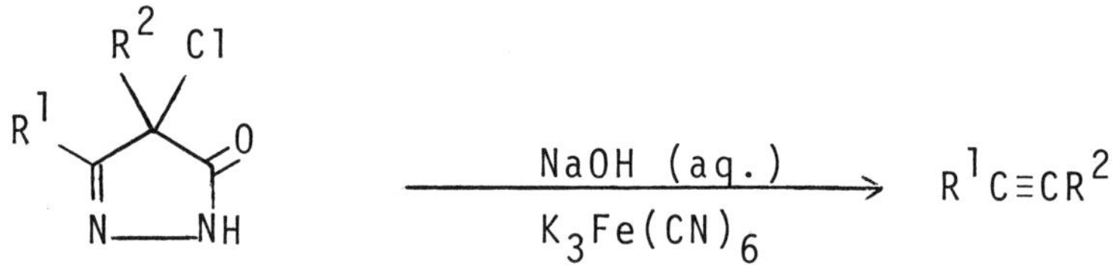

$$\underset{\substack{\\}}{\text{pyrazolinone}} \xrightarrow[K_3Fe(CN)_6]{NaOH\ (aq.)} R^1C \equiv CR^2$$

$R^1 = R^2 = Ph, 80\%$
$R^1 = Ph, R^2 = Me, 60\%$
$R^1 = Me, R^2 = Ph, 12\%$

I.C-7 C. Wentrup and W. Reichen, Helv. Chim. Acta, 59, 2615 (1976).

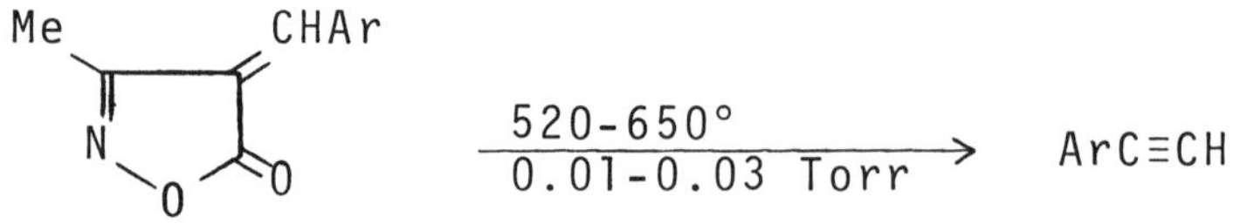

$$\underset{\substack{\\}}{\text{isoxazolone}} \xrightarrow[0.01-0.03\ Torr]{520-650°} ArC \equiv CH$$
$$45-95\%$$

I.C-8 J. Levisalles and H. Rudler, *J. Organomet. Chem.*, **122**, C17 (1976).

$$\text{n-BuCOCl} \xrightarrow{\underset{\text{HMPA}}{Na_2W(CO)_5}} \text{n-}C_5H_{12}$$

26%

+

n-BuC≡CH

15%

I.C-9 K. Okuhara, *J. Org. Chem.*, **41**, 1487 (1976).

$$\text{ArLi} \xrightarrow[\substack{\text{2) n-BuLi, 60°}\\ \text{3) H}_2\text{O}}]{\text{1) CF}_2\text{=CCl}_2} \text{ArC≡CH}$$

52-78%

I.D.1-1 E. V. Dehmlow, Tetrahedron Lett., 1976, 91.

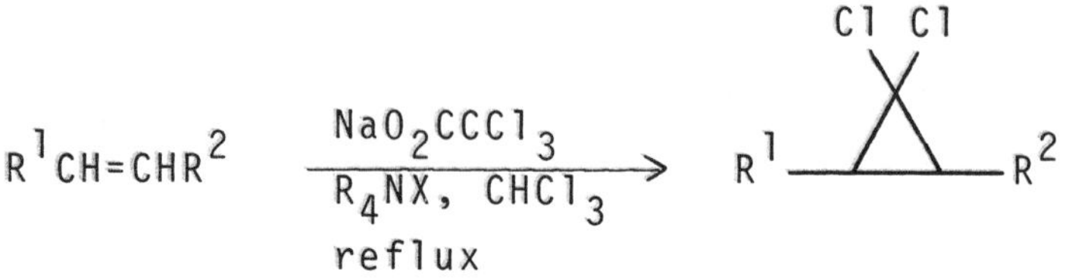

$$R^1CH=CHR^2 \xrightarrow[\text{reflux}]{NaO_2CCCl_3, R_4NX, CHCl_3}$$

[cyclopropane with gem-Cl,Cl and R^1, R^2 substituents]

R^1	R^2	% Yield
Ph	Ph	7
$-(CH_2)_4-$		57
$-(CH_2)_6-$		73

I.D.1-2 I. Tabushi, Y. Kuroda, and Z. Yoshida, Tetrahedron, 32, 997 (1976).

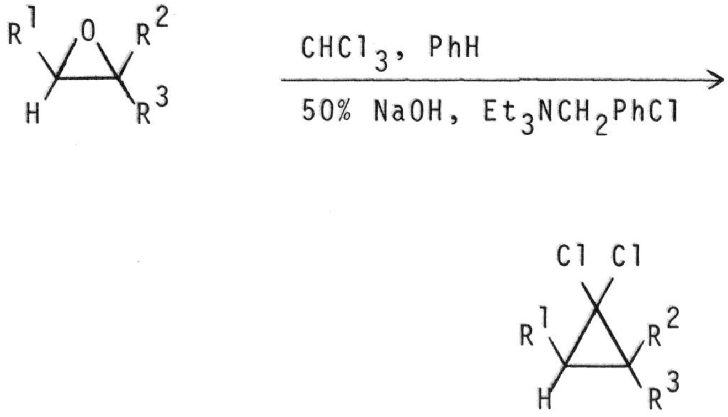

[epoxide with R^1, H on one carbon and R^2, R^3 on the other]

$$\xrightarrow[\text{50\% NaOH, Et}_3\text{NCH}_2\text{PhCl}]{CHCl_3, PhH}$$

[cyclopropane with gem-Cl,Cl, and R^1, H on one carbon, R^2, R^3 on the other]

15-39%

I.D.1-3 R. J. Kricks and A. A. Volpe, Synthesis, 1976, 313.

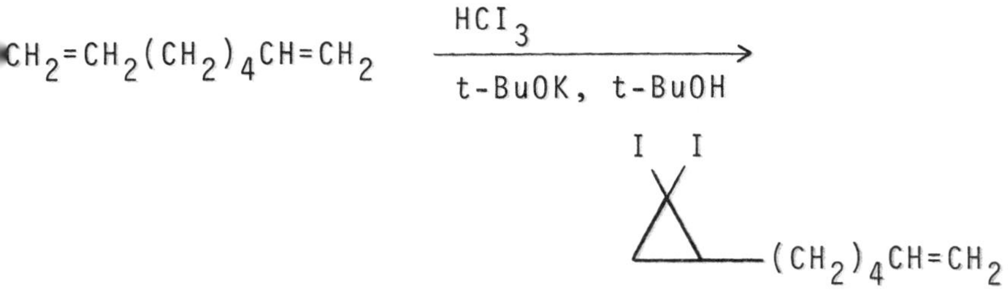

$$CH_2=CH_2(CH_2)_4CH=CH_2 \xrightarrow[\text{t-BuOK, t-BuOH}]{HCI_3}$$

[triangle with two I substituents]—$(CH_2)_4CH=CH_2$

29%

I.D.1-4 M. S. Baird, J. Chem. Soc., Perkin Trans. I, 1976, 54.

$(CH_2)_n$ [cyclic alkene] $\xrightarrow[\substack{\text{t-BuOH, -20° to 40°} \\ \text{light petroleum}}]{CHI_3, \text{ t-BuOK}}$

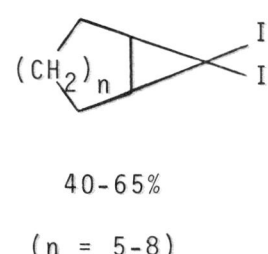

40-65%

(n = 5-8)

Many of the products are distillable contrary to earlier reports.

I.D.1-5 M. Kamel, W. Kimpenhaus, and J. Buddrus, Chem. Ber., 109, 2351 (1976).

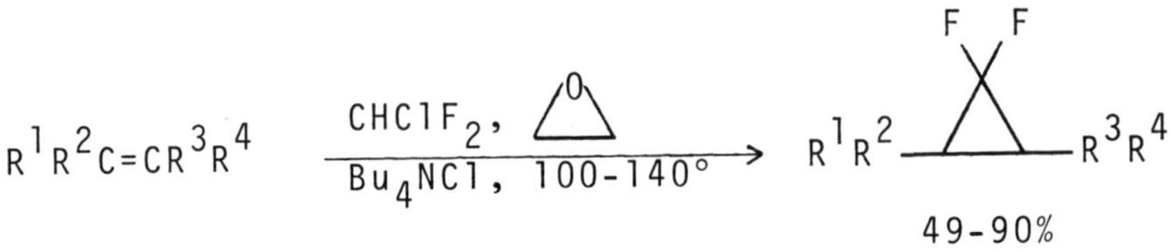

$$R^1R^2C=CR^3R^4 \xrightarrow[\text{Bu}_4\text{NCl, 100-140°}]{\text{CHClF}_2, \triangle^O} R^1R^2\underset{\text{49-90\%}}{\triangle}R^3R^4$$

with F, F at the apex of the product cyclopropane.

I.D.1-6 R. A. Olofson, K. D. Lotts, and G. N. Barber, Tetrahedron Lett., 1976, 3381 and 3379; see also: G. N. Barber and R. A. Olofson, *ibid*, 3783.

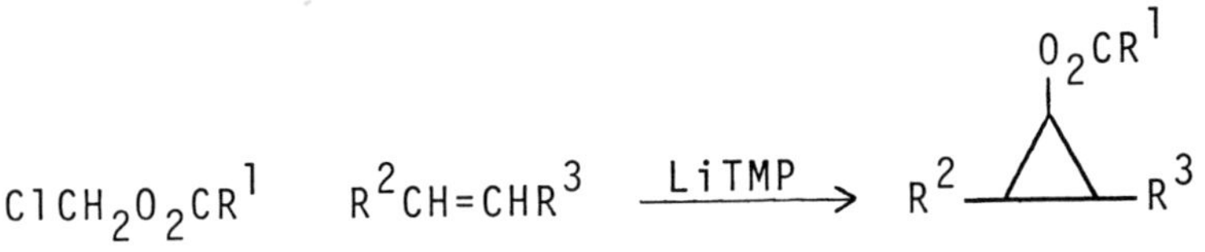

$$ClCH_2O_2CR^1 \quad R^2CH=CHR^3 \xrightarrow{\text{LiTMP}} R^2\underset{}{\overset{O_2CR^1}{\triangle}}R^3$$

R^1	R^2	R^3	% Yield
t-Bu	H	OEt	30
MeO	$-(CH_2)_4-$		10
Me_2N	$-(CH_2)_4-$		25

I.D.1-7 A. Jonczyk and M. Makosza, Synthesis, 1976, 387; see also: I. Artaud, J. Seyden-Penne, and P. Viout, Comptes Rendus, Ser. C, 285, 503 (1976).

$$R^1C(X)=CH_2 + R^2CHX^2Cl \xrightarrow[PhCH_2\overset{+}{N}Et_3Cl^-]{50\% \ NaOH}$$

R^1	R^2	X^1	X^2	% Yield
H	H	CN	CO_2t-Bu	45
H	Me	CO_2t-Bu	CN	67
Me	Me	CN	CN	75

The isomeric ratio was determined in all cases.

I.D.1-8 N. Kawabata, M. Naka, and S. Yamashita, J. Am. Chem. Soc., 98, 2676 (1976).

$$R^1CH=CHR^2 \xrightarrow[Cu, \ I_2 \ (cat.)]{R^3CHX_2, \ PhH}$$

R^1	R^2	R^3	X	% Yield
Ph	H	H	I	90
C_6H_{13}	H	H	I	86
$-(CH_2)_4-$		CO_2Me	Br	31
$-(CH_2)_4-$		Cl	I	48
$-(CH_2)_4-$		H	I	85-87

I.D.1-9 T. Sasaki, S. Eguchi, M. Ohno, and F. Nakata, *J. Org. Chem.*, **41**, 2408 (1976).

$$R^1_2C(Cl)C\equiv CH \xrightarrow[\text{crown ether (DC-18)}]{\underset{51\% \text{ KOH (aq.), PhH}}{R^2CH=CHR^3}} \underset{R^3}{\overset{R^2}{\triangle}}=C=CR^1_2$$

37-65%

I.D.1-10 A. Doutheau and J. Gore, *Bull. Soc. Chim. Fr.*, **1976**, 1189.

$$R^1CHBrCR^2=CR^3C\equiv CH \xrightarrow[-10°]{\underset{t\text{-BuOK, pentane}}{R^4R^5C=CR^6R^7}}$$

$$\underset{R^6 \quad R^7}{\overset{R^5 \quad R^4}{\triangle}}=C=CR^3CR^2=CHR^1$$

R^1	R^2	R^3	R^4	R^5	R^6	R^7	% Yield
Me	H	H	Ph	H	H	H	41
Me	H	H	Me	Me	Me	Me	43
H	H	Me	Ph	H	H	H	42
Me	Me	Me	Me	Me	Me	Me	21

Does not work well for the chloroenynes.

CARBON—CARBON BONDS FORMING REACTIONS

I.D.1-11 R. F. Cunico and Y.-K. Han, *J. Organomet. Chem.*, 105, C29 (1976).

$$Me_2C=C(Cl)SiMe_3 \xrightarrow[R^1CH=CHR^2]{Me_4NF} R^1\underset{35-66\%}{\overset{CMe_2}{\triangle}}R^2$$

I.D.1-12 D. Seyferth and D. Dagani, *J. Organomet. Chem.*, 104, 145 (1976).

$$R^1R^2C=CR^3R^4 \xrightarrow[Ph_2Hg, PhH, 150°]{Me_2C=C(Br)HgBr} R^1R^2\underset{24-87\%}{\overset{CMe_2}{\triangle}}R^3R^4$$

I.D.1-13 D. Seyferth, C. K. Haas, and D. Dagani, *J. Organomet. Chem.*, 104, 9 (1976).

$$RHgCXYBr \xrightarrow[PhH]{R^1CH=CHR^2} R^1\overset{X\ Y}{\underset{}{\triangle}}R^2$$

(R = cyclohexyl and $PhCH_2CH_2$, X,Y = Cl, Cl; Cl, Br; Br, Br)

Use of alkyl group R allows the reaction to occur slowly at room temperature.

I.D.1-14 A. J. Hubert, A. F. Noels, A. J. Anciaux, and P. Teyssie, Synthesis, 1976, 600.

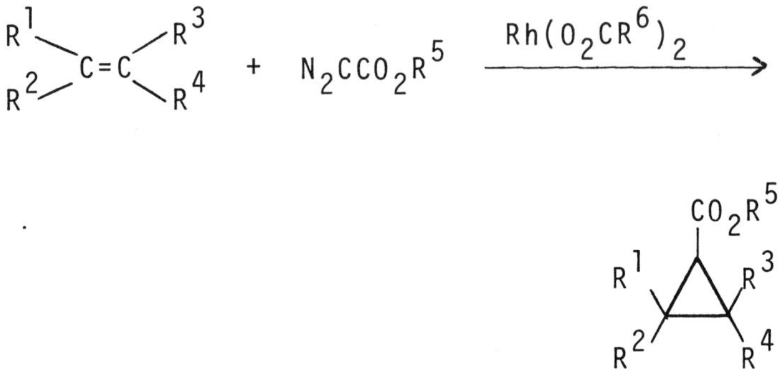

$$R^1R^2C=CR^3R^4 + N_2CCO_2R^5 \xrightarrow{Rh(O_2CR^6)_2}$$

[cyclopropane product with R^1, R^2, R^3, R^4, and CO_2R^5 substituents]

A comparison of various rhodium(II) carboxylates with copper(II) triflate and palladium(II) acetate was made.

I.D.1-15 S. H. Goh, K. C. Chan, and H. L. Chong, Aust. J. Chem., 29, 1699 (1976).

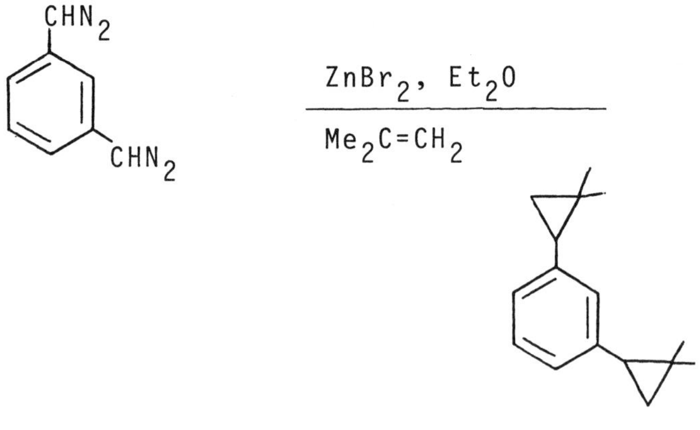

[1,3-bis(diazomethyl)benzene] $\xrightarrow[Me_2C=CH_2]{ZnBr_2,\ Et_2O}$ [1,3-bis(2,2-dimethylcyclopropyl)benzene]

62%

I.D.2-1 S. Cohen and A. Vogev, *J. Am. Chem. Soc.*, **98**, 2013 (1976).

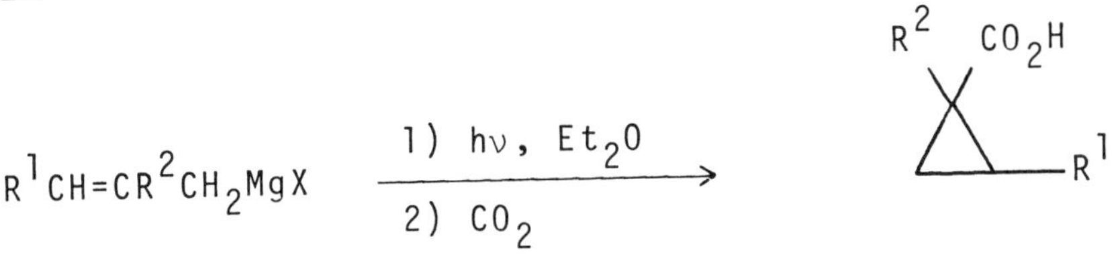

$$R^1CH=CR^2CH_2MgX \xrightarrow[\text{2) }CO_2]{\text{1) }h\nu, Et_2O}$$

[triangle with R^2, CO_2H at top and R^1 at right]

$R^1 = R^2 = H$, 45%
$R^1 = H$, $R^2 = Me$, 70%
$R^1 = Ph$, $R^2 = H$, 75%

I.D.2-2 A. Couture, M. Hoshino, and P. deMayo, *J. Chem. Soc., Chem. Commun.*, **1976**, 131.

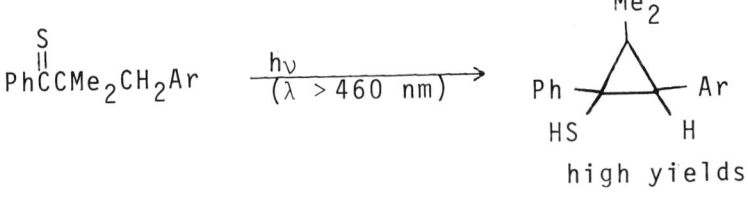

high yields

I.D.2-3 P. Baeckstrom, *J. Chem. Soc., Chem. Commun.*, **1976**, 476; see also: M. J. Bullivant and G. Pattenden, *J. Chem. Soc., Perkin Trans. I*, **1976**, 256.

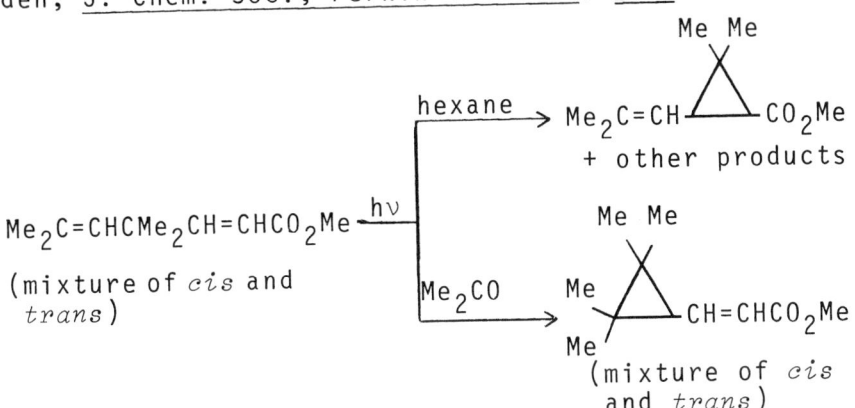

I.D.2-4 G. Jones, II and L. P. McDonnell, J. Chem. Soc., Chem. Commun., 1976, 18.

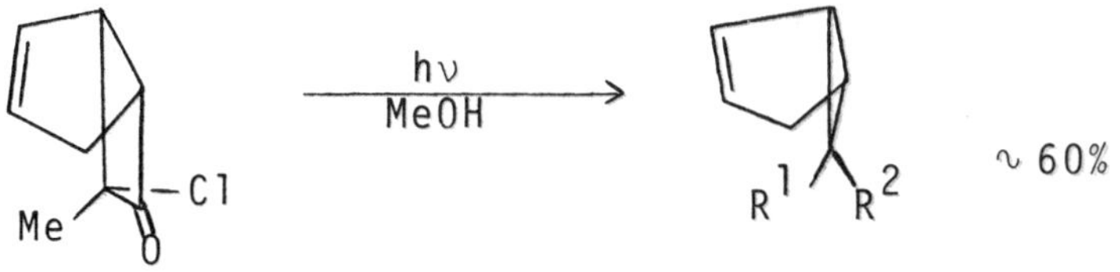

[Structure diagram: bicyclic ketone with Me and Cl substituents] →(hν / MeOH)→ [bicyclic product with R^1, R^2] ~60%

(R^1 = Me, R^2 = CO_2Me / R^1 = CO_2Me / R^2 = Me = 6/2)

Other stereoisomer at α-carbonyl position gives mainly openchain product upon photolysis.

I.D.2-5 M. Braun and D. Seebach, Chem. Ber., 109, 669 (1976).

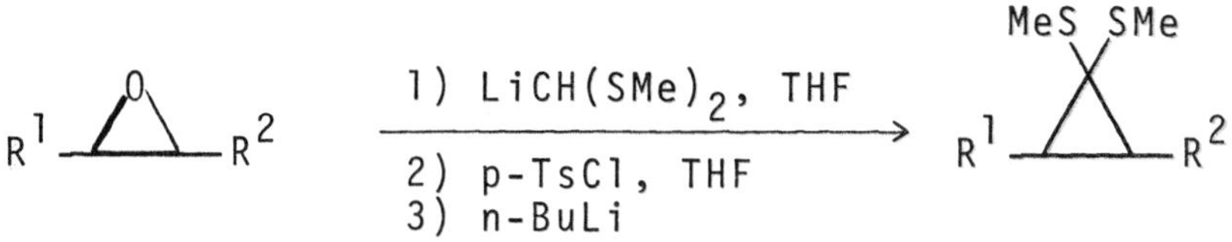

R^1—[epoxide]—R^2 →(1) LiCH(SMe)$_2$, THF; 2) p-TsCl, THF; 3) n-BuLi)→ R^1—[cyclopropane with MeS, SMe]—R^2

R^1	R^2	% Yield
Me	Me	66
n-$C_{10}H_{21}$	H	77
CH_2=CH	H	60
‒(CH$_2$)$_4$‒		81

An "improved one-pot-modification" is described.

I.D.2-6 J.-L. Burgot, J. Masson, P. Metzner, and J. Vialle, <u>Tetrahedron Lett.</u>, <u>1976</u>, 4297; see also: J.-L. Burgot, J. Masson, and J. Vialle, *ibid*, 4775.

MeC(S)CMe$_2$COMe $\xrightarrow{\text{RMgBr, THF}}$ [cyclopropane with HO, Me, Me, Me, Me, SR substituents]

R = Et, 60%

I.D.2-7 A. Hercouet and M. LeCorre, <u>Tetrahedron Lett.</u>, <u>1976</u>, 825; see also: A. Turcant and M. LeCorre, *ibid*, 1277.

Ph$_3$$\overset{+}{P}$-$\overset{-}{C}HCO\overset{-}{C}H\overset{+}{S}Me_2$ $\xrightarrow{\text{XCH=CH}_2}$ Ph$_3$$\overset{+}{P}$$\overset{-}{C}$HCO—△—X

X = CO$_2$Me, 60%
X = COMe, 53%

I.D.2-8 H. Braun and G. Huber, <u>Tetrahedron Lett.</u>, <u>1976</u>, 2121.

CH$_2$=CHCH=$\overset{+}{C}$HSMe$_2$ $\xrightarrow{\text{KCHR}^1\text{R}^2}$ CH$_2$=CH—△—$\overset{R^1}{\underset{R^2}{}}$
Cl$^-$

R^1 = R^2 = CO$_2$Et, 45%
R^1 = R^2 = COMe, 60%
R^1 = CO$_2$Me, R^2 = COMe, 60%

I.D.2-9 C. W. Jefford and A. F. Boschung, Helv. Chim. Acta, 59, 962 (1976).

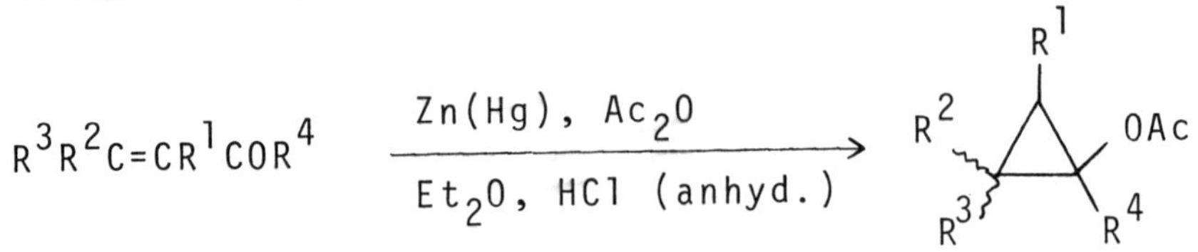

$$R^3R^2C=CR^1COR^4 \xrightarrow[\text{Et}_2\text{O, HCl (anhyd.)}]{\text{Zn(Hg), Ac}_2\text{O}}$$

(cyclopropane product with R^1, R^2, R^3, R^4, OAc, and Me substituents)

R^1	R^2	R^3	R^4	% Yield
H	Me	Me	Me	65
H	H	H	Me	32
H	H	$-(CH_2)_4-$		40
Me	H	$-(CH_2)_3-$		40

I.D.2-10 W. Tam and M. F. Rettig, J. Organomet. Chem., 108, C1 (1976).

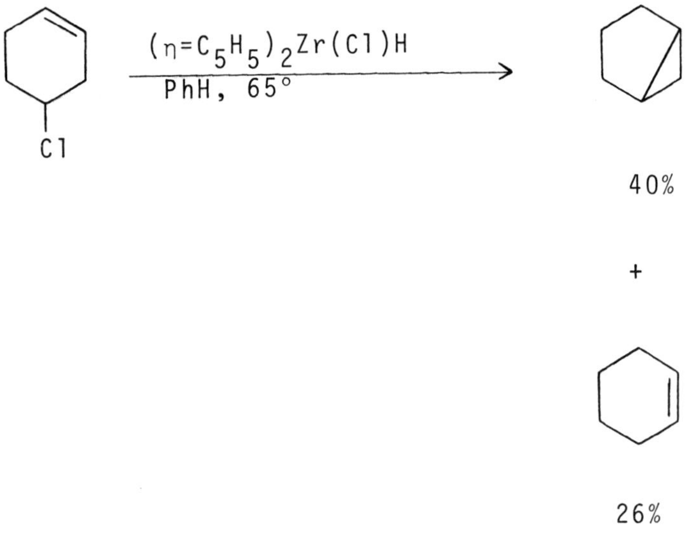

(4-chlorocyclohexene) $\xrightarrow[\text{PhH, 65°}]{(\eta = C_5H_5)_2Zr(Cl)H}$ (norbornane/bicyclic product)

40%

+

(cyclohexene)

26%

I.D.2-11   H.-G. Heine and D. Wendisch, <u>Justus Liebigs Ann. Chem.</u>, <u>1976</u>, 463.

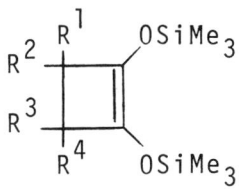

1) Br_2, CH_2Cl_2, $-10°$
2) H_2O, r.t.

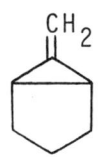

∿ 90%

I.D.2-12   U. Langer and H. Musso, <u>Justus Liebigs Ann. Chem.</u>, <u>1976</u>, 1180.

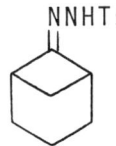

NaOMe / diglyme, 130°

39%

I.E.1-1 M. Kakushima, J. Espinosa, and Z. Valenta, Can. J. Chem., 49, 3304 (1976).

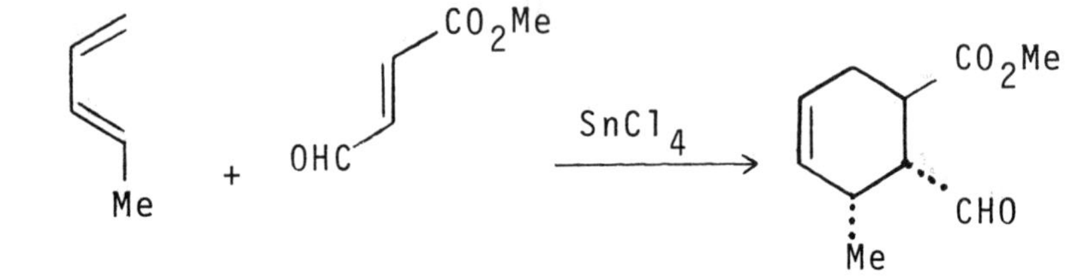

Lewis acid catalyzed Diels-Alder reaction of simple dienophiles exhibits clean orientation reversal.

I.E.1-2 M. E. Jung and C. A. McCombs, Tetrahedron Lett., 1976, 2935.

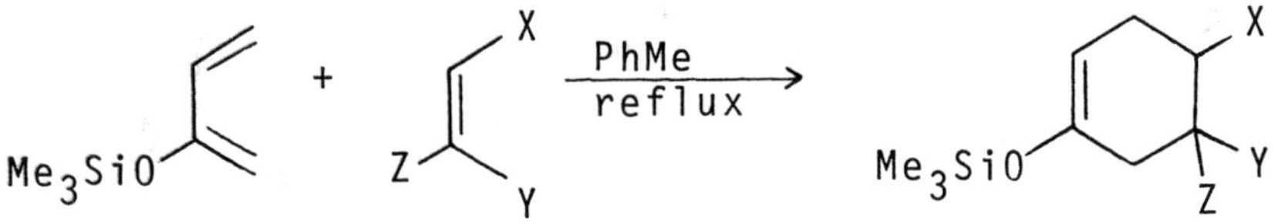

X	Y	Z	% Yield
COMe	H	H	60
CO_2Et	H	CO_2Et	77
CO_2Et	CO_2Et	H	39

I.E.1-3 B. M. Trost and A. J. Bridges, *J. Am. Chem. Soc.*, **98**, 5017 (1976).

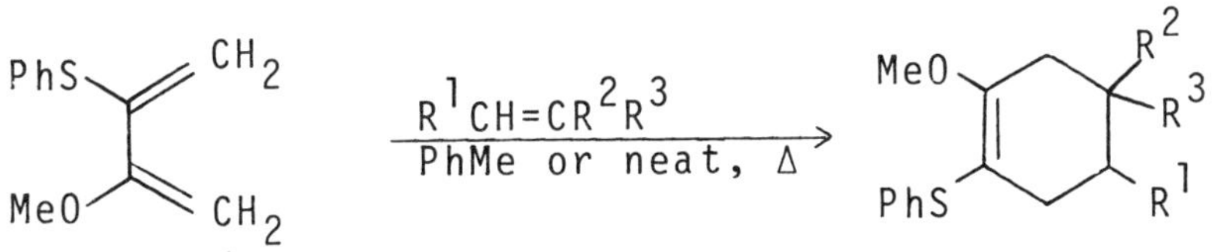

[Diagram: PhS/MeO-substituted bis-methylenecyclohexane diene + $R^1CH=CR^2R^3$, PhMe or neat, Δ → cyclohexene with MeO, PhS, R^1, R^2, R^3 substituents]

R^1	R^2	R^3	% Yield
H	CN	H	63
H	COMe	H	75
H	CHO	Me	72
-C(O)OCO-		H	48

I.E.1-4 I. Fleming and A. Percival, *J. Chem. Soc., Chem. Commun.*, **1976**, 681.

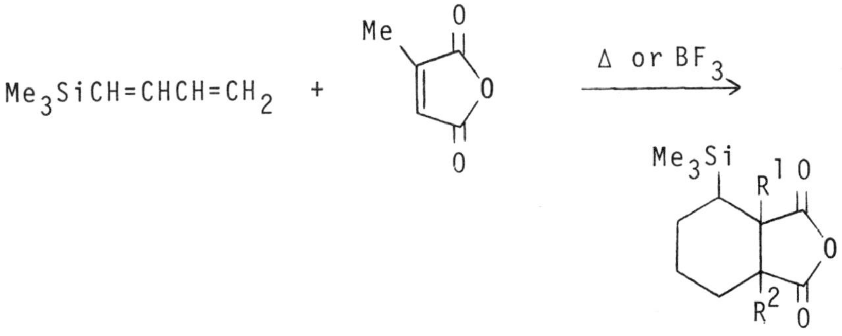

$Me_3SiCH=CHCH=CH_2$ + [methyl-substituted furandione] $\xrightarrow{\Delta \text{ or } BF_3}$ [bicyclic product with Me_3Si, R^1, R^2]

R^1 = Me, R^2 = H, 28%

R^1 = H, R^2 = Me, 24%

Little regioselectivity shown with other dienophiles.

I.E.1-5 L. C. Dunn, Y.-M. Chang, and K. N. Houk, J. Am. Chem. Soc., **98**, 7095 (1976).

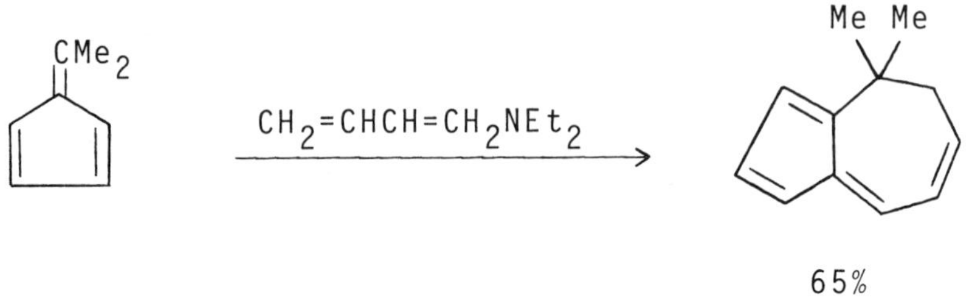

[cyclopentadiene with =CMe$_2$ substituent] + CH$_2$=CHCH=CH$_2$NEt$_2$ → [bicyclic Me,Me product] 65%

I.E.1-6 J. S. Baum and H. G. Viehe, J. Org. Chem., **41**, 183 (1976).

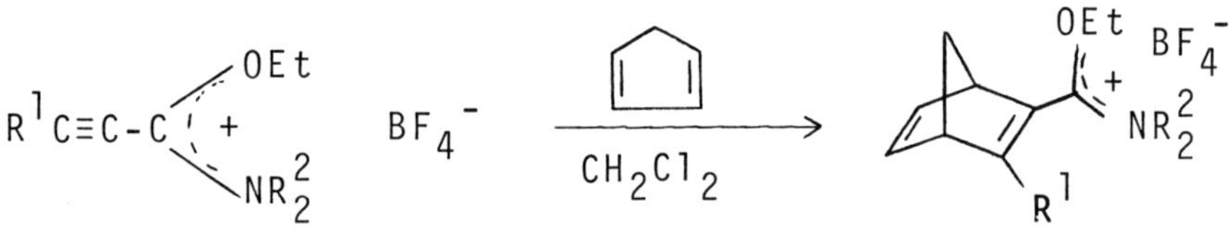

R^{1}C≡C-C(OEt)(+)(NR$_2^2$) BF$_4^-$ + [cyclopentadiene] →(CH$_2$Cl$_2$) [norbornene product with OEt, NR$_2^2$, R^1, BF$_4^-$]

R^1	R^2	% Yield
Ph	Me	85
t-Bu	Me	32
H	Me	87
H	H	71

The acetylenic iminium compounds are more reactive in Diels-Alder cycloadditions than the corresponding esters, acid chlorides or nitriles.

I.E.1-7 R. G. Salomon, J. R. Burns, and W. J. Dominic, J. Org. Chem., 41, 2918 (1976).

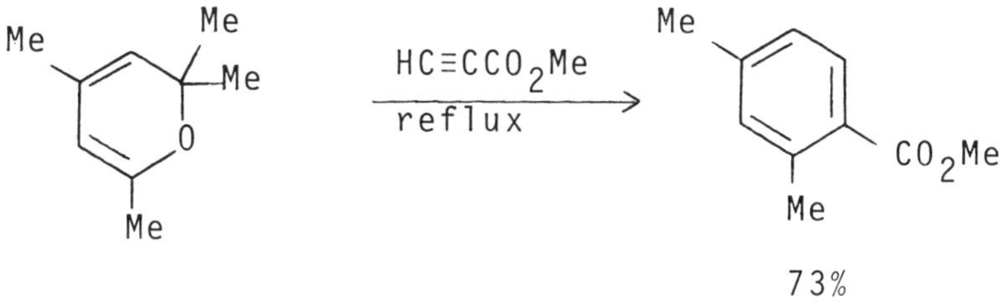

[Reaction: 2,2,4,6-tetramethyl-2H-pyran + HC≡CCO$_2$Me, reflux → methyl 2,4-dimethylbenzoate derivative, 73%]

Other 2H-pyrans were studied and found to undergo similar regiospecific reactions.

I.E.1-8 J. H. Lukas, F. Baardman, and A. P. Kouwenhoven, Angew. Chem. Int. Ed. Engl., 15, 369 (1976).

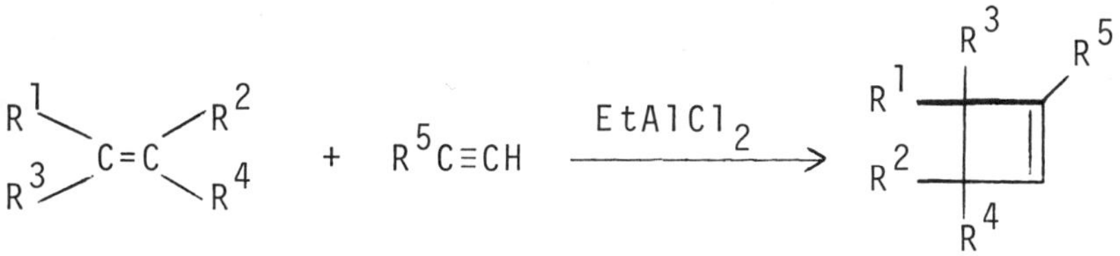

$$R^1R^3C=CR^2R^4 + R^5C\equiv CH \xrightarrow{EtAlCl_2} \text{cyclobutene with } R^1, R^2, R^3, R^4, R^5$$

R^1	R^2	R^3	R^4	R^5	% Yield
Me	Me	Me	Me	Me	27
Me	Me	Me	Me	C_5H_{11}	50
H	Me	H	Me	Et	30

I.E.1-9 P.H.J. Ooms, M. A. Bertisen, H. W. Scheeren, and R.J.F. Nivard, J. Chem. Soc., Perkin Trans I, 1976, 1538.

$$(R^1O)_2C=C(OR^1)_2 + R^2CH=CHCH=C(X)CN \xrightarrow{25-70°}$$

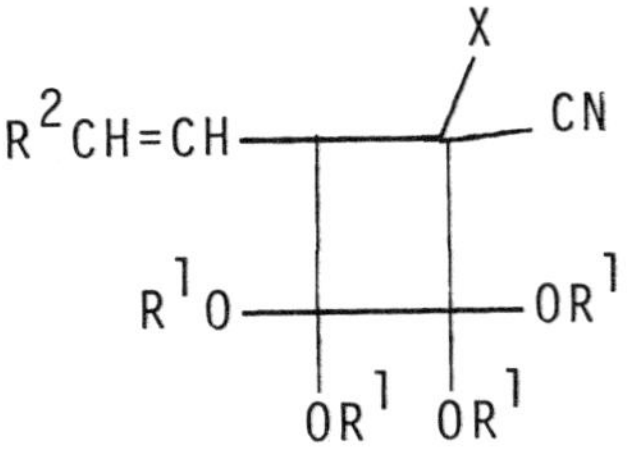

[structure: cyclobutane ring with $R^2CH=CH$ and X, CN on one carbon; R^1O, OR^1, OR^1, OR^1 on the others]

R^1	R^2	X	% Yield
Me	i-Pr	CN	70
Me	Ph	CN	88
Et	p-$NO_2C_6H_4$	CN	80
Me	Ph	CO_2Et	82
Et	Ph	SO_2Ph	80

I.E.1-10 R. Huisgen, R. Schug, and G. Steiner, Bull. Soc. Chim. Fr., 1976, 1813.

Review: Cycloadditions 2 + 2 Par Des Intermediaires Dipolaires-1,4.

I.E.1-11 E. M. Mil'vitskaya, A. V. Tarakanova, and
A. F. Plate, Russ. Chem. Rev., 45, 469 (1976).

Review: Thermal Rearrangements of Vinylcyclo-
propanes.

I.E.1-12 E. Piers, C. K. Lau, and I. Nagakura,
Tetrahedron Lett., 1976, 3233.

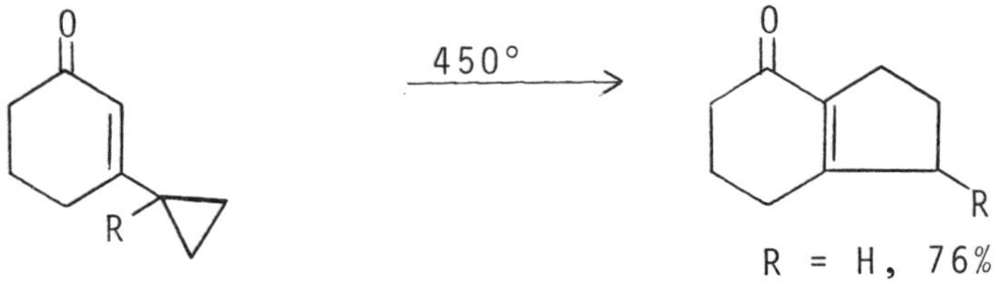

$\xrightarrow{450°}$

R = H, 76%
R = Me, 79%

I.E.1-13 E. Piers and I. Nagakura, Tetrahedron
Lett., 1976, 3237; see also: J. P. Marino and L. J.
Browne, *ibid*, 3245; P. A. Wender and M. P. Filosa,
J. Org. Chem., 41, 3490 (1976).

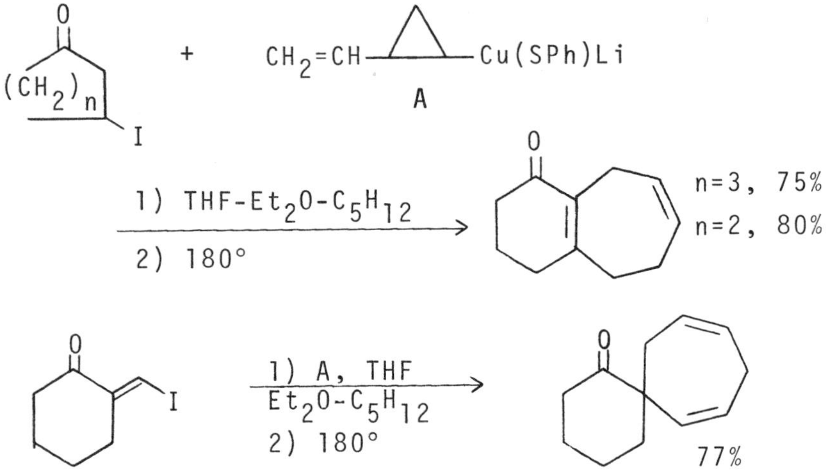

$(CH_2)_n$ + $CH_2=CH-\triangle-Cu(SPh)Li$
            A

1) THF-Et$_2$O-C$_5$H$_{12}$
2) 180°

n=3, 75%
n=2, 80%

1) A, THF
   Et$_2$O-C$_5$H$_{12}$
2) 180°

77%

I.E.2-1 J. Haslouin and F. Rouessac, <u>Tetrahedron Lett.</u>, <u>1976</u>, 4651.

$$\text{[bicyclic lactone with } R^1, R^2\text{]} \xrightarrow[\text{0.01 Torr}]{550°} \text{[lactone with } R^1, R^2, =CH_2\text{]}$$

quantitative yield

I.E.2-2 J. Haslouin and F. Rouessac, <u>Bull. Soc. Chim. Fr.</u>, <u>1976</u>, 1122; see also: O. Gringore, J. Haslouin, and F. Rouessac, *ibid*, 1523.

$$\text{[norbornene-OSiMe}_3\text{, R]} \xrightarrow{500°} CH_2=CROSiMe_3$$

quantitative yield

I.E.2-3 F. Jung, <u>J. Chem. Soc., Chem. Commun.</u>, <u>1976</u>, 525.

$$\text{[cyclopropane with } R^1, R^2, R^3\text{, O-S=O]} \xrightarrow[\text{reflux}]{CHCl_3} CH_2=CHCH_2\underset{|}{\overset{R^1}{C}}=\underset{}{\overset{R^3}{C}}R^2$$

100% yield

(>99.5% isomeric purity)

I.E.2-4 H. Krapf, P. Riedl, and J. Sauer, Chem. Ber., 109, 576 (1976); see also: M. Hirth, H. Krapf, P. Riedl, J. Sauer, and E. Oeser, *ibid*, 562.

$$\text{Ph-cyclopropene-X/Ph} + CH_2=C(Y)(NMe_2) \xrightarrow{\Delta} CH_2=C(Y)CPh=CPhC(X)NMe_2$$

X = O and S
Y = NMe_2, OEt, and SMe

I.E.2-5 H. Wolfers, U. Kraatz, and F. Korte, Chem. Ber., 109, 1061 (1976).

$$R^2, R^1\text{-butenolide} \xrightarrow[140-160°]{R^3CH_2C(OEt)_3} R^2, R^1\text{-butenolide with }=C(OEt)CH_2R^3$$

30-90%

I.E.2-6 O. Achmatowicz, Jr. And M. Pietraszkiewicz, J. Chem. Soc., Chem. Commun., 1976, 484.

$$R^1CH=CR^3CH_2R^3 \xrightarrow[\text{Lewis acid, r.t.}]{TsN=CHCO_2Bu, 120° or}$$

$$R^2CH=CR^3CHR^1CH(NHTs)CO_2Bu$$

R^1	R^2	R^3	% Yield
H	H	H	70
H	Et	H	90
H	H	Ph	92
$\ce{-(CH_2)_3-}$		H	81

I.E.2-7 D. S. Karanewsky and Y. Kishi, J. Org. Chem., 41, 3026 (1976).

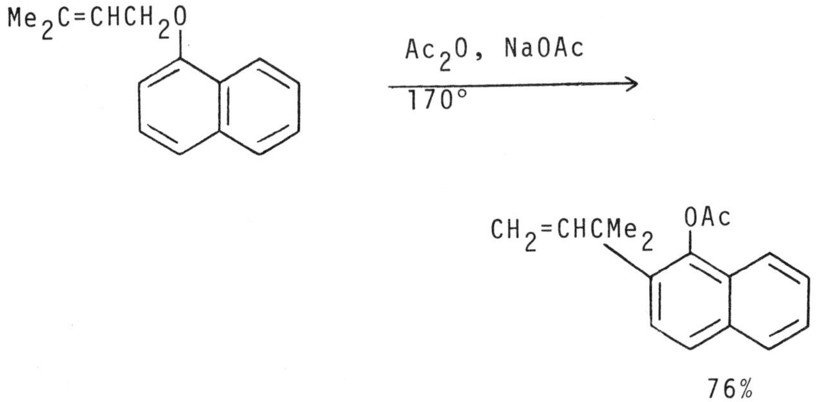

Me$_2$C=CHCH$_2$O—[1-naphthyl] $\xrightarrow{\text{Ac}_2\text{O, NaOAc}}_{170°}$ CH$_2$=CHCMe$_2$—[2-position]-naphthyl-1-OAc

76%

Conditions for formation of normal Claisen rearrangement products have been developed for cases where standard conditions give the abnormal product.

I.E.2-8 R. E. Ireland, R. H. Mueller, and A. K. Willard, J. Am. Chem. Soc., 98, 2868 (1976).

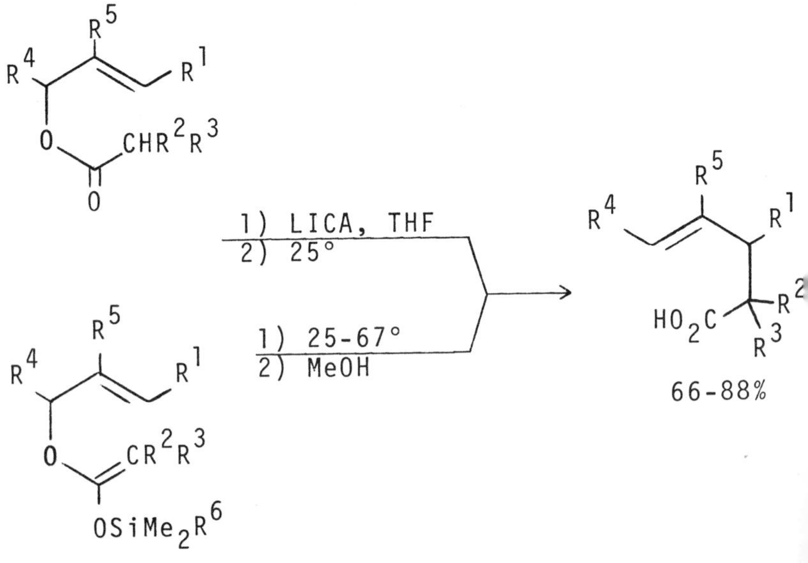

$\xrightarrow[\text{2) 25°}]{\text{1) LICA, THF}}$

$\xrightarrow[\text{2) MeOH}]{\text{1) 25-67°}}$

66-88%

I.E.2-9 K. A. Parker and R. W. Kosley, Jr., Tetrahedron Lett., 1976, 341.

$HC\equiv CC(OH)Me_2$ $\xrightarrow{\begin{array}{c}1)\ (EtO)_2C(NMe_2)Me\\ 115°\\ 2)\ H_3O^+\end{array}}$

$Me_2CHCO(CH_2)_2CO_2Et$

73%

$Me_3SiC\equiv CC(OH)R^1R^2$ $\xrightarrow{\begin{array}{c}(EtO)_2C(NMe_2)Me\\ \text{xylene, reflux}\end{array}}$

$R^1R^2C=C=C(SiMe_3)CH_2CONMe_2$

59-86%

I.E.2-10 R. W. Thies and E. P. Seitz, J. Chem. Soc., Chem. Commun., 1976, 846.

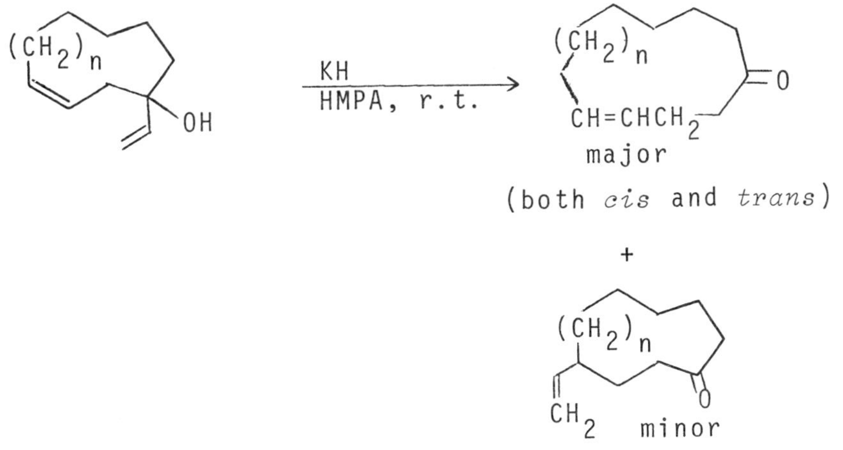

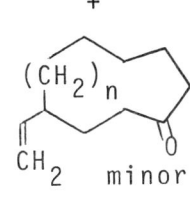

(both *cis* and *trans*)

Use of the K salt allows the [1,3] shift of oxy-Cope system to occur at room temperature.

I.E.2-11 J. F. King and D.R.K. Harding, J. Am. Chem. Soc., 98, 3312 (1976).

$$CH_2=CRCH_2SO_2CH=CH_2 \xrightarrow[\text{py, EtOH}]{165-175°} CH_2=CR(CH_2)_3SO_3^-$$

R = H, 70%
R = Me, 96%
R = Ph, 81%

I.E.2-12 F. Vögtle and E. Goldschmitt, Chem. Ber., 109, 1 (1976).

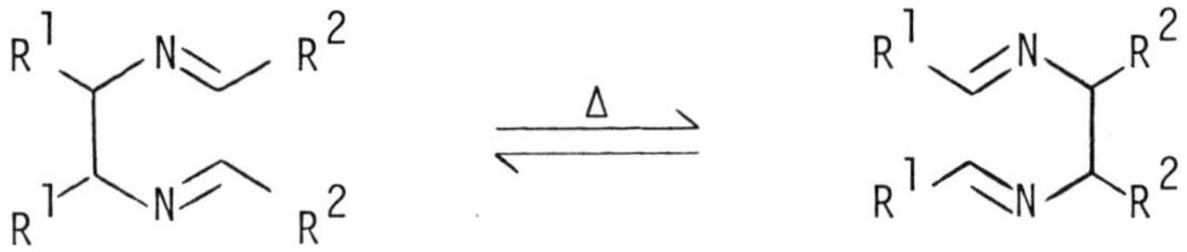

A full paper describing the general diaza-Cope rearrangement and the use of an o-hydroxy group to determine the equilibrium of the valence isomerization.

I.E.2-13   C. W. Spangler, Chem. Rev., **76**, 187 (1976).

Review: Thermal [1,j]Sigmatropic Rearrangements

I.E.2-14   L. E. Overman and L. A. Clizbe, J. Am. Chem. Soc., **98**, 2352 (1976).

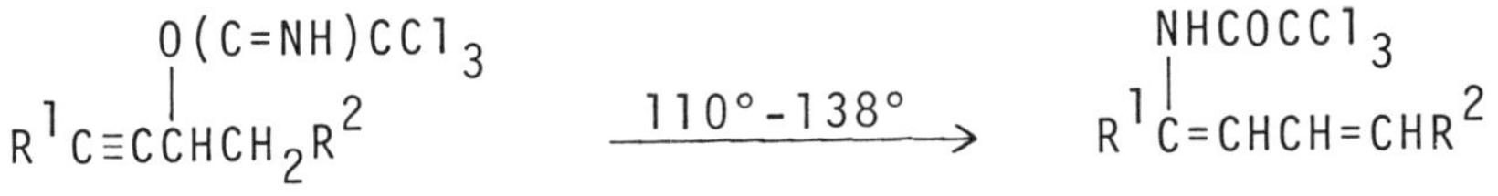

$$R^1C\equiv CCH(O(C=NH)CCl_3)CH_2R^2 \xrightarrow{110°-138°} R^1C(NHCOCCl_3)=CHCH=CHR^2$$

$R^1 = R^2 = H$, 38%
$R^1 = Ph$, $R^2 = Me$, 45%
$R^1 = R^2 = t\text{-Bu}$, 85%

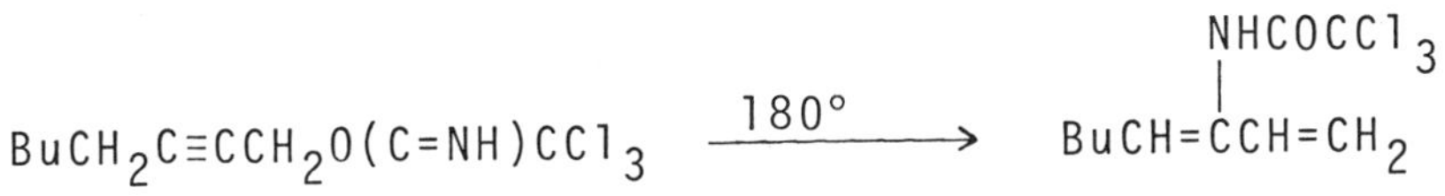

$$BuCH_2C\equiv CCH_2O(C=NH)CCl_3 \xrightarrow{180°} BuCH=C(NHCOCCl_3)CH=CH_2$$

78%

The trichloroacetamido-1,3-dienes are useful in Diels-Alder reactions.

I.F.1-1 G. P. Schiemenz and U. Schmidt, Justus Liebigs Ann. Chem., 1976, 1514.

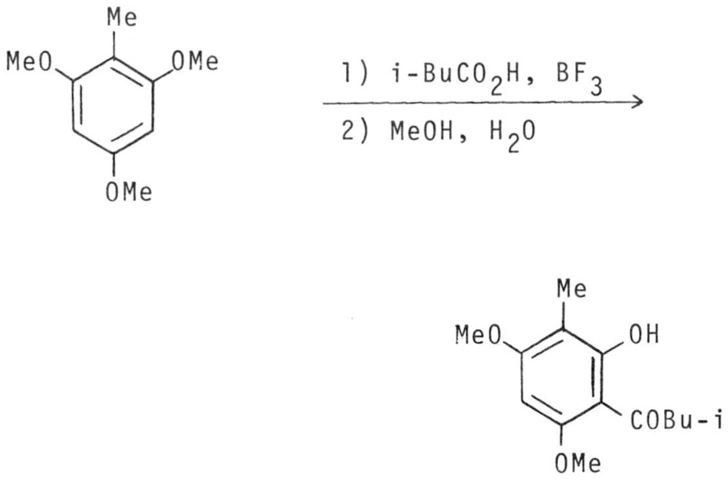

1,3,5-trimethoxy-2-methylbenzene + 1) i-BuCO$_2$H, BF$_3$ 2) MeOH, H$_2$O → 2-hydroxy-3-isobutyryl-4,6-dimethoxy-toluene derivative (MeO, Me, OH, COBu-i, OMe substituted benzene)

73%

I.F.1-2 E. P. Papadopoulos, J. Org. Chem., 41, 962 (1976).

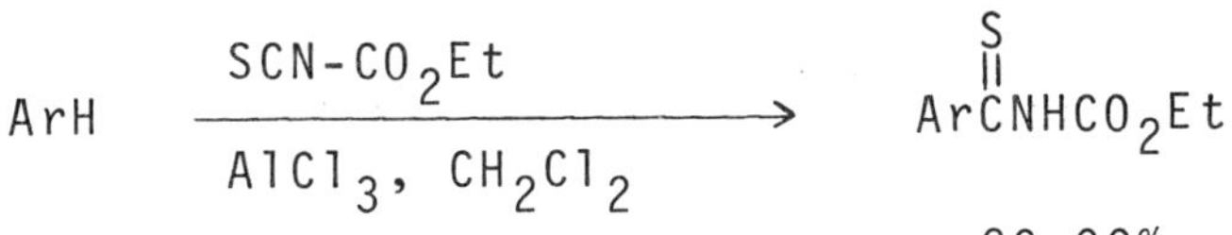

$$\text{ArH} \xrightarrow[\text{AlCl}_3, \text{CH}_2\text{Cl}_2]{\text{SCN-CO}_2\text{Et}} \text{Ar}\overset{\overset{\text{S}}{\|}}{\text{C}}\text{NHCO}_2\text{Et}$$

29-90%

By changing the workup procedure the aromatic thioamides can be isolated directly in approximately the same yield.

I.F.1-3   A. E. Feiring, J. Org. Chem., 41, 148 (1976).

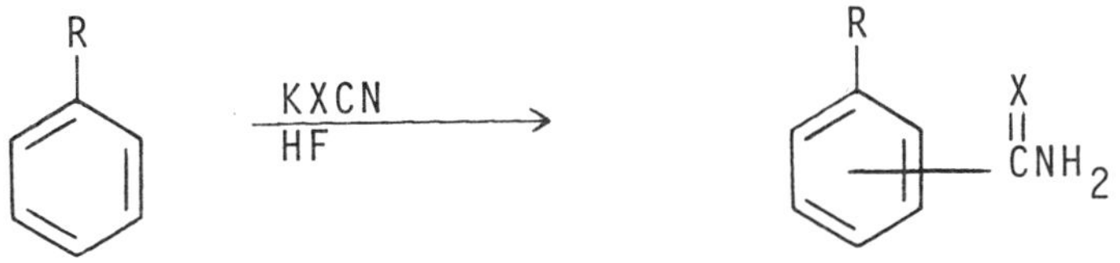

ArR + KXCN / HF → Ar(R)(C(=X)NH$_2$)

X	R	% Yield
O	H	63
O	Me	50 (o/p = 26/74)
S	Me	36 (only p-)

I.F.1-4   J. Zavada, M. Pankova, and Z. Arnold, Coll. Czech. Chem. Commun., 41, 1777 (1976).

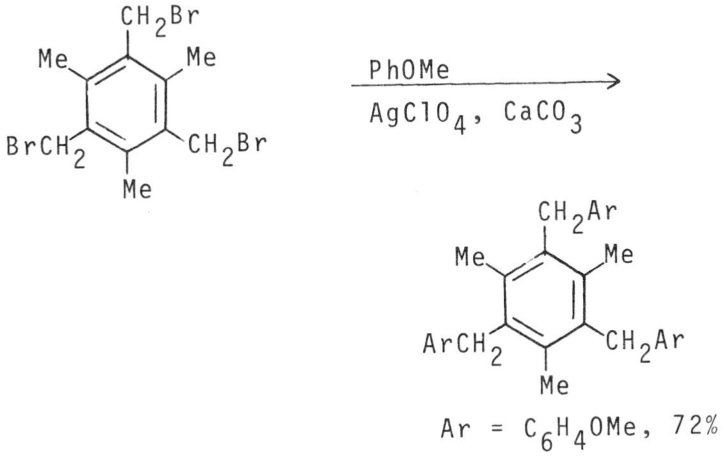

Hexasubstituted benzene [1,3,5-Me$_3$-2,4,6-(CH$_2$Br)$_3$-C$_6$] + PhOMe / AgClO$_4$, CaCO$_3$ → 1,3,5-Me$_3$-2,4,6-(CH$_2$Ar)$_3$-C$_6$

Ar = C$_6$H$_4$OMe, 72%

Use of AgClO$_4$-CaCO$_3$ allows the synthesis of systems that are unstable to normal Friedel-Crafts catalysts.

I.F.1-5 D. Ben-Ishai, I. Sataty, and Z. Bernstein, Tetrahedron, 32, 1571 (1976).

$$ArH + RCONHCH(OH)CO_2H \xrightarrow{H_2SO_4} ArCHCO_2H$$
$$\qquad\qquad\qquad\qquad\qquad\qquad\qquad\quad | $$
$$\qquad\qquad\qquad\qquad\qquad\qquad\qquad NHCOR$$

Ar	R	% Yield
Ph	Ph	91
$C_{10}H_7$	OMe	41
p-MeOC_6H_4	Ph	72
C_4H_3S	Ph	84

I.F.1-6 J.-P. Gesson and J.-C. Jacquesy, J. Chem. Soc., Chem. Commun., 1976, 652.

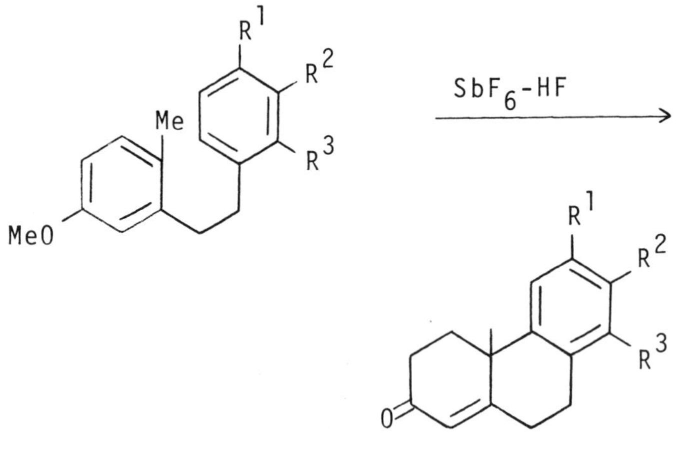

45-78%

I.F.1-7   J. Hocker, H. Giesecke, and R. Merten, Angew. Chem. Int. Ed. Engl., 15, 169 (1976).

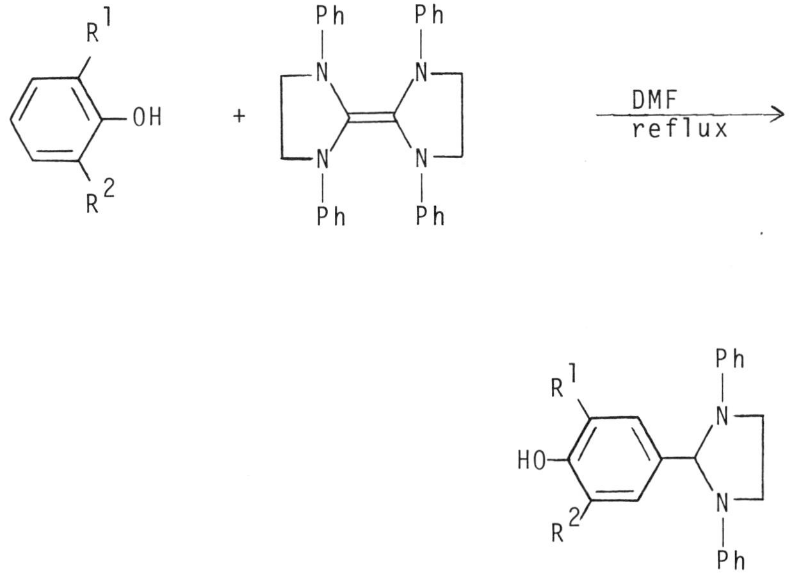

R^1	R^2	% Yield
H	H	55
Me	Me	58
t-Bu	t-Bu	84
Me	Cl	74

I.F.2-1 J. B. Melpolder and R. F. Heck, J. Org. Chem., 41, 265 (1976); see also: A. J. Chalk and S. A. Magennis, *ibid*, 273 and 1206.

$$PhI + CH_2=CHCHROH \xrightarrow[Et_3N, MeCN, 100°]{Pd(OAc)_2 \text{ (cat.)}}$$

$$PhCH_2CH_2COR$$

Study of the palladium-catalyzed arylation of allylic alcohols.

I.F.2-2 A. Sekiya and N. Ishikawa, J. Organomet. Chem., 118, 349 (1976).

$$ArX + RMgBr \xrightarrow[Et_2O \text{ or } THF]{(Ph_3P)_2Pd(Ph)I \text{ (cat.)}} ArR$$

ArX	R	% Yield
PhI	PhC≡C	84
PhI	Et	32
p-ClC$_6$H$_4$Br	Ph	73

I.F.2-3 R. A. Kretchmer and R. Glowinski, *J. Org. Chem.*, 41, 2661 (1976).

$$\text{ArHgX} \xrightarrow[\text{py, 113°}]{\text{Cu, PdCl}_2 \text{ (cat.)}} \text{Ar-Ar}$$
$$47-95\%$$

I.F.2-4 F. E. Ziegler, K. W. Fowler, and S. Kanfer, *J. Am. Chem. Soc.*, 98, 8282 (1976).

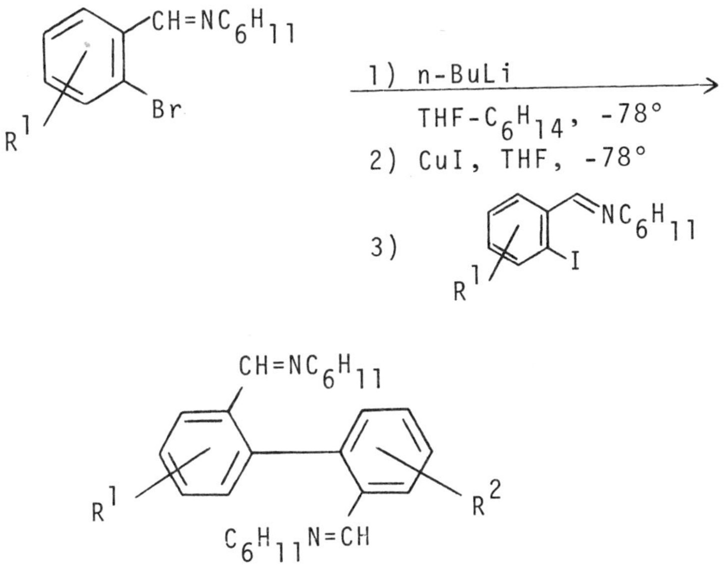

1) n-BuLi, THF-C_6H_{14}, -78°
2) CuI, THF, -78°
3) [aryl iodide with CH=NC$_6$H$_{11}$ and R^1, I substituents]

44-63%

I.F.2-5 E. Negishi and S. Baba, J. Chem. Soc., Chem. Commun., 1976, 596.

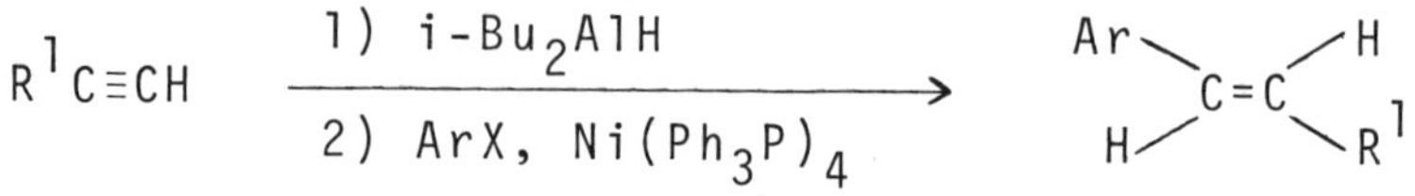

$$R^1C\equiv CH \xrightarrow[\text{2) ArX, Ni(Ph}_3\text{P})_4]{\text{1) i-Bu}_2\text{AlH}} \begin{array}{c}Ar\\ \end{array}\!\!\!C=C\!\!\!\begin{array}{c}H\\ R^1\end{array}$$

R^1	ArX	% Yield
n-Bu	PhI	89
n-Bu	PhBr	85
cyclohexyl	p-MeC$_6$H$_4$Br	75
n-Bu	p-NCC$_6$H$_4$Br	64

I.F.3-1 E. Havinga and J. Cornelisse, Pure and Appl. Chem., 47, 1 (1976).

Plenary Lecture: Aromatic Photosubstitution Reactions.

I.F.3-2 F. Minisci, Fortschr. Chem. Forsch., 62, 1 (1976).

Review: Recent Advances of Homolytic Aromatic Substitutions.

I.F.3-3   O. N. Chupakhin and I. Ya. Postovskii, Russ. Chem. Rev., $\underline{45}$, 454 (1976).

Review: Nucleophilic Substitution of Hydrogen in Aromatic Systems.

I.F.3-4   T. Sakakibara and Y. Odaira, J. Org. Chem., $\underline{41}$, 2049 (1976).

$$ArH \xrightarrow[Ac_2O \text{ or } CCl_4, \text{ HOAc}]{Na\ Pd[CH_2(CO_2)_2]_2} ArCO_2H$$

R = H, 98%
R = OMe, 55%

I.F.3-5   M. E. Kurz and R.T.Y. Chen, J. Chem. Soc., Chem. Commun., $\underline{1976}$, 968.

$$ArH \xrightarrow[Mn(OAc)_3, \text{ HOAc}]{MeNO_2} ArCH_2NO_2$$

Ar	% Yield
C_6H_5	78
MeC_6H_4	77
ClC_6H_4	20
$O_2NC_6H_4$	0

I.F.3-6   S. Uemura, S. Tanaka, and M. Okano, _J. Chem. Soc., Perkin Trans. I_, 1976, 1966.

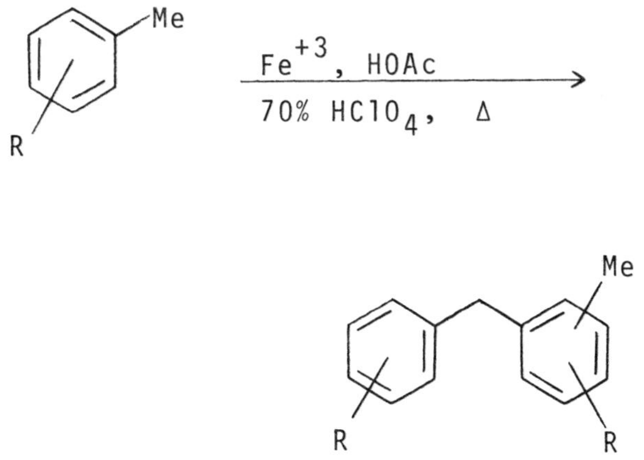

R-C6H4-Me $\xrightarrow{\text{Fe}^{+3}, \text{HOAc}}_{70\% \text{ HClO}_4, \Delta}$

[diarylmethane product with Me and R substituents]

good yields

I.F.3-7   K. Tsujimoto, K. Miyake, and M. Ohashi, _J. Chem. Soc., Chem. Commun._, 1976, 386.

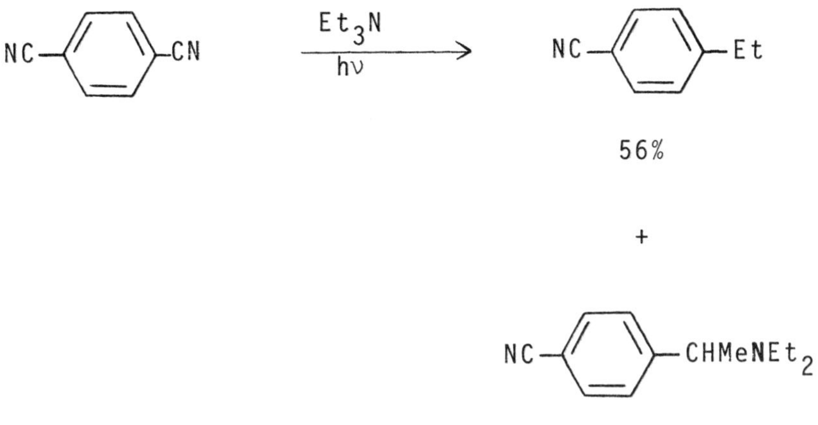

NC-C6H4-CN $\xrightarrow{\text{Et}_3\text{N}}_{h\nu}$ NC-C6H4-Et

56%

+

NC-C6H4-CHMeNEt$_2$

18%

I.G.1-1 J. A. Sinclair and H. C. Brown, J. Org. Chem., 41, 1078 (1976).

$$Sia_2BC \equiv CR^1 \xrightarrow[\text{2) } I_2, \text{ THF, } -78°]{\text{1) } LiC \equiv CR^2, \text{ THF, } -78°} R^1C \equiv CC \equiv CR^2$$

$R^1 = n-C_6H_{13}$, $R^2 =$ cyclohexyl, 73%
$R^1 =$ cyclohexyl, $R^2 =$ t-Bu, 70%

A "one-pot" procedure is described.

I.G.1-2 A. Pelter, R. Hughes, K. Smith, and M. Tabata, Tetrahedron Lett., 1976, 4385.

$$(cyclohexyl)_2BSMe \xrightarrow[\substack{\text{2) } LiC \equiv CR^2 \\ \text{3) } I_2}]{\text{1) } LiC \equiv CR^1} R^1C \equiv CC \equiv CR^2$$

$R^1 = n-C_6H_{13}$, $R^2 = n-Bu$, 61%
$R^1 = t-Bu$, $R^2 = Ph$, 51%

I.G.1-3 K. Utimoto, Y. Yabuki, K. Okada, and H. Nozaki, Tetrahedron Lett., 1976, 3969.

$$R^1_3B \xrightarrow[\substack{\text{1) LiC≡CH} \\ \text{2) n-BuLi} \\ \text{3) } R^2X \\ \text{4) } I_2}]{} R^1C≡CR^2$$

$$R^1 = C_6H_{13}, R^2 = n\text{-}Bu, 72\%$$
$$R^1 = Bu, R^2 = PhCH_2, 87\%$$

I.G.1-4 A. Pelter, T. W. Bentley, C. R. Harrison, C. Subrahmanyam, and R. J. Laub, J. Chem. Soc., Perkin Trans. 1, 1976, 2419; see also: A. Pelter, K. J. Gould, and C. R. Harrison, *ibid*, 2428.

$$R^1_3\bar{B}C≡CR^2 \; Li^+ \xrightarrow{R^3X} R^1_2BCR^1=CR^2R^3 \longrightarrow$$

mixture of *cis* and *trans*

$$\xrightarrow{ox.} R^1COCHR^2R^3$$
$$\xrightarrow{H^+} R^1CH=CR^2R^3$$

Full paper describes conditions favoring formation of dialkylvinylboranes and subsequent reactions; stereoselectivity in alkene formation is obtained using alkynyldialkyl(thexyl)borates.

I.G.1-5 K.-W. Chiu, E.-I. Negishi, M. S. Plante, and A. Silveira, Jr., *J. Organomet. Chem.*, 112, C3 (1976).

$$(\text{C}_6\text{H}_{11})_2\text{B}-\underset{\underset{\text{Bu}}{|}}{\text{C}}=\text{C}\overset{\text{Bu}}{\underset{\text{H}}{}} \quad \text{Li}^+ \xrightarrow{\text{1N HCl}} \text{C}_6\text{H}_{10}=\text{C}\overset{\text{Bu}}{\underset{\text{CH}_2\text{Bu}}{}}$$

73%

Other systems studied and the migrating group is the more highly substituted group.

I.G.1-6 Y. Yamamoto, H. Yatagai, A. Sonoda, and S.-I. Murahashi, *J. Chem. Soc., Chem. Commun.*, 1976, 452.

$$R^1C\equiv CR^2 \xrightarrow[\substack{2)\ 3\ \text{eq. MeCu, } -30° \\ 3)\ R^3X}]{1)\ H_2BCl} \overset{R^1}{\underset{H}{}}C=C\overset{R^2}{\underset{R^3}{}}$$

R^1	R^2	R^3X	% Yield
n-Bu	H	CH_2=$CHCH_2Br$	95
n-Bu	n-Bu	CH_2=$CHCH_2Br$	70
n-Bu	H	n-BuI	47

For the non-activated alkyl halides (R^3 = n-Bu) a phosphite was added.

I.G.1-7 J. Hooz and R. Mortimer, <u>Tetrahedron Lett.</u>, <u>1976</u>, 805.

R^1_3B $\xrightarrow{\begin{array}{l}1)\ R^2C{\equiv}CLi\\2)\ n\text{-}Bu_3SnCl\\3)\ HCO_2H\end{array}}$

$\underset{R^1}{H}\diagdown C=C\diagup\underset{R^2}{H}$

73-89%

$\underset{Et}{Et_2B}\diagdown C=C\diagup\underset{Bu}{SnBu_3}$ $\xrightarrow{\begin{array}{l}1)\ MeSCH_2Li,\ TMEDA\\2)\ MeI\\3)\ HCO_2H\end{array}}$ $\underset{Et}{Me}\diagdown C=C\diagup\underset{Bu}{H}$

52%

I.G.1-8 P. Jacob, III and H. C. Brown, <u>J. Am. Chem. Soc.</u>, <u>98</u>, 7832 (1976).

$R^1C{\equiv}CR^2$ $\xrightarrow{\begin{array}{l}1)\ 9\text{-}BBN,\ THF\\2)\ CH_2=CHCOMe\\ \quad reflux\\3)\ oxidative\ workup\end{array}}$ $\underset{H}{R^1}\diagdown C=C\diagup\underset{(CH_2)_2COMe}{R^2}$

R^1	R^2	% Yield
n-Bu	H	87
t-Bu	H	85
Ph	H	93
t-Bu	Me	69
Et	Et	35

I.G.1-9 R. Köster, H.-J. Zimmerman, and W. Fenzl, Justus Liebigs Ann. Chem., 1976, 1116.

$$CH_2=CMeCHO \xrightarrow{R^1_2BR^2} R^2CH_2CMe=CHOBR^1_2$$
$$\text{A}$$

$$\xrightarrow{PhCOCOPh} R^2CHMeCHO$$
$$\text{B}$$

R^1	R^2	% Yield A	B
Et	Et	89	-
Et	MeCH=CHCH$_2$	65	-
Et	PhCH$_2$	86	-
i-Pr	i-Pr	79	98

I.G.1-10 R. Murphy and R. H. Prager, Tetrahedron Lett., 1976, 463.

$$CH_2=CR^1(CH_2)_nCR^2=CH_2 \xrightarrow[\substack{2) \ KOH, \ MeOH \\ aq. \ AgNO_3}]{1) \ BH_3 \cdot THF} \underset{CHR^1}{\overset{HCR^1}{(CH_2)_n \quad (CH_2)_2}}$$

n	R^1	R^2	% Yield
0	Me	Me	79
1	Me	H	85
2	H	H	66 (17% methylcyclopentane)
3	H	H	67
4	H	H	42

I.G.1-11 M. W. Rathke, E. Chao, and G. Wu, J. Organomet. Chem., 122, 145 (1976).

$$Cl_2CHB(i\text{-}PrO)_2 \xrightarrow[\text{2) }H_2O_2,\text{ pH 8.8}]{\text{1) RM}} RCHO$$

RM = n-BuLi, 60%
RM = t-BuLi, 35%
RM = PhMgBr, 70%

$$Cl_2CHB(i\text{-}PrO)_2 \xrightarrow[\text{2) }H_2O_2,\text{ pH 8.8}]{\text{1) 2 eq. n-BuLi}} (n\text{-}Bu)_2CHOH$$
40%

I.G.1-12 R. J. Hughes, A. Pelter, K. Smith, E. Negishi, and T. Yoshida, Tetrahedron Lett., 1976, 87.

$$R^1R^2_2B \xrightarrow[\substack{\text{2) }HgCl_2\text{ (3 eq.)}\\ \text{3) }H_2O_2,\text{ NaOH}}]{\text{1) }LiCH(SPh)_2} R^2_2CHOH$$

R^1	R^2	% Yield
n-C_6H_{13}	n-C_6H_{13}	85
cyclohexyl	cyclohexyl	88
thexyl	cyclopentyl	80

I.G.1-13 A. B. Levy and S. J. Schwartz, *Tetrahedron Lett.*, 1976, 2201.

$$R_3B \xrightarrow[\substack{2)\ 2\underline{N}\ HCl \\ 3)\ H_2O_2,\ NaOH}]{1)\ LiC(OMe)=CH_2} R_2C(OH)Me$$

R	% Yield
$n\text{-}C_6H_{13}$	100
sec-Bu	92
cyclopentyl	87

$$(RCH_2)_3B \xrightarrow[2)\ H_2O_2,\ NaOH]{1)\ LiC(OMe)=CH_2} RCH_2COMe$$

$R = C_5H_{11}$, 82%

I.G.1-14 N. Miyaura, M. Itoh, and A. Suzuki, *Synthesis*, 1976, 618.

$$n\text{-}Bu_3B \xrightarrow[\substack{2)\ CuCN,\ THF \\ 3)\ PhCH_2Br}]{1)\ MeLi} PhC_5H_{11} + PhCH_2CH_2Ph$$

68%   16%

I.G.1-15   Y. Takahashi, M. Tokuda, M. Itoh, and A. Suzuki, Synthesis, 1976, 616.

$$R_3B + MeNO_2 \xrightarrow{electrolysis} RCH_2NO_2$$

R	% Yield
n-Pr	84
n-C_8H_{17}	92
sec-Bu	76
cyclopentyl	150

I.G.1-16   K. Utimoto, N. Sakai, M. Obayashi, and H. Nozaki, Tetrahedron, 32, 769 (1976).

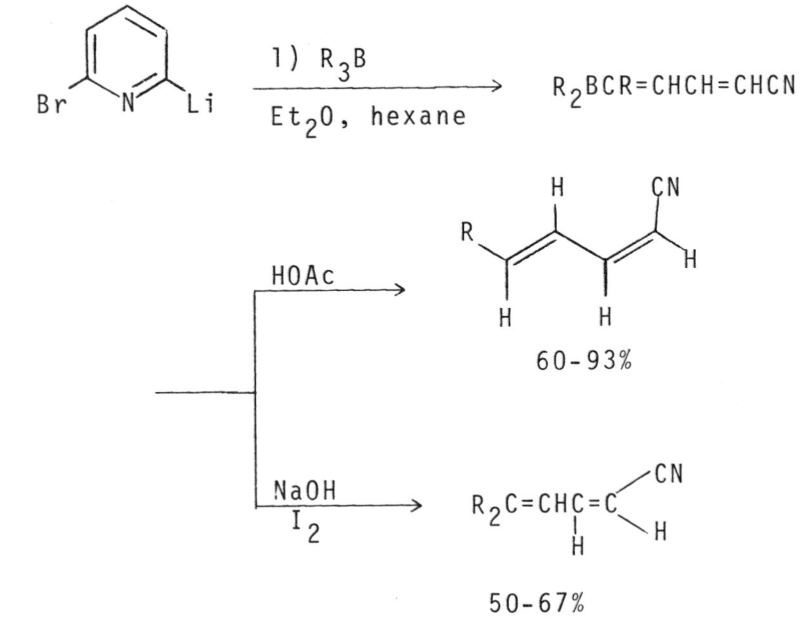

$$\underset{Br}{\underset{|}{\text{pyridine}}}-Li \xrightarrow[Et_2O, hexane]{1)\ R_3B} R_2BCR=CHCH=CHCN$$

$\xrightarrow{HOAc}$ (R, H, H, H, H, CN on diene) 60-93%

$\xrightarrow[I_2]{NaOH}$ $R_2C=CHC=C\begin{smallmatrix}CN\\H\end{smallmatrix}$ with H on middle C

50-67%

I.G.2-1 C. A. Bertelo and J. Schwartz, J. Am. Chem. Soc., 98, 262 (1976).

$$R^1R^2C=CR^3CH=CH_2 \xrightarrow{\begin{array}{l}1)\ (Cp)_2Zr(H)Cl,\ PhH\\ 2)\ CO\ (50\ psi),\ PhH\\ 3)\ H_3O^+\end{array}}$$

$$R^1R^2C=CR^3(CH_2)_2CHO$$

R^1	R^2	R^3	% Yield (based on RZr)
Me	H	H	81
H	H	Me	98
Me	H	Me	91
Me	Me	H	98

I.G.2-2 T. Yamamoto, T. Kohara, and A. Yamamoto, Chem. Lett., 1976, 1217.

$$R_2Ni\ (bpy) \xrightarrow[-78°\ to\ rt]{CO,\ THF} RCOR$$

R = Me, Et, Pr (80-90%)

$$PhMgBr\ +\ PhBr \xrightarrow[NiEt_2\ (bpy)]{CO,\ Et_2O} PhCOPh\ +\ Ph_3COH$$

(bpy) = 2,2'-bipyridine

I.G.2-3 W. Best, B. Fell, and G. Schmitt, <u>Chem. Ber.</u>, <u>109</u>, 2914 (1976).

RC≡CR $\xrightarrow{\text{Ni(CO)}_4 \atop \text{HCl}}$ [cyclopentenone with 4 R groups]

R = H, alkyl, or Ph

15-70%

I.G.2-4 J. X. McDermott, M. E. Wilson, and G. M. Whitesides, <u>J. Am. Chem. Soc.</u>, <u>98</u>, 6529 (1976).

Cp$_2$Ti[cyclopentane ring] $\xrightarrow{\text{CO (1 atm.)} \atop \text{Et}_2\text{O}}$ [cyclopentanone]

80-90%

[cyclohexane-CH$_2$Li, CH$_2$Li] $\xrightarrow{\text{1) Cp}_2\text{TiCl}_2 \atop \text{2) CO}}$ [bicyclic ketone]

20%

I.G.2-5 R. Aumann and J. Knecht, <u>Chem. Ber.</u>, <u>109</u>, 174 (1976).

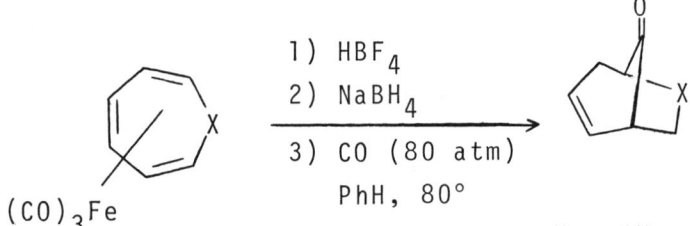

1) HBF$_4$
2) NaBH$_4$
3) CO (80 atm)
   PhH, 80°

X = CH$_2$, 57%
X = CH=CH, 62%

I.G.2-6 R. F. Heldeweg and H. Hogeveen, J. Am. Chem. Soc., 98, 6040 (1976).

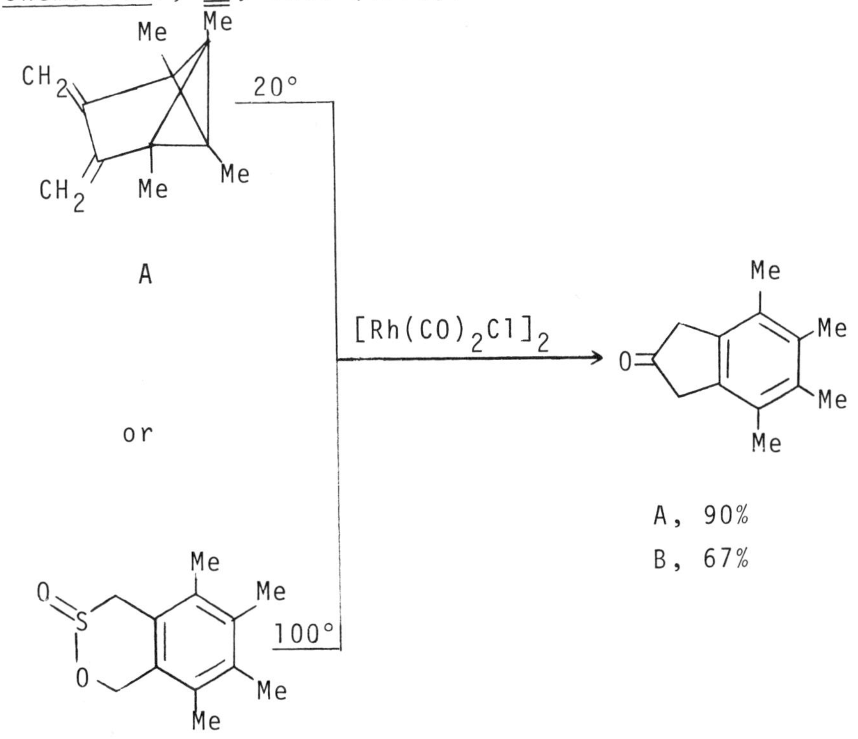

A (tetracyclic diene with Me groups and =CH$_2$ groups)

or

B (cyclic sulfoxide/sulfinate with tetramethylbenzene)

$[Rh(CO)_2Cl]_2$, 20° (from A) or 100° (from B) → tetramethyl-indanone

A, 90%
B, 67%

I.G.2-7 L. Cassar, M. Foa, and A. Gardano, J. Organomet. Chem., 121, C55 (1976).

$$RX \xrightarrow[\text{30\% NaOH, xylene, 95\%}]{\text{CO (5 atm), Pd(Ph}_3\text{P)}_4\text{, Bu}_4\text{NI}} RCO_2Na$$

R = PhCH$_2$, X = Cl, 83%
R = p-BrC$_6$H$_4$, X = Cl, 90%

The use of phase-transfer catalysis allows selectivity in polyhalogenated aromatic compounds.

I.G.3-1 S. Danishefsky, T. Kitahara, M. Tsai, and J. Dynak, J. Org. Chem., 41, 1669 (1976).

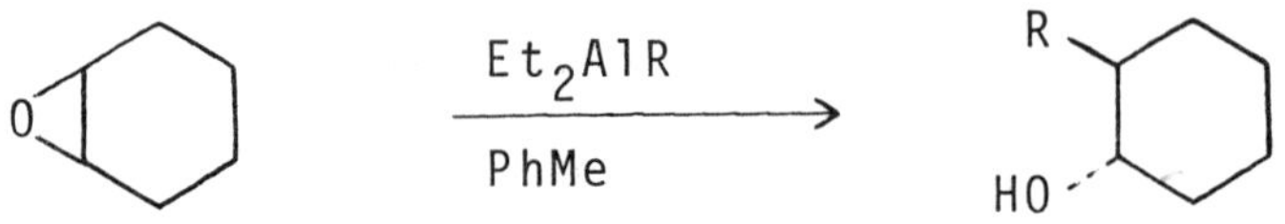

R = CH_2CO_2t-Bu, 68%
R = C≡COEt, 95%

The RLi derivatives give no or very poor yields of product.

I.G.3-2 Y. Kitagawa, S. Hashimoto, S. Iemura, H. Yamamoto, and H. Nozaki, J. Am. Chem. Soc., 98, 5030 (1976).

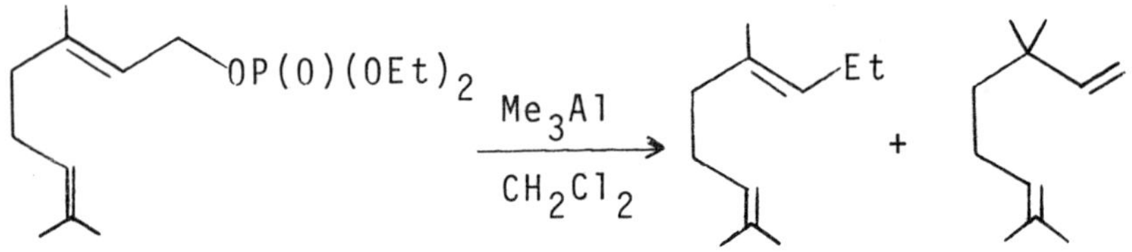

9 : 1

> 90%

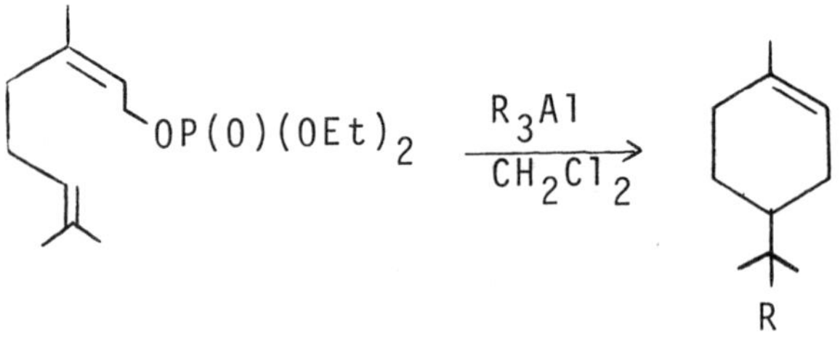

R = Me, 68%
R = Et, 70%
R = i-Bu, 68%

I.G.3-3 S. Hashimoto, Y. Kitagawa, S. Iemura, H. Yamamoto, and H. Nozaki, Tetrahedron Lett., 1976, 2615.

$$Me_2C=CH(CH_2)_2CMe=CHCH_2OR^1 \xrightarrow{R^2_3Al}{hexane}$$

$$Me_2C=CH(CH_2)_2CMe=CHCH_2R^2 + Me_2C=CH(CH_2)_2CR^2MeCH=CH_2$$
$$\qquad\qquad A \qquad\qquad\qquad\qquad\qquad B$$

R^1	R^2	% Yield	A:B
Ac	Me	74	92:8
Ac	i-Bu	66	97:3
CO_2Et	Et	69	85:13
CO_2Et	i-Bu	73	97:3
OTHP	Me	80	96:4
OTHP	i-Bu	<3	-

I.G.3-4 P. W. Chum and S. E. Wilson, Tetrahedron Lett., 1976, 1257.

$$CH_2=CH(CH_2)_2CH=CH_2 \xrightarrow[\text{mineral oil, }115°]{i-Bu_2AlH\ (10\ mol\ \%)}$$

81%

$$[CH_2=CH(CH_2)_2]_2C=CH \xrightarrow[\text{mineral oil, }110°]{i-Bu_2AlH\ (cat.)}$$

78%

I.G.3-5 E. Negishi, S. Baba, and A. O. King, J. Chem. Soc., Chem. Commun., 1976, 17.

$$R^1C\equiv CH \xrightarrow{\begin{array}{l}1)\ i\text{-BuAlH, }n\text{-}C_6H_{14},\ 50°\\ 2)\ n\text{-BuLi, }n\text{-}C_6H_{14},\ 25°\\ 3)\ \triangle\!-\!R^2 \text{ (epoxide)}\end{array}} \begin{array}{c}R^1\\ \diagdown\\ H\end{array}C=C\begin{array}{c}H\\ \diagdown\\ CH_2CHR^2OH\end{array}$$

R^1	R^2	% Yield
n-Bu	H	75
n-Bu	Me	74
t-Bu	Me	82

I.G.3-6 S. Baba and E. Negishi, J. Am. Chem. Soc., 98, 6729 (1976).

$$R^1C\equiv CH \xrightarrow{\begin{array}{l}1)\ i\text{-Bu}_2AlH,\ THF\\ 2)\ R^2R^3C=CHX\\ \text{cat. PdL}_n \text{ or NiL}_n\end{array}} \begin{array}{c}R^1\\ \diagdown\\ H\end{array}C=C\begin{array}{c}H\\ \diagdown\\ CH=CR^2R^3\end{array}$$

R^1	R^2	R^3	X	% Yield
$n\text{-}C_5H_{11}$	n-Bu	H	I	74
t-Bu	n-Bu	H	I	36
n-Bu	t-Bu	H	I	82
n-Bu	CO_2Me	Me	Br	75

The reaction is stereospecific in terms of both the vinylalane and the vinyl halide.

I.G.3-7 R. A. Coleman, C. M. O'Doherty, H. E. Tweedy, T. V. Harris, and D. W. Thompson, J. Organomet. Chem., 107, C15 (1976).

$$RC \equiv C(CH_2)_nOH \xrightarrow[\text{2) Et}_2\text{AlCl, CH}_2\text{Cl}_2, -78°]{\text{1) TiCl}_2(C_5H_7O_2)_2, \text{py}}$$

$$\underset{R}{\overset{Et}{\diagdown}} C = C \underset{(CH_2)_nOH}{\overset{H}{\diagup}}$$

R	n	% Yield
H	2	53
Me	2	53
H	3	20
H	1	0
H	4	0

I.G.3-8 K. Tamao, K. Sumitani, Y. Kiso, M. Zembayashi, A. Fujioka, S. Kodama, I. Nakajima, A. Minato, and M. Kumada, Bull. Chem. Soc., Japan, 49, 1958 (1976).

$$R^1MgX + R^2X' \xrightarrow{L_2NiX_2 \text{ (cat.)}} R^1-R^2$$

(R = aryl or alkenyl)

Full paper covering the general scope and limitations of the nickel-phosphine complex —catalyzed coupling.

I.G.3-9 P. W. Jennings, D. G. Pillsbury, J. L. Hall, and V. T. Brice, J. Org. Chem., **41**, 719 (1976).

$$RX \xrightarrow[\substack{Ni(acac)_2 \text{ or } Fe(acac)_3 \\ DMF, Ph_3P, Et_4NBr}]{electrolysis} R\text{-}R$$

R = $PhCH_2$, 87%
R = Ph, 65%
R = Bu, 28%

I.G.3-10 S. Takahashi, Y. Suzuki, K. Sonogashira, and N. Hagihara, J. Chem. Soc., Chem. Commun., **1976**, 839.

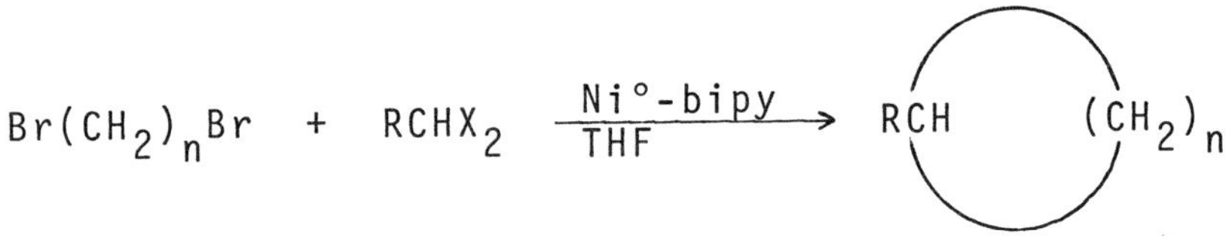

$$Br(CH_2)_nBr + RCHX_2 \xrightarrow[THF]{Ni°\text{-}bipy} \text{cyclo-}RCH(CH_2)_n$$

n	$RCHX_2$	% Yield
4	CH_2Br	70
4	CH_2Cl_2	51
4	$MeCHBr_2$	59
4	$CHCl_3$	27
5	CH_2Br_2	37

I.G.3-11 Y. Ito, S. Fujii, T. Konoike, and T. Saegusa, <u>Synth. Commun.</u>, <u>6</u>, 429 (1976).

$$\text{RCHXY} \xrightarrow[\text{DMSO}]{\text{Ag}_2\text{O}} \text{RCXYCXYR}$$

R	X	Y	% Yield
$PhCH_2$	CO_2Me	CO_2Me	74
Ph	H	CO_2Me	92
Me	Me	NO_2	40
-(CH$_2$)$_4$CO-		CO_2Et	90

I.G.3-12 R. C. Larock, <u>J. Org. Chem.</u>, <u>41</u>, 2241 (1976).

$$\underset{H}{\overset{R}{>}}C=C\underset{HgCl}{\overset{H}{<}} \xrightarrow{\text{Li}_2\text{PdCl}_4} \underset{H}{\overset{R}{>}}C=C\underset{H}{\overset{H}{<}}\underset{H}{\overset{}{>}}C=C\underset{R}{\overset{H}{<}}$$

75-100%

I.G.3-13 G. A. Olah and G. K. Surya Prakash, <u>Synthesis</u>, <u>1976</u>, 607.

$$R^1R^2\text{CHCl} \xrightarrow[\text{THF}]{\text{Ti(II)}} R^1R^2\text{CHCHR}^1R^2$$

R^1	R^2	% Yield
Ph	Ph	85
Ph	H	76
CH_2=CMe	H	76

I.G.3-14 E. J. Corey, R. L. Danheiser, and S. Chandrasekaran, J. Org. Chem., 41, 260 (1976).

$$R^1COR^2 + R^3COR^4 \xrightarrow[\text{THF}]{\text{TiCl}_4\text{-Mg(Hg)}} R^1R^2\underset{|}{\overset{OH}{C}}-\underset{|}{\overset{OH}{C}}R^3R^4$$

R^1	R^2	R^3	R^4	% Yield
Ph	H	Ph	H	84
-(CH$_2$)$_5$-		-(CH$_2$)$_5$-		93
-(CH$_2$)$_4$-		Me	Me	65
-(CH$_2$)$_5$-		Me	H	72

Other titanium chloride species were studied. The CpTiCl$_3$-LiAlH$_4$ reagent was efficiently employed in many cases.

I.G.3-15 N. E. Glushkova and N. P. Kharitonov, J. Gen. Chem. USSR, 45, 1981 (1975).

$$p\text{-}XC_6H_4CHO + Et_3SiH \xrightarrow[110-140°]{\text{Ni(colloidal)}}$$

$p\text{-}XC_6H_4CH(OSiEt_3)CH(OSiEt_3)C_6H_4X\text{-}p$
A
+
$p\text{-}XC_6H_4CH_2OSiEt_3$
B

X	% Yield A	B
H	59	24
Me	38	43
Me$_2$N	41	55

I.G.3-16 T. N. Mitchell and B. Kleine, *Tetrahedron Lett.*, 1976, 2173.

$$\underset{R^1}{\overset{R^2}{\text{isoxazole}}} \xrightarrow{(Me_3Si)_2Hg} [(Me_3Si)_2NCR^1=CHCHR^2\overset{OSiMe_3}{|}]_2$$

"almost quantitative"

I.G.3-17 G. Cahiez, D. Bernard, and J. F. Normant, *J. Organomet. Chem.*, 113, 99 (1976).

$$\underset{R^1}{\overset{R^2}{>}}C=C\underset{I}{\overset{R^3}{<}} \xrightarrow[RLi, Et_2O]{MnCl_2, LiCl} \underset{R^1}{\overset{R^2}{>}}C=C\underset{R^3}{\overset{R^3}{<}}C=C\underset{R^2}{\overset{R^1}{<}}$$

R^1, R^2, R^3, = H or alkyl
70-91%

$$\underset{Et}{\overset{Bu}{>}}C=C\underset{I}{\overset{H}{<}} \xrightarrow[RLi, -60°]{Mn^{+2}, THF} \underset{Et}{\overset{Bu}{>}}C=C\underset{R}{\overset{H}{<}}$$

R = Me or n-Bu, 85-90%

I.G.3-18 R. F. Cunico and F. J. Clayton, J. Org. Chem., **41**, 1480 (1976).

$$\underset{H}{\overset{Me_3Si}{\diagdown}}C=C\underset{SnR_3}{\overset{H}{\diagup}} \xrightarrow[-70°]{BuLi}$$

n-BuBr → $\underset{H}{\overset{Me_3Si}{\diagdown}}C=C\underset{n-Bu}{\overset{H}{\diagup}}$ 81%

CO_2 → $\underset{H}{\overset{Me_3Si}{\diagdown}}C=C\underset{CO_2H}{\overset{H}{\diagup}}$ 86%

I.G.3-19 W. E. Truce, A. W. Borel, and P. J. Marek J. Org. Chem., **41**, 401 (1976).

$$\underset{H}{\overset{R^1SO_2}{\diagdown}}C=C\underset{I}{\overset{R^2}{\diagup}} \xrightarrow{R^3Cu} \underset{H}{\overset{R^1SO_2}{\diagdown}}C=C\underset{R^3}{\overset{R^2}{\diagup}}$$

R^1	R^2	R^3	% Yield
Et	n-Bu	Me	96
Et	Ph	n-Bu	90
Et	t-Bu	Me	47
Ph	Ph	Ph	50
Ph	H	PhC≡C	63
Me	Me	PhC≡C	53

I.G.3-20 K. Tamao, M. Zembayashi, and M. Kumada, Chem. Lett., 1976, 1237; see also: *ibid*, 1239.

$$\text{RMgBr} + \text{BrCH=CHOEt} \xrightarrow[\text{Et}_2\text{O or THF}]{\text{Ni(dppp)Cl}_2 \text{ (cat.)}}$$

RCH=CHOEt

R = n-Bu, 66%
R = Ph, 70%

I.G.3-21 F. Derguini-Boumechal and G. Linstrumelle, Tetrahedron Lett., 1976, 3225.

$$\underset{R^2}{\overset{H}{\diagdown}}C=C\underset{MgBr}{\overset{R^1}{\diagup}} \xrightarrow[\text{THF}]{R^3X, \text{CuI}} \underset{R^2}{\overset{H}{\diagdown}}C=C\underset{R^3}{\overset{R^1}{\diagup}}$$

R^1	R^2	R^3X	% Yield
H	H	n-C_8H_{17}I	82
Me	H	n-C_8H_{17}OTs	80
Me	Me	n-PrI	97 (88% Z, 12% E)

I.G.3-22 F. Pochat and E. Levas, Tetrahedron Lett., 1976, 1491.

$$R^1R^2C(SEt)_2 \xrightarrow[\text{MeCN, 60°}]{\text{Hg(CN)}_2, \text{I}_2} R^1R^2C(SEt)CN$$
$$47\text{-}80\%$$

I.G.3-23 I. Ojima, Y. Miyazawa, and M. Kumagai, J. Chem. Soc., Chem. Commun., 1976, 927.

$$MeCOCO_2R \xrightarrow[\text{TiCl}_4, \text{CH}_2\text{Cl}_2]{\text{CH}_2=\text{CHCH}_2\text{SiMe}_3} CH_2=CHCH_2CMe(OH)CO_2R$$
$$82\text{-}95\%$$

I.G.3-24 A. Hosomi and H. Sakurai, Tetrahedron Lett., 1976, 1295; see also: G. Deleris, J. Dunoguès, and R. Calas, *ibid*, 2449.

$$Me_3SiCHR^1CH=CHR^2 \xrightarrow[\text{TiCl}_4, \text{CH}_2\text{Cl}_2]{R^3COR^4} R^1CH=CHCHR^2C(OH)R^3R^4$$

R^1	R^2	R^3	R^4	% Yield
H	H	n-Pr	H	87
H	H	$-(CH_2)_5-$		83
Me	H	Me	Me	72
Ph	H	n-Pr	H	54
H	Me	Me	Me	45
H	Ph	n-Pr	H	87

I.G.3-25 A. Hosomi, M. Endo, and H. Sakurai, Chem. Lett., 1976, 941.

$$R^1R^2C(OMe)_2 \xrightarrow[-78°]{CH_2=CHCH_2SiMe_3, TiCl_4, CH_2Cl_2} CH_2=CHCH_2C(OMe)R^1R^2$$

R^1	R^2	% Yield
n-Bu	H	77
n-PrCHBr	H	91
Me	Me	98
$-(CH_2)_5-$		71

I.G.3-26 R. C. Larock and J. C. Bernhardt, Tetrahedron Lett., 1976, 3097.

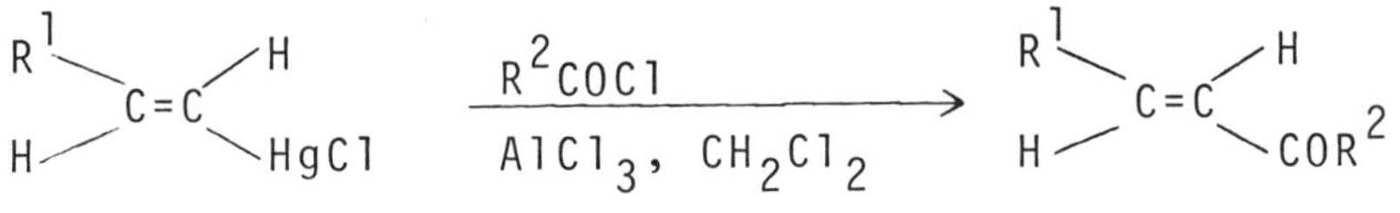

$$\underset{H}{\overset{R^1}{\diagdown}}C=C\underset{HgCl}{\overset{H}{\diagup}} \xrightarrow{R^2COCl, AlCl_3, CH_2Cl_2} \underset{H}{\overset{R^1}{\diagdown}}C=C\underset{COR^2}{\overset{H}{\diagup}}$$

72-100%

I.G.3-27 J. Meijer, H. Westmijze and P. Vermeer, Recl. Trav. Chim. Pays-Bas, **95**, 102 (1976).

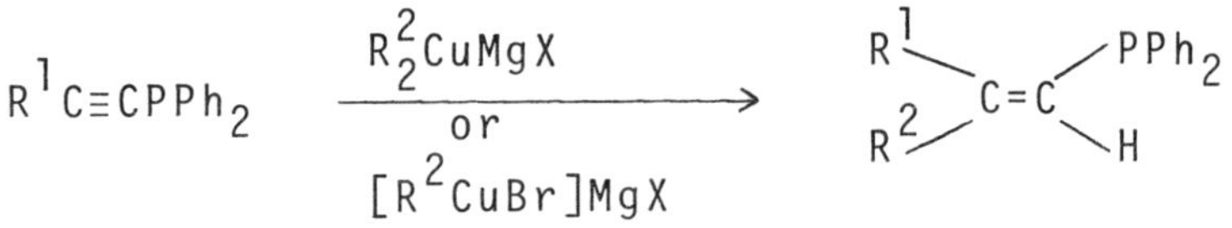

$$R^1C\equiv CPPh_2 \xrightarrow[\text{or } [R^2CuBr]MgX]{R^2_2CuMgX} \underset{R^2}{\overset{R^1}{>}}C=C\underset{H}{\overset{PPh_2}{<}}$$

Excellent yields obtained but are dependent on the copper reagent employed.

I.G.3-28 A. Alexakis, A. Commercon, J. Villieras, and J. F. Normant, Tetrahedron Lett., **1976**, 2313.

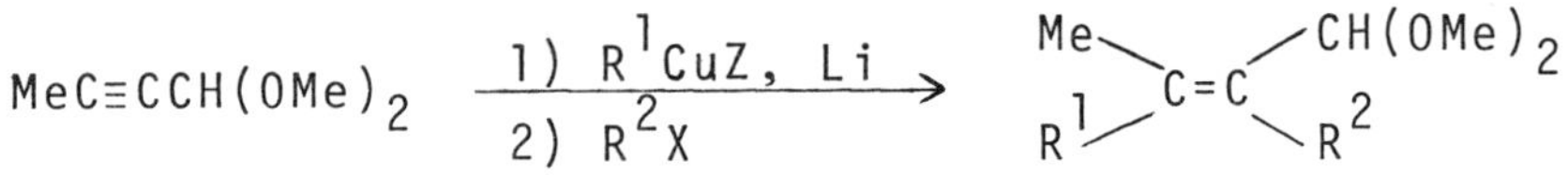

$$MeC\equiv CCH(OMe)_2 \xrightarrow[\text{2) } R^2X]{\text{1) } R^1CuZ, Li} \underset{R^1}{\overset{Me}{>}}C=C\underset{R^2}{\overset{CH(OMe)_2}{<}}$$

$R^1CuZ, Li = n\text{-}BuCu(t\text{-}BuO)Li, R^2X = Me, 87\%$

$R^1CuZ, Li = Et_2CuLi, R^2X = CH_2=CHCH_2Br, 85\%$

I.G.3-29 K. Kaneda, T. Uchiyama, M. Terasawa, T. Imanaka, and S. Teranishi, Chem. Lett., 1976, 449.

$$CH_2=CHCH_2X + R^1C\equiv CR^2 \xrightarrow{\text{Pd-polymer cat.}}$$

$$CH_2=CHCH_2CR^2=CR^1X \quad (A) \quad + \quad CH_2=CHCH_2CR^1=CR^2X \quad (B)$$

X	R^1	R^2	% Yield A	% Yield B
Cl	Ph	H	61	23
Br	Ph	H	79	13
Cl	n-Pr	H	78	-
Cl	Et	Et	54	
Cl	H	H	75	

I.G.3-30 Y. Frangin and M. Gaudemar, Bull. Soc. Chim. Fr., 1976, 1173.

$$RC\equiv CY \xrightarrow[\substack{\text{THF} \\ \text{2) } H_3O^+}]{\text{1) 2.5 eq. } CH_2=CHCH_2ZnBr} CH_2=CHCH_2CR=CH_2 \quad A$$

$$+ \quad (CH_2=CHCH_2)_2CRMe \quad B$$

R	Y	% Yield	A/B
$n-C_8H_{17}$	H	72	61/39
$n-C_7H_{15}$	MgBr	64	0/100
Ph	H	73	92/8
Ph	H	80	6/94

Other systems also studied.

I.G.3-31 H. Klein, H. Eijsinga, H. Westmijze, J. Meijer, and P. Vermeer, Tetrahedron Lett., 1976, 947; see also: P.H.M. Schreurs, J. Meijer, P. Vermeer, and L. Brandsma, *ibid*, 2387.

$$H_2C=C=CHOMe \xrightarrow[THF]{RCu(Br \text{ or } R)MgX} RCH_2CH=CHOMe$$

$$80-90\%$$

The E/Z ratio of the product is influenced by the type of organocopper reagent used.

I.G.3-32 L. M. Zubritskii, L. N. Cherkasov, T. N. Fomina, and K. V. Bal'yan, J. Gen. Chem., USSR, 46, 440 (1976).

$$Me_3SiC\equiv CCH=CH_2 \xrightarrow[2) H_2O]{1) PhMgBr, NiCl_2} Me_3SiC\equiv CCH_2CH_2Ph$$

$$30-35\%$$

I.G.3-33 H. Matsumoto, T. Nikado, and Y. Nagai, J. Org. Chem., 41, 396 (1976).

$$R^1CH=CH_2 + Cl_2CR^2CO_2R^3 \xrightarrow[PhH, 120°]{RuCl_2(PPh_3)_3 \text{ (cat.)}}$$

$$R^1CHR^2CH_2CCl_2CO_2R^3$$

R^1	R^2	R^3	% Yield
C_6H_{13}	Cl	Me	88
Ph	Cl	Et	82
CN	Cl	Et	71
MeCO	Cl	Et	64
C_6H_{13}	H	Me	89

I.G.3-34 N. Genco, D. Marten, S. Raghu, and M. Rosenblum, J. Am. Chem. Soc., 98, 848 (1976).

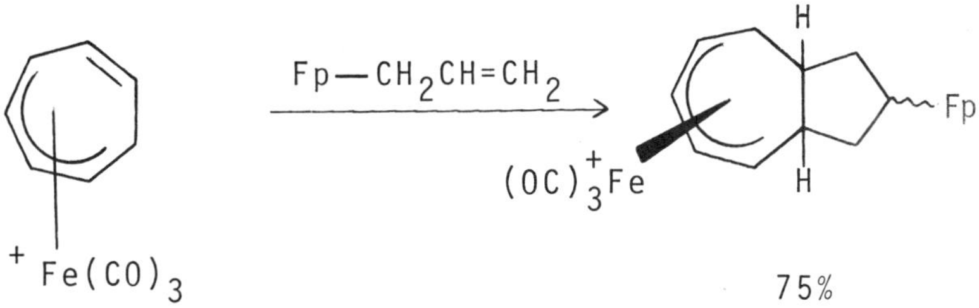

[Reaction: cycloheptatriene–Fe(CO)$_3$ cation + Fp—CH$_2$CH=CH$_2$ → bicyclic product with (OC)$_3$Fe and Fp substituents, 75%]

I.G.3-35 A. J. Pearson, Aust. J. Chem., 29, 1101 (1976).

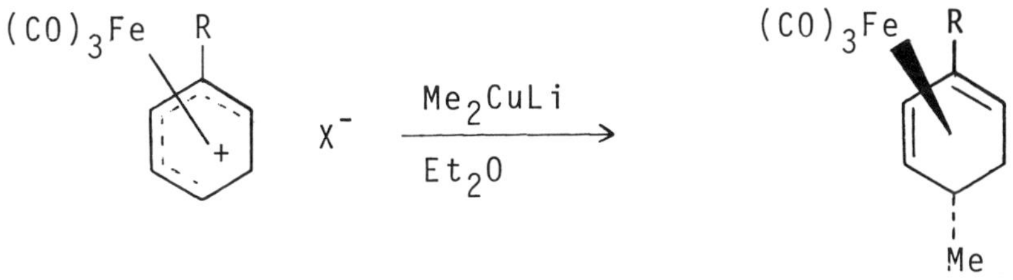

[Reaction: (CO)$_3$Fe-cyclohexadienyl cation with X$^-$ counterion, Me$_2$CuLi / Et$_2$O → (CO)$_3$Fe-cyclohexadiene with Me substituent]

R	X$^-$	% Yield
H	BF$_4^-$	80
Me	PF$_6^-$	60
MeO	BF$_4^-$	50

I.G.3-36   A. J. Pearson, *Aust. J. Chem.*, **29**, 1841 (1976).

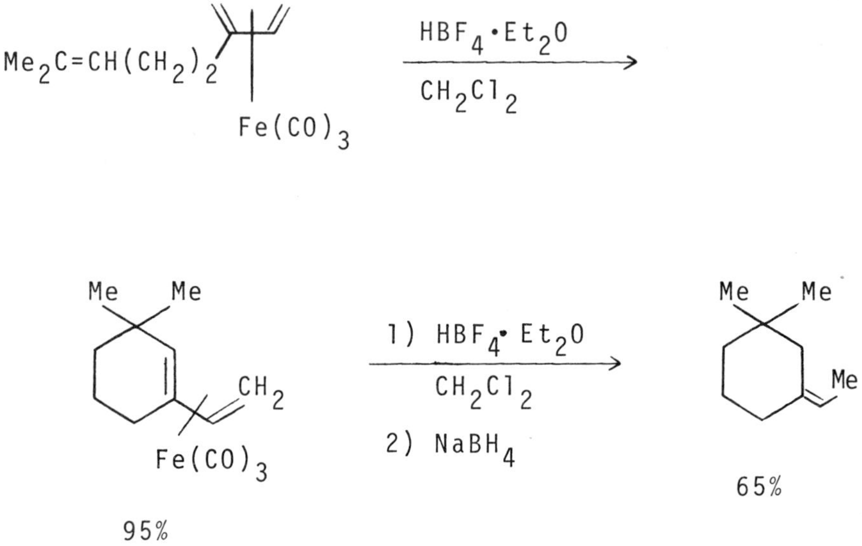

$Me_2C=CH(CH_2)_2$—[diene]—$Fe(CO)_3$ $\xrightarrow[CH_2Cl_2]{HBF_4 \cdot Et_2O}$

[cyclohexene with gem-dimethyl and vinyl-Fe(CO)$_3$ substituent] $\xrightarrow[\text{2) NaBH}_4]{\text{1) HBF}_4 \cdot Et_2O,\ CH_2Cl_2}$ [4,4-dimethyl-(methylene)cyclohexane]

95%   65%

I.G.3-37   A. J. Birch and A. J. Pearson, *J. Chem. Soc., Chem. Commun.*, **1976**, 601.

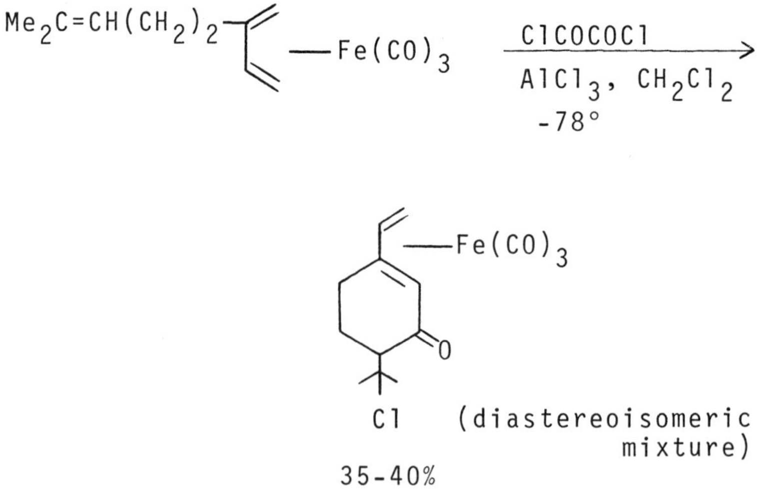

$Me_2C=CH(CH_2)_2$—[diene]—$Fe(CO)_3$ $\xrightarrow[\substack{AlCl_3,\ CH_2Cl_2 \\ -78°}]{ClCOCOCl}$

[cyclohexenone with vinyl-Fe(CO)$_3$ and gem-Me,Cl substituents]   (diastereoisomeric mixture)

35-40%

I.G.3-38 D.-J. Collins, W. R. Jackson, and R. N. Timms, Tetrahedron Lett., 1976, 495.

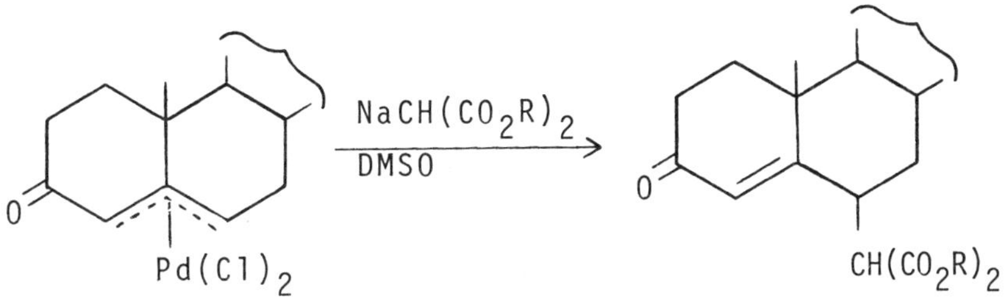

"up to 90% yield"

I.G.3-39 R. C. Larock and M. A. Mitchell, J. Am. Chem. Soc., 98, 6718 (1976).

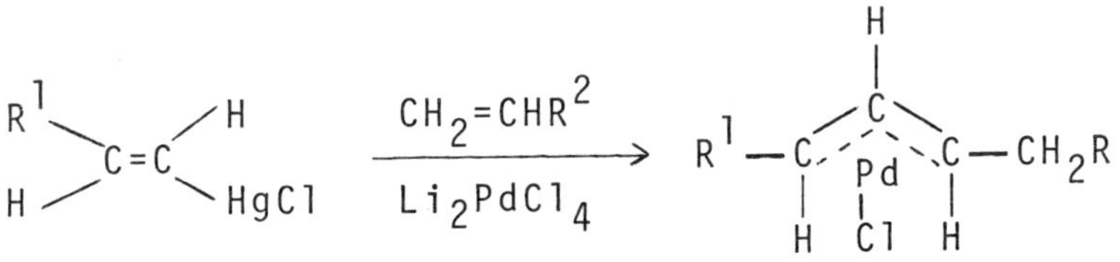

R^1	R^2	% Yield
Ph	CO_2Et	100
t-Bu	CO_2Et	90
t-Bu	H	92
t-Bu	CN	87
t-Bu	Me	59

I.G.3-40   R. L. Funk and K.P.C. Vollhardt, J. Chem. Soc., Chem. Commun., 1976, 833.

HC≡CCH$_2$CH(OTMS)C≡CH $\xrightarrow[\text{CpCo(CO)}_2]{\text{Me}_3\text{SiC≡CSiMe}_3}$

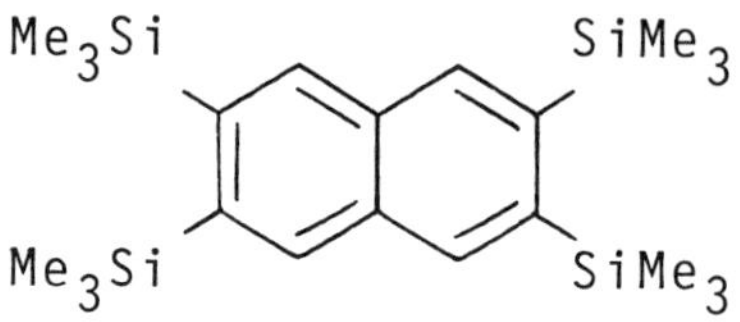

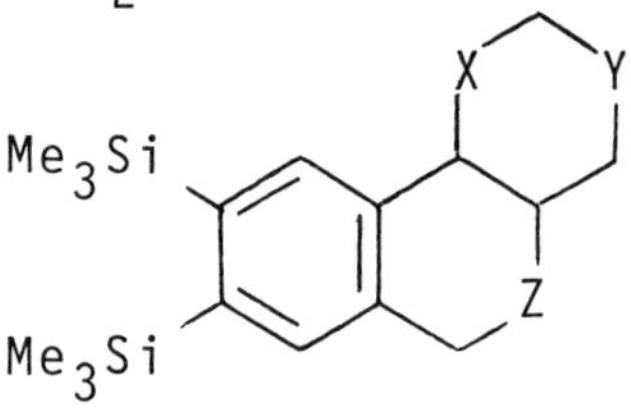

2,3,6,7-tetrakis(trimethylsilyl)naphthalene

30%

I.G.3-41   R. L. Funk and K.P.C. Vollhardt, J. Am. Chem. Soc., 98, 6755 (1976).

HC≡CCHXCH$_2$YCH$_2$CH=Z
  |
HC≡CCH$_2$

$\xrightarrow[\text{CpCo(CO)}_2]{\text{Me}_3\text{SiC≡CSiMe}_3}$

X	Y	Z	% Yield
O	CH$_2$	CH$_2$	60
CH$_2$	O	CH$_2$	90
CH$_2$	CH$_2$	CH$_2$	80

I.G.3-42   W. Münzenmaier and H. Straub, *Synthesis*, 1976, 49.

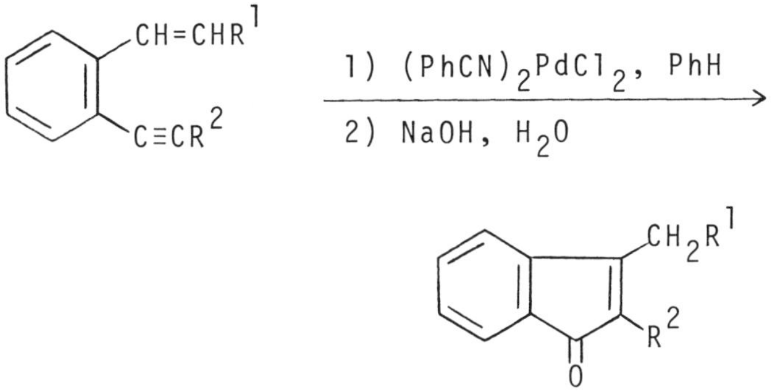

[benzene ring with CH=CHR^1 and C≡CR^2 substituents] $\xrightarrow{\text{1) (PhCN)}_2\text{PdCl}_2\text{, PhH} \atop \text{2) NaOH, H}_2\text{O}}$ [indanone with CH$_2$R^1 and R^2 substituents]

R^1	R^2	% Yield
Ph	Ph	54
t-Bu	Ph	35
Me	Me	30

I.G.3-43   C. F. Lochow and R. G. Miller, *J. Am. Chem. Soc.*, **98**, 1281 (1976).

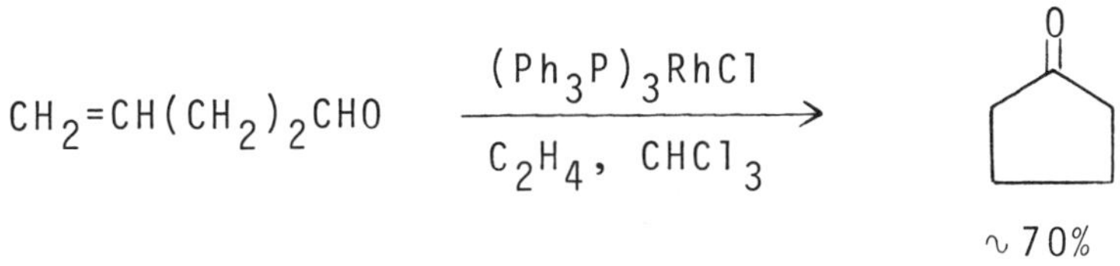

$CH_2\text{=}CH(CH_2)_2CHO$ $\xrightarrow{\text{(Ph}_3\text{P)}_3\text{RhCl} \atop \text{C}_2\text{H}_4\text{, CHCl}_3}$ [cyclopentanone]

∿ 70%

I.G.3-44 A. Baba, Y. Ohshiro, and T. Agawa, J. Organomet. Chem., 110, 121 (1976).

[Ph—C(=O)—C(Ph)=Ph cyclopropenone] + PhCR=C=O $\xrightarrow{\text{Ni(CO)}_4, \text{DMF}}$ [cyclopentenedione product with Ph, Ph, Ph, R substituents]

R = Ph or Et, 85%

I.G.e-45 T. Mitsudo, K. Kokuryo, and Y. Takegami, J. Chem. Soc., Chem. Commun., 1976, 722.

[norbornene] $\xrightarrow[\text{PhH, 80°}]{\text{MeO}_2\text{CC} \equiv \text{CCO}_2\text{Me}, \text{H}_2\text{Ru(PPh}_3)_4}$ [bicyclic product with CO$_2$Me groups]

52%

I.G.3-46 R. Baker, M. S. Nobbs, and D. T. Robinson, J. Chem. Soc., Chem. Commun., 1976, 723.

$CH_2=CHCH=CH_2$ + $R^1R^2C=NNHPh$ $\xrightarrow{\text{(Ph}_3\text{P)}_4\text{Pd or (cod)}_2\text{NiPPh}_3}$

$CH_2=CH(CH_2)_3CH=CHCH_2CR^1R^2N=NPh$ +

$CH_2=CH(CH_2)_3\underset{\underset{CR^1R^2N=NPh}{|}}{CHCH=CH_2}$

Product distribution is dependent on the catalyst used.

I.G.3-47 S. Akutagawa, Bull. Chem. Soc. Japan, 49, 3646 (1976).

$$CH_2=CHCH=CH_2 \xrightarrow[\text{NiP(O-C}_6\text{H}_4\text{Me)}_3 \text{ (cat.)}]{\text{RCOMe, 70°}} CH_2=CH(CH_2)_3\overset{H}{C}=\overset{H}{C}CH_2C(OH)RMe$$

R	% Yield
Me	62
i-Pr	55
Ph	52

I.G.3-48 P. A. Grieco, M. Nishizawa, N. Marinovic, and W. J. Ehmann, J. Am. Chem. Soc., 98, 7102 (1976).

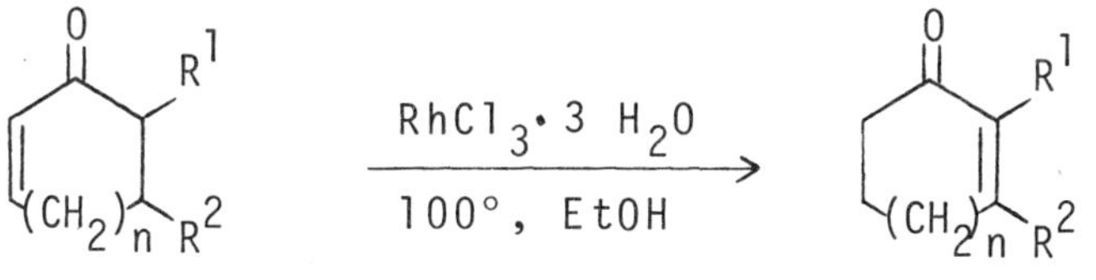

n = 0-3, 62-95%

I.G.3-49 E. C. Taylor, R. L. Robey, K.-T. Liu,
B. Favre, H. T. Bozimo, R. A. Conley, C.-S. Chiang,
A. McKillop and M. E. Ford, J. Am. Chem. Soc., 98,
3037 (1976).

$$ArCH=CR^1COR^2 \xrightarrow[MeOH, HC(OMe)_3]{Tl(NO_3)_3} ArCR^1XCH(OMe)_2$$

Ar	R^1	R^2	X	% Yield
Ph	H	H	$CH(OMe)_2$	79
p-$O_2NC_6H_4$	Me	H	$CH(OMe)_2$	50
H	H	OMe	CO_2Me	96
p-ClC_6H_4	H	OMe	CO_2Me	96

I.G.3-50 J. Levisalles, H. Rudler, and D. Villemin,
J. Organomet. Chem., 122, C15 (1976).

$$(RCH_2)_2O \xrightarrow{PhLi, WCl_6} RPhCHOCH_2R$$

R = Me, 80%
R = -CH_2CH_2-, 90%

$$Et_3N \xrightarrow{PhLi, WCl_6} Et_2NCHPhMe$$

35%

CARBON—CARBON BONDS FORMING REACTIONS

I.G.4-1   H. C. Brown, Pure and Appl. Chem., 47, 49 (1976).

Plenary Lecture: Organoboranes — The Modern Miracle.

I.G.4-2   C. F. Lane and G. W. Kabalka, Tetrahedron, 32, 981 (1976).

Review: Catecholborane. A New Hydroboration Agent.

I.G.4-3   J. Weill-Raynal, Synthesis, 1976, 633.

Review: Formation of Carbon-Carbon Bonds by Using Organoboranes.

I.G.4-4   B. M. Mikhailov, Russ. Chem. Rev., 45, 557 (1976).

Review: Methods of Synthesis and Properties of Allylboranes.

I.G.4-5 E.-I. Negishi, *J. Organometal. Chem.*, **108**, 281 (1976).

Review: Chemistry of Organoborates.

I.G.4-6 S. S. Washburne, *J. Organomet. Chem.*, **123**, 1 (1976).

Review: Silicon-Application to Organic Synthesis. Annual Survey Covering the Year 1974.

I.G.4-7 P. J. Smith, *Chem. Ind. (London)*, **1976**, 1025.

Review: Organic Synthesis *via* Organictin Intermediates — Some Recent Developments.

I.G.4-8 J. Schwartz and J. A. Labinger, *Angew. Chem. Int. Ed. Engl.*, **15**, 333 (1976).

Review: Hydrozirconation: A New Transition Metal Reagent for Organic Synthesis.

I.G.4-9 A. P. Kozikowski and H. F. Wetter, Synthesis, 1976, 561.

Review: Transition Metals in Organic Synthesis.

I.G.4-10 G. Henrici-Olive' and S. Olive', Fortschr. Chem. Forsh., 67, 107 (1976).

Review: Olefin Insertion in Transition Metal Catalysis.

I.G.4-11 K. C. Bishop, III, Chem. Rev., 76, 461 (1976).

Review: Transition Metal Catalyzed Rearrangements of Small Ring Organic Molecules.

I.G.4-12 I. P. Beletskaya, G. A. Artamkina, and O. A. Reutov, Russ. Chem. Rev., 45, 330 (1976).

Review: The Interaction of Organimetallic Derivatives with Organic Halides.

II.A.1-1   G. A. Lee and H. H. Freedman, <u>Tetrahedron Lett</u>., 1641 (1976).

$$Ar-CH_2OH \xrightarrow[Bu_4N^{\oplus} \; HSO_4^{\ominus}]{NaOCl} Ar-CHO$$

47-100%

$$\underset{R \quad R'}{CH-OH} \xrightarrow[TBAB]{NaOCl} \underset{R \quad R'}{C=O}$$

R,R' = alkyl, aryl

II.A.1-2   K. Omura, A. K. Sharma, and D. Swern, J. Org. Chem., <u>41</u>, 957 (1976).

$$\underset{R^2}{\overset{R^1}{CHOH}} \xrightarrow[Et_3N]{DMSO \; TFAA} \underset{R^2}{\overset{R^1}{C=O}}$$

R^1, R^2 = alkyl, aryl, H   44-98%

Eight examples given.

II.A.1-3   R. W. Binkley, J. Org. Chem., <u>41</u>, 3030 (1976).

$$\underset{R'}{\overset{R}{CHOH}} \xrightarrow[2. \; h\nu]{1. \; CH_3\overset{O}{\overset{\|}{C}}-\overset{O}{\overset{\|}{C}}-Cl, \; pyridine} \underset{R'}{\overset{R}{C=O}}$$

R,R' = H, alkyl, aryl, cyclic   76-100%

OXIDATIONS

II.A.1-4   S. L. Huang, K. Omura, and D. Swern, *J. Org. Chem.*, **41**, 3329 (1976).

$$\begin{array}{c} R \\ R' \end{array}\!\!C\!\!\begin{array}{c} H \\ OH \end{array} \xrightarrow{\text{DMSO, TFAA}} \begin{array}{c} R \\ R' \end{array}\!\!C\!=\!O$$

R,R' = alkyl, aryl, cyclic, H

Yields generally 70-95%

Especially useful for very hindered alcohols.

II.A.1-5   G. Cardillo, M. Orena, and S. Sandri, *Synthesis*, 394 (1976).

$$\begin{array}{c} R^1 \\ R^2 \end{array}\!\!CHOH \xrightarrow{CrO_3/HMPT} \begin{array}{c} R^1 \\ R^2 \end{array}\!\!C\!=\!O$$

R^1, R^2 = H, alkyl, aryl, cyclic

80-100% for aldehydes, aryl ketones
40-49% for dialkyl ketones

II.A.1-6   G. Cainelli *et al.*, *J. Am. Chem. Soc.*, **98**, 6737 (1976).

Use of CrO_3 bound to an ion-exchange resin to effect alcohol oxidations:

$$\begin{array}{c} R \\ R' \end{array}\!\!C\!\!\begin{array}{c} H \\ OH \end{array} \xrightarrow[\text{resin}]{CrO_3} \begin{array}{c} R \\ R' \end{array}\!\!C\!=\!O$$

R,R' = H, alkyl, aryl

Yields generally 75-95%

II.A.1-7   G. H. Posner, R. B. Perfetti, and A. W. Runquist, <u>Tetrahedron Lett</u>., 3499 (1976).

$$\underset{R-CH-R'}{\overset{OH}{|}} + Cl_3CCHO \xrightarrow{Al_2O_3} R-\overset{O}{\underset{\|}{C}}-R'$$

R,R' = alkyl, aryl, cyclic   63-90%

II.A.1-8   M. Fetizon, F. Gomez-Parra, and J.-M. Louis, <u>J. Het. Chem</u>., <u>13</u>, 525 (1976).

$$R-CH_2OH \xrightarrow[celite]{Ag_2CO_3} R-CHO$$

R = heterocyclic   50-97%

II.A.1-9   W. L. Waters <u>et al</u>., <u>J. Org. Chem</u>., <u>41</u>, 889 (1976).

$$\underset{R^2}{\overset{R^1}{\diagdown}}\underset{OH}{\overset{H}{C\diagup}} \xrightarrow{O_3} \underset{R^2}{\overset{R^1}{\diagdown}}C=O$$

R^1, R^2 = alkyl, cyclic   57-83% (VPC)

II.A.1-10   T. Ogawa and M. Matsui, <u>J. Am. Chem. Soc</u>., <u>98</u>, 1629 (1976).

$$\underset{R^2}{\overset{R^1}{\diagdown}}CHOSnBu_3 \xrightarrow[CCl_4]{NBS} \underset{R^2}{\overset{R^1}{\diagdown}}C=O$$

R's = H, alkyl, aryl   60-90%

II.A.1-11 K. Saigo, A. Morikawa, and T. Mukaiyama, Bull. Chem. Soc. Japan, 49, 1656 (1976).

$$\begin{array}{c} R^1 \\ R^2 \end{array}\!\!\!CHOH + Et_3SnOMe \longrightarrow \begin{array}{c} R^1 \\ R^2 \end{array}\!\!\!CHOSnEt_3$$

R^1, R^2 = H, alkyl, aryl, etc.

$$\downarrow \text{Base} \mid Br_2$$

$$\begin{array}{c} R^1 \\ R^2 \end{array}\!\!\!C=O$$

generally 60-80%

II.A.1-12 Y. Ueno and M. Okawara, Tetrahedron Lett., 4597 (1976).

$$\underset{\displaystyle RCH-(CH_2)_nOH}{\overset{\displaystyle OH}{|}} \xrightarrow[CH_2Cl_2]{(Bu_3Sn)_2OBr_2} R-\overset{\displaystyle O}{\underset{\displaystyle \|}{C}}-(CH_2)_nOH$$

R = Me, Et, Ph
n = 0, 1

66-86%

II.A.1-13 G. A. Olah, J. Welch, and T.-L. Ho, J. Am. Chem. Soc., 98, 6717 (1976).

$$\underset{\displaystyle R'}{\overset{\displaystyle H}{\underset{\displaystyle |}{R-\underset{|}{C}-OCH_3}}} \xrightarrow{UF_6} \begin{array}{c} R \\ R' \end{array}\!\!\!C=O$$

R,R' = H, alkyl, aryl

57-86%

II.A.1-14   M. E. Jung, J. Org. Chem., 41, 1479 (1976).

$$\underset{R-C-R'}{\overset{H\diagup\diagdown OH}{}} \xrightarrow[\text{pyridine}]{\underset{(Me_3Si)_2NH}{Me_3SiCl}} \underset{R-C-R'}{\overset{H\diagup\diagdown OTMS}{}}$$

R,R' = H, alkyl, aryl, vinyl

$\downarrow Ph_3C^{\oplus}\ BF_4^{\ominus}$
CH_2Cl_2

$$\underset{R\diagup\diagdown R'}{\overset{O}{\|}} \xleftarrow{H_2O} \underset{R\diagup\diagdown R'}{\overset{OTMS}{\overset{|}{C^{\oplus}}}} \ BF_4^{\ominus}$$

38-100% (VPC)

II.A.1-15   G. A. Olah and T.-L. Ho, Synthesis, 609 (1976).

$$\underset{R}{\overset{Ar-CH-OH}{|}} \xrightarrow[CH_2Cl_2]{NO^{\oplus}\ BF_4^{\ominus}} \underset{R}{\overset{Ar-C=O}{|}}$$

R = H, Me, Ph   60-78%

II.A.1-16 J. J. Kaminski and N. Bodor, Tetrahedron, 32, 1097 (1976).

Use of I as a brominating and oxidizing agent, similar to NBS.

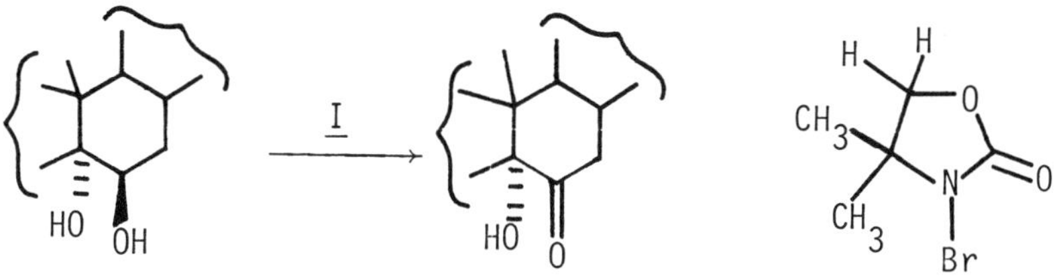

[Structure: diol → hydroxy ketone, 83%]

[Structure of I: 4,4-dimethyl-3-bromo-oxazolidin-2-one]

83%   I

II.A.1-17 M. Al Neirabeyeh et al., Synthesis, 811 (1976).

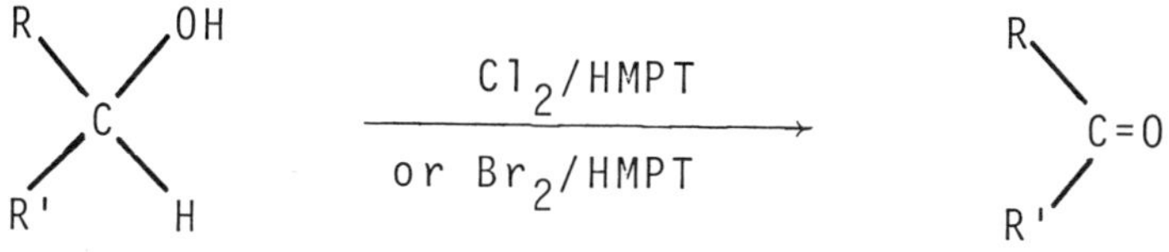

$$\underset{R'}{\overset{R}{>}}C\underset{H}{\overset{OH}{<}} \xrightarrow[\text{or } Br_2/HMPT]{Cl_2/HMPT} \underset{R'}{\overset{R}{>}}C=O$$

R,R' = H, alky, aryl

Oxidizes 2° alcohols more rapidly than 1°, e.g.:

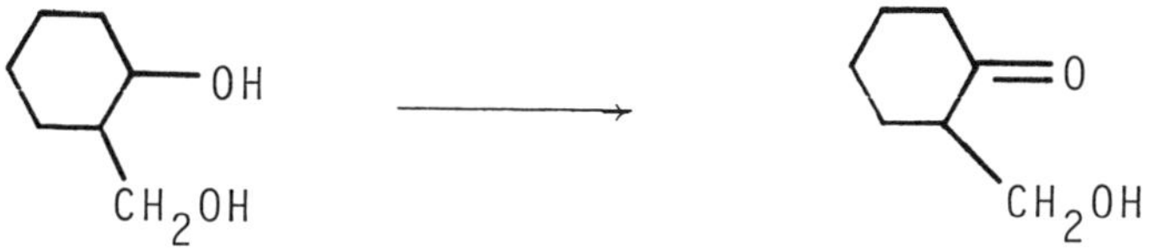

[Structure: 2-(hydroxymethyl)cyclohexanol → 2-(hydroxymethyl)cyclohexanone]

93%

II.A.1-18   M. E. Jung and L. M. Speltz, J. Am. Chem. Soc., 98, 7882 (1976).

A method for the selective oxidation of secondary alcohols, e.g.:

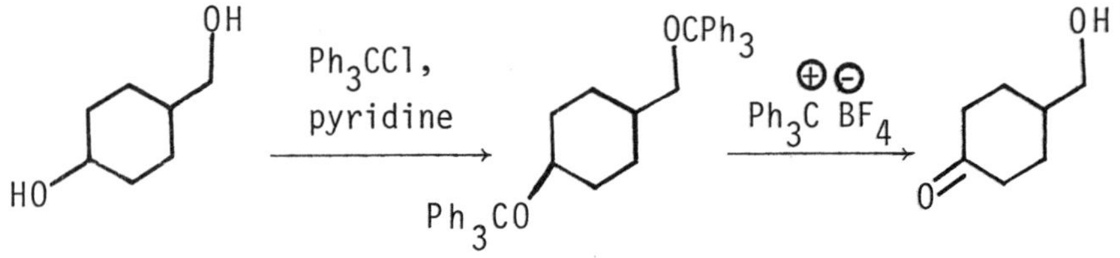

79% overall

II.A.1-19   T. C. Sharma and V. Saksena, Indian J. Chem., 14B, 143 (1976).

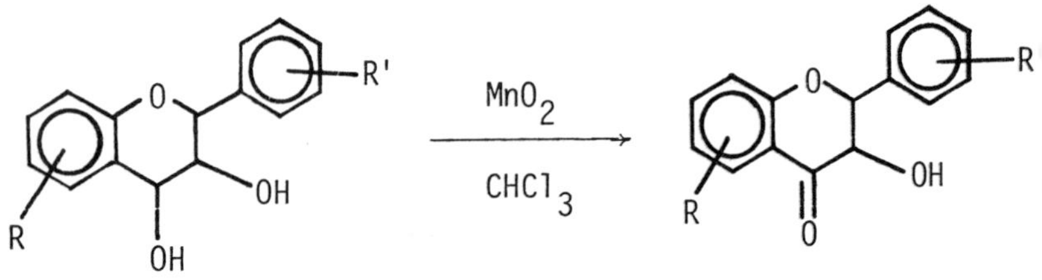

No yields given.

II.A.2-1   S. A. Maslov and E. A. Blyumberg, Russ. Chem. Rev., 45, 155 (1976).

Review: "Liquid-Phase Oxidation of Aldehydes

OXIDATIONS

II.B.1-1 Z. Cohen et al., J. Org. Chem., 41, 2651 (1976).

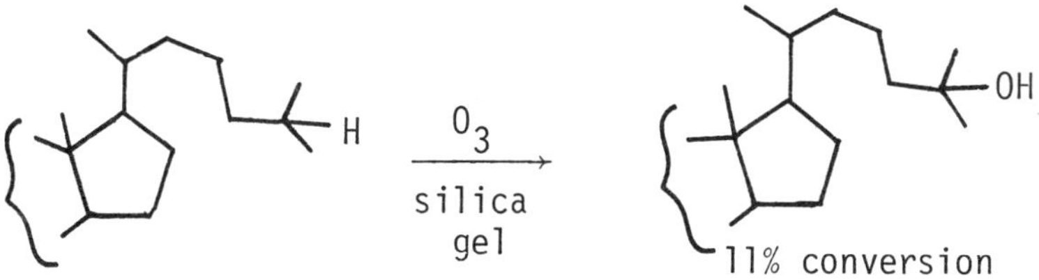

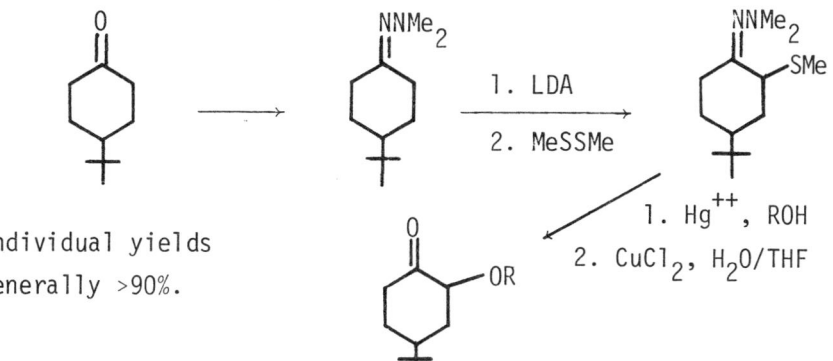

11% conversion

51% yield

II.B.1-2 W. Adam, O. Cueto, and V. Ehrig, J. Org. Chem., 41, 370 (1976).

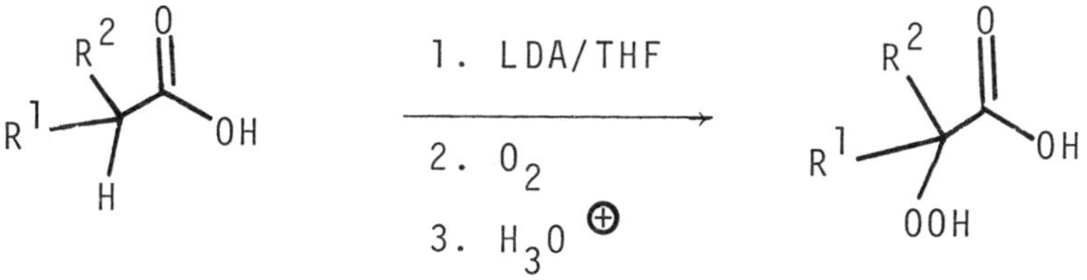

R^2R^1CH-COOH →
1. LDA/THF
2. O_2
3. $H_3O^{\oplus}$
→ $R^2R^1C(OOH)$-COOH

Yields >70%

II.B.1-3 E. J. Corey and S. Knapp, Tetrahedron Lett., 4687 (1976)

Individual yields generally >90%.

Cyclohexanone (4-t-Bu) → NNMe$_2$ hydrazone →
1. LDA
2. MeSSMe
→ α-SMe NNMe$_2$ hydrazone
→
1. Hg^{++}, ROH
2. CuCl$_2$, H$_2$O/THF
→ 2-OR cyclohexanone

II.B.1-4   H. H. Wasserman and J. L. Ives, J. Am. Chem. Soc., 98, 7868 (1976).

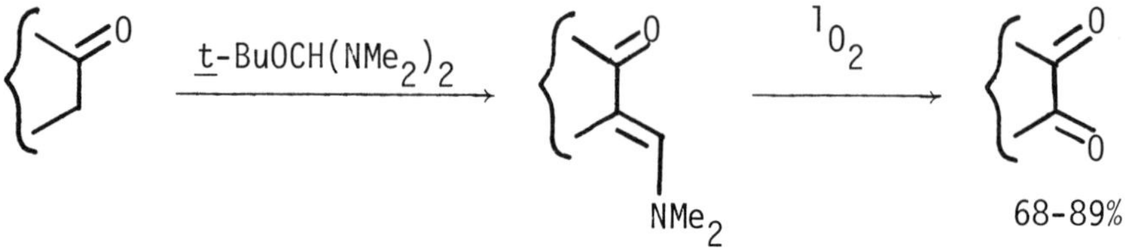

[cyclic ketone] →(t-BuOCH(NMe$_2$)$_2$)→ [cyclic ketone with =CH-NMe$_2$] →($^1O_2$)→ [cyclic 1,2-diketone]

68-89% overall

II.B.1-5   E. Vedejs and J. E. Telschow, J. Org. Chem., 41, 740 (1976).

$$R-CH_2-CN \xrightarrow[\text{2. MoO}_5 \cdot \text{Py} \cdot \text{HMPA}]{\text{1. LDA}} R-\underset{\underset{\text{OH}}{|}}{CH}-CN$$

35-57%

R = alkyl, aryl

II.B.1-6   G. Ortar, M. P. Arpiani, and A. Romeo, Steroids, 27, 197 (1976).

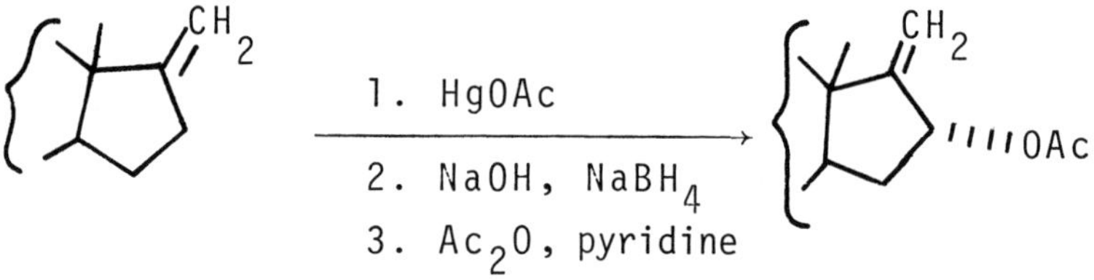

[methylenecyclopentane fused system] →(1. HgOAc; 2. NaOH, NaBH$_4$; 3. Ac$_2$O, pyridine)→ [same with OAc substituent]

57%

II.B.1-7 D. S. Fullerton and C.-M. Chen, Synth. Commun., 6, 217 (1976).

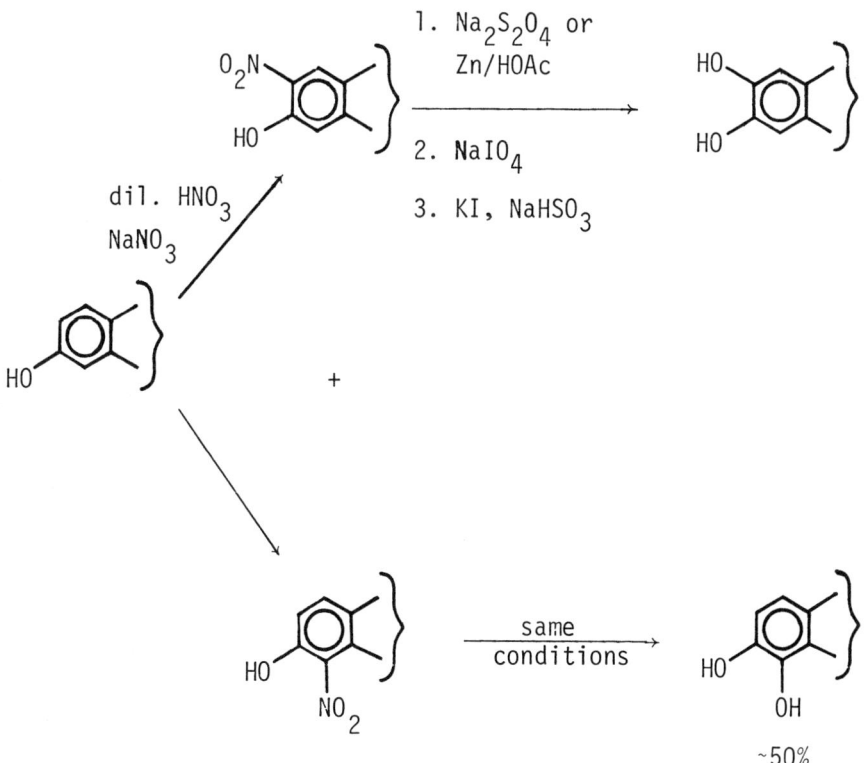

60-70%

II.B.1-8 G. Stubenrauch and R. Knuppen, Steroids, 28, 733 (1976).

~50%
(total for 2 isomers)

II.B.1-9 J. Buddrus and H. Plettenberg, *Angew. Chem. Int. Ed.*, 15, 436 (1976).

$$\text{tetrahydropyran} + 2\ I(OCOCF_3)_3 \longrightarrow \text{2,3-bis(trifluoroacetoxy)tetrahydropyran}$$

42%

II.B.1-10 R. W. Hoffman and N. Maak, *Tetrahedron Lett.*, 2237 (1976).

Formation of allyl alcohols from α,β-unsaturated sulfoxides, e.g.:

$$H_3C\text{-}C_6H_4\text{-}S\text{-}CH=\text{cyclohexylidene} \xrightarrow[\substack{2.\ CH_3OH,\ (CH_3O)_3P,\\ (CH_3)_2NH_2Cl}]{1.\ NaH} \text{2-(methylenyl)cyclohexanol}$$

75% yield

(Up to 60% optical purity can be obtained using optically active starting materials.)

II.B.1-11 P.R.O. deMontellano and C. K. Hsu, *Tetrahedron Lett.*, 4215 (1976).

$$\underset{CH_3\quad COOMe}{\overset{R\quad H}{\diagdown C=C\diagup}} \xrightarrow[\text{2. methyl methane-thiosulfonate}]{1.\ \text{tetramethylpiperidine}} \underset{SMe}{\overset{R\quad COOMe}{}}$$

$$\xrightarrow[\text{2. phosphate buffer}]{1.\ NaIO_4} \underset{R\quad COOMe}{CH_2OH}$$

~30-40% overall

II.B.2-1 N. Inukai et al., Chem. Pharm. Bull., 24, 820 (1976).

2-carbethoxycyclohexanone →
- LiCl, (PhCO)$_2$
- MgBr$_2$, H$_2$O$_2$
- LiI, (PhCO)$_2$

→ 2-X-2-carbethoxycyclohexanone

62%, X = Cl
100%, X = Br
72%, X = I

II.B.2-2 A. Rahman, A. Basha, and V. U. Ahmad, Experientia, 32, 1491 (1976).

N-acetylpiperidine $\xrightarrow{PCl_5}$ N-(1-chlorovinyl)piperidine $\xrightarrow{Br_2}$ N-(bromoacetyl)piperidine

80%

II.B.2-3 L. Blanco, P. Amice, and J. M. Conia, Synthesis, 194, 196 (1976).

$$R^1-\underset{\underset{R^3}{|}}{\overset{\overset{R^2}{|}}{C}}-\overset{O}{\overset{\|}{C}}-R^3 \xrightarrow[Et_3N]{ClSiMe_3} R^1-\overset{R^2}{\overset{|}{C}}=\overset{OTMS}{\overset{|}{C}}-R^3 \xrightarrow[CCl_4 \text{ or THF}]{Br_2 \text{ or NBS}} R^1-\underset{\underset{Br}{|}}{\overset{\overset{R^2}{|}}{C}}-\overset{O}{\overset{\|}{C}}-R^3$$

R's = H, alkyl, cyclic   60-95%

II.B.2-4 F. Kamper, H. J. Schöfer, and
H. Luftmann, Angew. Chem. Int. Ed., 15, 306 (1976).

Regioselective chlorination of centered methylene groups:

$$CH_2\text{-}[(CH_2)_nCOOH]_2 \xrightarrow[H_2SO_4]{i\text{-}Pr_2NCl} CHCl\text{-}[(CH_2)_nCOOH]_2$$

n = 2-4

58-96% specificity

2-96% yields

II.B.2-5 M.F.G.-Loustalot et al., Synthesis, 33 (1976).

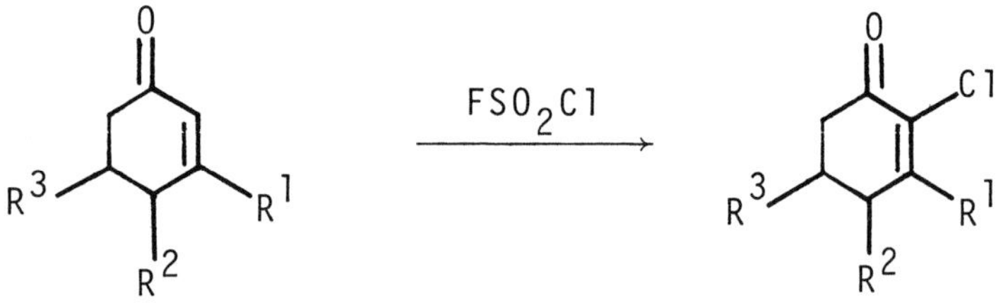

R's = H, alkyl

FSO_2Cl

50-75%

II.B.2-6 R. C. Cambie et al., J. Chem. Soc. Perkin I, 1161 (1976).

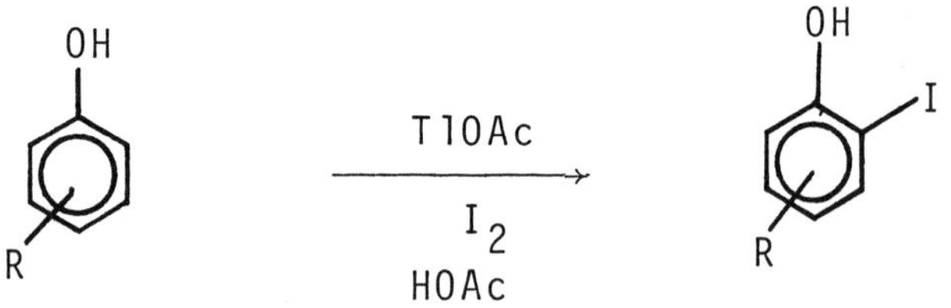

R = H, alkyl, aryl

TlOAc
I_2
HOAc

yields generally 50-80%

OXIDATIONS

II.B.2-7 L. L. Miller and B. F. Watkins, J. Am. Chem. Soc., 98, 1515 (1976).

$$Ar-H \xrightarrow[\text{Iodine(I)}]{\text{Electrogenerated}} Ar-I$$

Yields generally 50-90%

II.B.2-8 I. Agranat et al., Experientia, 15, 417 (1976).

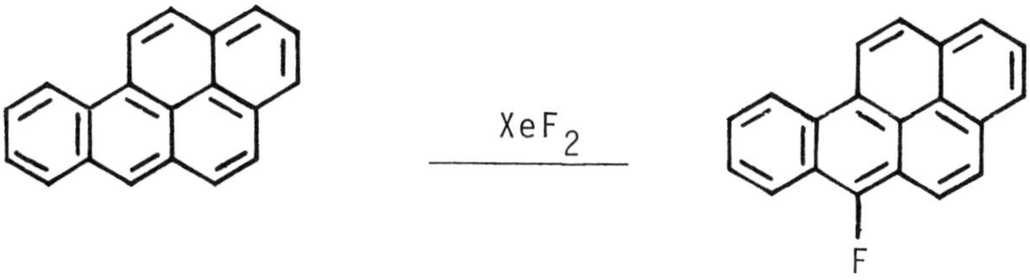

[benzo[a]pyrene] $\xrightarrow{XeF_2}$ [6-fluorobenzo[a]pyrene]

26%

II.B.2-9 D. E. Pearson et al., Synthesis, 621 (1976).

Use of trimethyl phosphate as a reaction medium for aromatic halogenation of substrates sensitive to HX. No HX is generated because of a fast reaction to form methyl halide.

II.B.3-1   P. A. Zoretic and P. Soja, J. Org. Chem., 41, 3587 (1976).

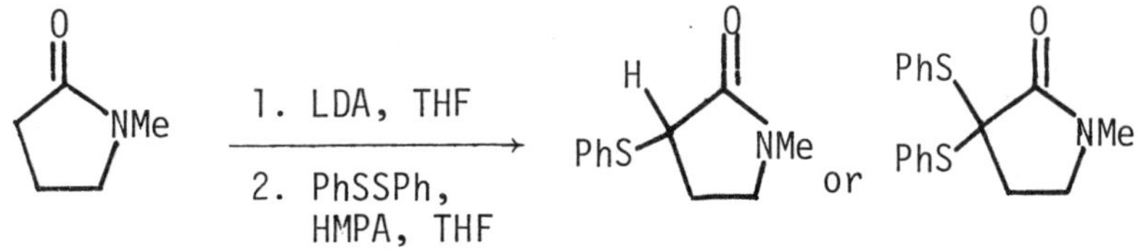

[Reaction: N-methylpyrrolidinone + 1. LDA, THF; 2. PhSSPh, HMPA, THF → mono-PhS product (60%) or bis-PhS product (73%)]

60%   73%

depending on conditions

II.C-1   G. A. Lee and H. H. Freedman, Tetrahedron Lett., 1641 (1976).

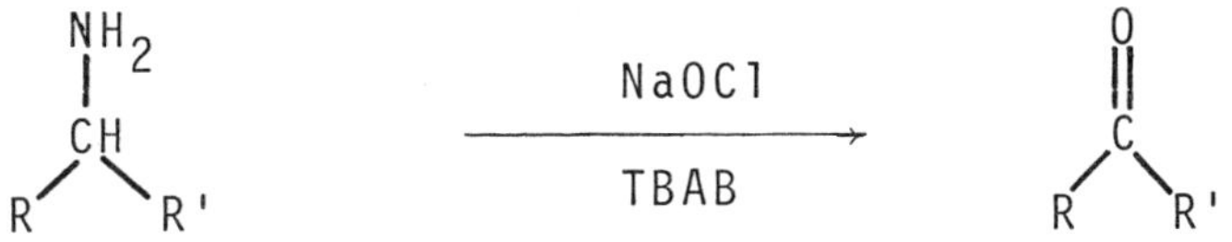

[Reaction: R(R')CH-NH$_2$ → (NaOCl, TBAB) → R(R')C=O]

R,R' = alkyl, aryl   84-98%

II.C-2   R. Kirchoff, Tetrahedron Lett., 2533 (1976).

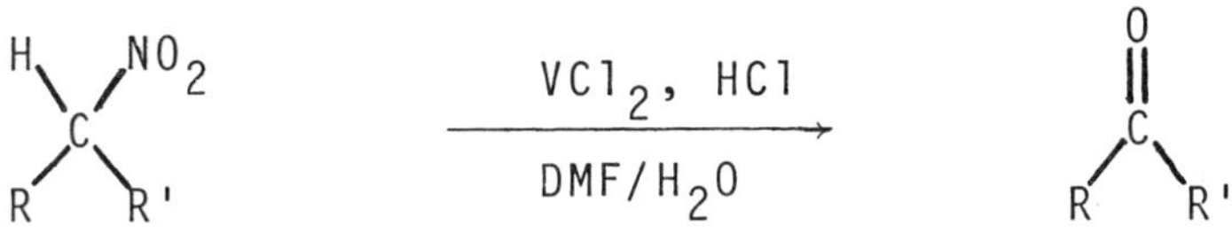

[Reaction: R(R')CH-NO$_2$ → (VCl$_2$, HCl, DMF/H$_2$O) → R(R')C=O]

R = H, alkyl   24% (heptaldehyde)
R' = alkyl, benzyl   53-71% (ketones)

OXIDATIONS

II.D-1 M. R. Czarny, Synth. Commun., 6, 285 (1976).

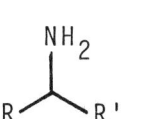

$$\underset{RR'}{NH_2} \xrightarrow{Ph-\overset{O}{\underset{\|}{Se}}-Cl} \underset{RR'}{NH}$$

R,R' = Ph;   97%
also 2-adamantylamine

$$Ph\frown NH_2 \xrightarrow{Ph-\overset{O}{\underset{\|}{Se}}-Cl} PhCN$$

85%

II.D-2 K. J. Chapman and L. K. Dyall, Aust. J. Chem., 29, 367 (1976).

$$ArNH_2 \; + \; Cl_2O \longrightarrow ArNHCl$$

"high yields"

II.E-1 T. Aida et al., Bull. Chem. Soc. Japan, 49, 1441 (1976).

$$2RSH \xrightarrow[DMSO]{I_2, \; HI} R-S-S-R$$

R = alkyl, aryl   generally >90%

II.E-2   M. Hojo and R. Masuda, Tetrahedron Lett., 613 (1976).

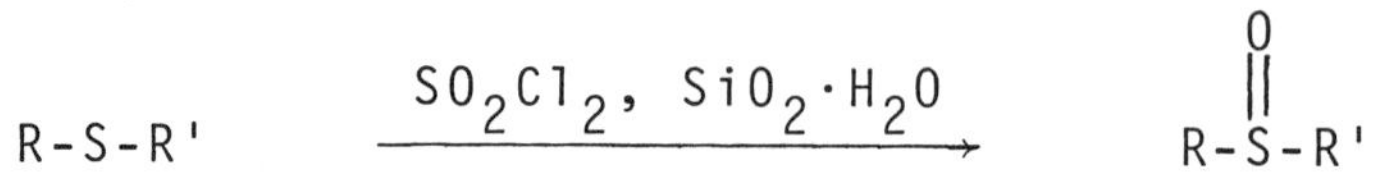

$$R-S-R' \xrightarrow{SO_2Cl_2, \ SiO_2 \cdot H_2O} R-S(=O)-R'$$

R,R' = alkyl, aryl, vinyl, CH_2Cl 82-100% (crude)

II.E-3   R. Louw et al., J. Chem. Soc., Chem. Commun., 496 (1976).

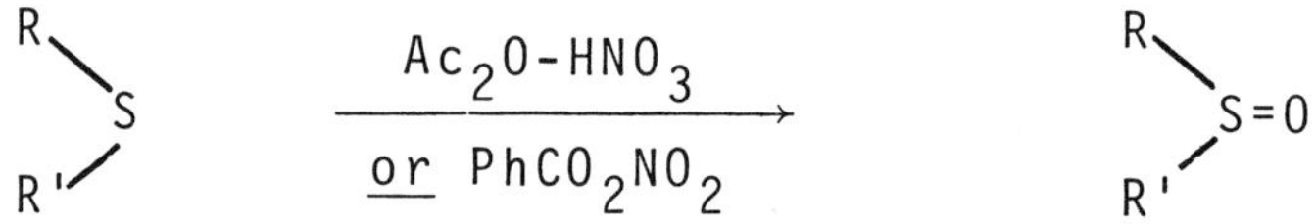

$$R_2S \xrightarrow[\text{or } PhCO_2NO_2]{Ac_2O-HNO_3} R_2S=O$$

R,R' = 1° alkyl, aryl   83-100%

II.E-4   C. R. Harrison and P. Hodge, J. Chem. Soc. Perkin I, 2252 (1976).

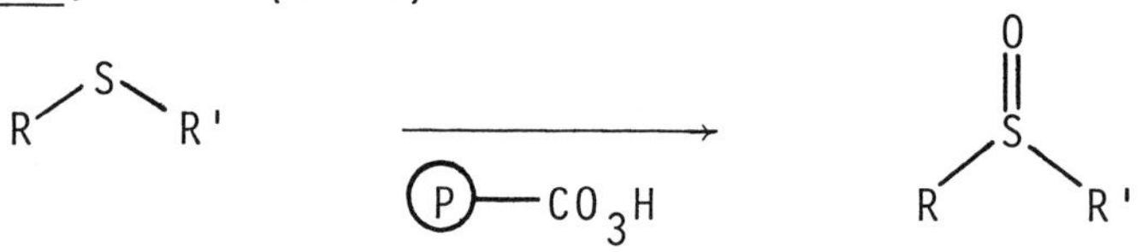

$$R-S-R' \xrightarrow{\text{P}-CO_3H} R-S(=O)-R'$$

Wide variety of substrates including penicillins, etc.   ~70-100%

II.E-5   O. G. Lowe, J. Org. Chem., 41, 2061 (1976).

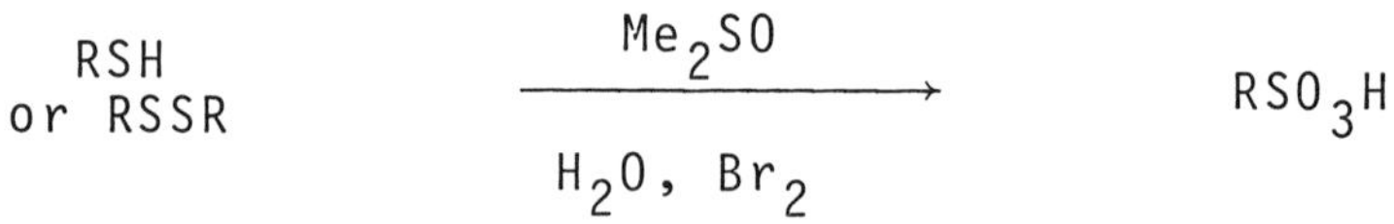

$$\text{RSH or RSSR} \xrightarrow[H_2O, \ Br_2]{Me_2SO} RSO_3H$$

OXIDATIONS

II.F.1-1   M. Parrilli et al., Tetrahedron Lett., 207 (1976).

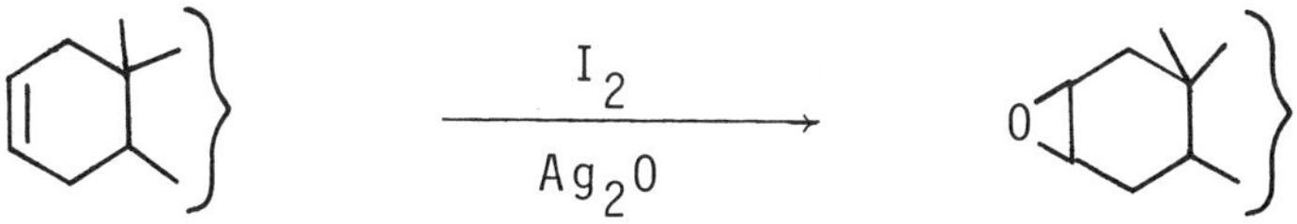

$$\text{alkene} \xrightarrow[\text{Ag}_2\text{O}]{\text{I}_2} \text{epoxide}$$

70-86%
Mixture of α and β.

II.F.1-2   N. Shimiza and P. D. Bartlett, J. Am. Chem. Soc., 98, 4193 (1976).

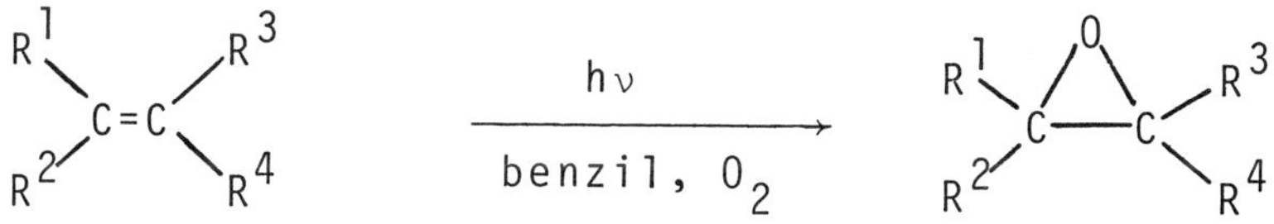

$$R^1R^2C=CR^3R^4 \xrightarrow[\text{benzil, O}_2]{h\nu} R^1R^2C\underset{O}{-}CR^3R^4$$

R's = alkyl, aryl, H

5-98%
trans product preferred

II.F.1-3   C. R. Harrison and P. Hodge, J. Chem. Soc. Perkin I, 605 (1976).

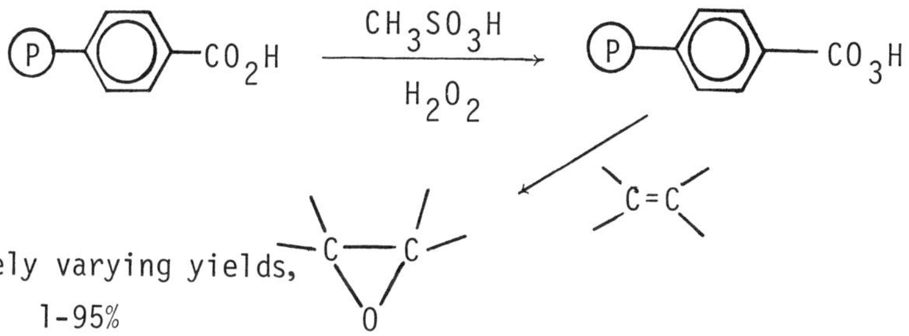

$$\text{P}-\text{C}_6\text{H}_4-\text{CO}_2\text{H} \xrightarrow[\text{H}_2\text{O}_2]{\text{CH}_3\text{SO}_3\text{H}} \text{P}-\text{C}_6\text{H}_4-\text{CO}_3\text{H}$$

$$\searrow \quad \nearrow \;\; \text{C=C}$$

widely varying yields, epoxide
1-95%

II.F.2-1   V. G. Atabekyan, M. V. Mukhina, and
A. S. Sopova, J. Org. Chem. (USSR), 12, 1234 (1976).

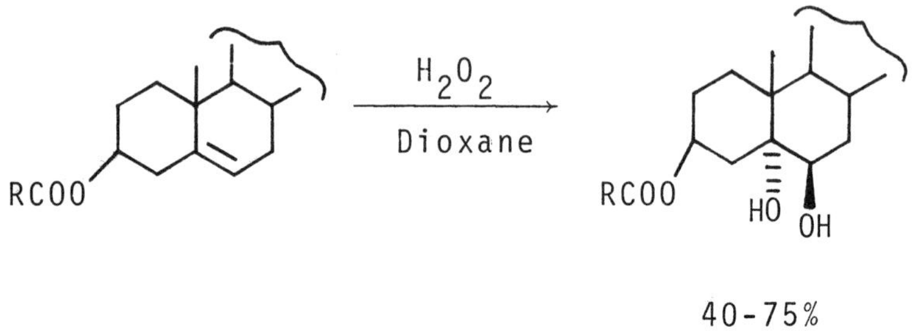

RCOO― → [H$_2$O$_2$ / Dioxane] → RCOO― (HO, OH)

40-75%

II.F.2-2   E. Glotter and A. Schwartz, J. Chem. Soc. Perkin I, 1660 (1976).

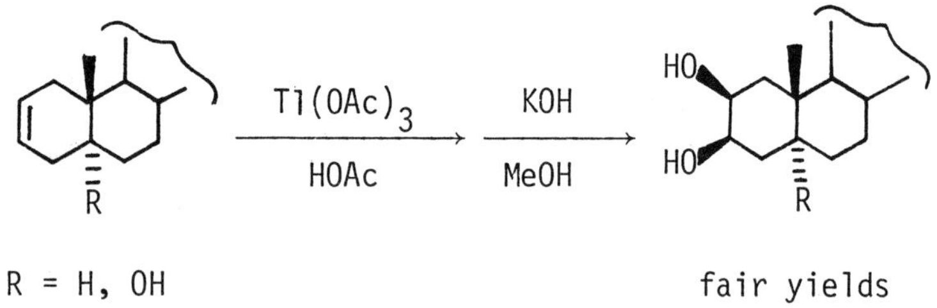

→ Tl(OAc)$_3$ / HOAc → KOH / MeOH →

R = H, OH   fair yields

II.F.2-3   V. Van. Rheenen, R. C. Kelly, and D. Y. Cha, Tetrahedron Lett., 1973 (1976).

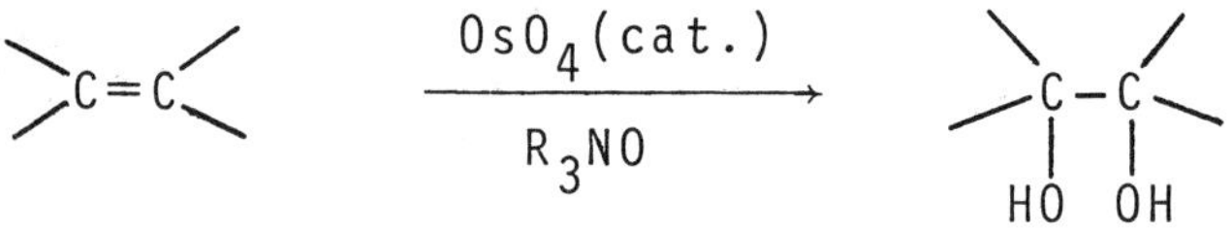

>C=C< → OsO$_4$(cat.) / R$_3$NO → >C(HO)-C(OH)<

Use of tertiary amine oxides as the oxidants in catalytic oxidations of olefins to cis-1,2-glycols. Yields are comparable to those obtained using a full mole of O$_5$O$_4$.

II.F.3-1 V. F. Mel'nikov, V. I. Koshutin, and V. A. Smirnov, J. Gen. Chem. (USSR), 45, 2072 (1975).

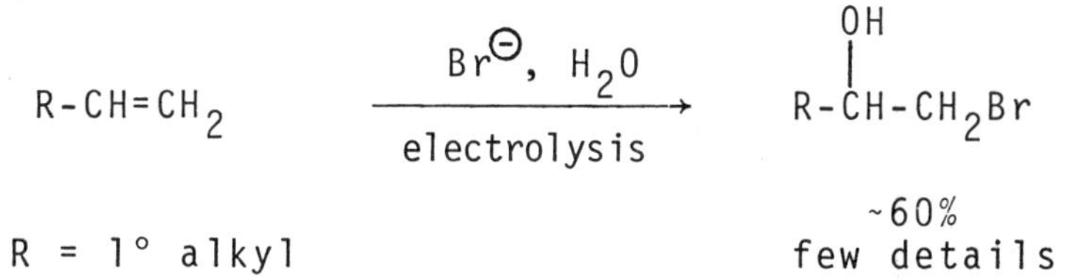

$$R-CH=CH_2 \xrightarrow[\text{electrolysis}]{Br^\ominus, H_2O} R-CH(OH)-CH_2Br$$

R = 1° alkyl

~60%
few details

II.F.3-2 A. Zwierzak and K. Osowska, Angew. Chem. Int. Ed., 15, 302 (1976).

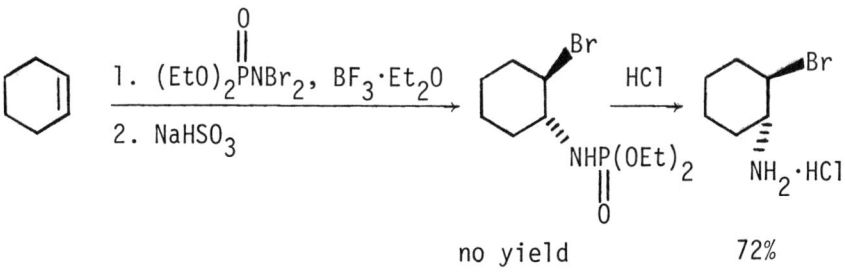

II.F.3-3 M. Adinolfi et al., Tetrahedron Lett., 3661 (1976).

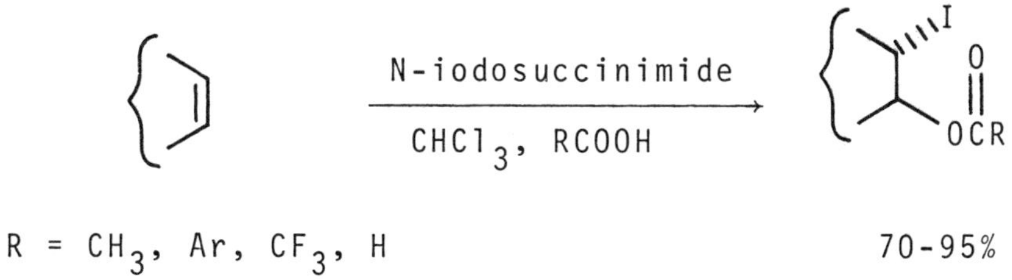

$$\text{cyclopentene} \xrightarrow[\text{CHCl}_3, \text{RCOOH}]{\text{N-iodosuccinimide}} \text{trans-2-iodo-1-acyloxycyclopentane}$$

R = CH_3, Ar, CF_3, H

70-95%

II.F.3-4 K. B. Sharpless and S. P. Singer, *J. Org. Chem.*, 41, 2504 (1976).

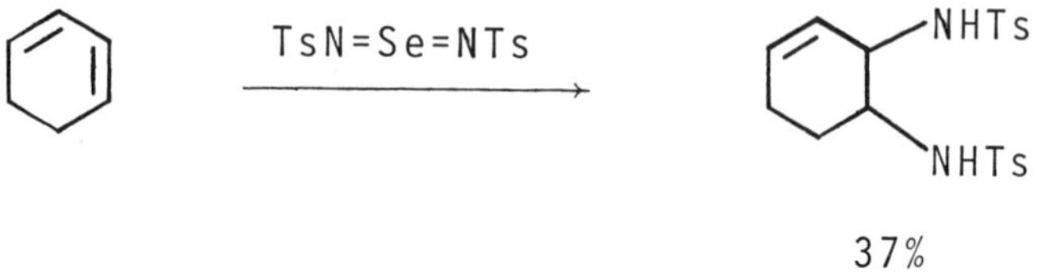

cyclohexene + TsN=Se=NTs → 3-cyclohexene-1,2-bis(NHTs) 37%

II.F.3-5 K. B. Sharpless, A. O. Chong, and K. Oshima, *J. Org. Chem.*, 41, 177 (1976).

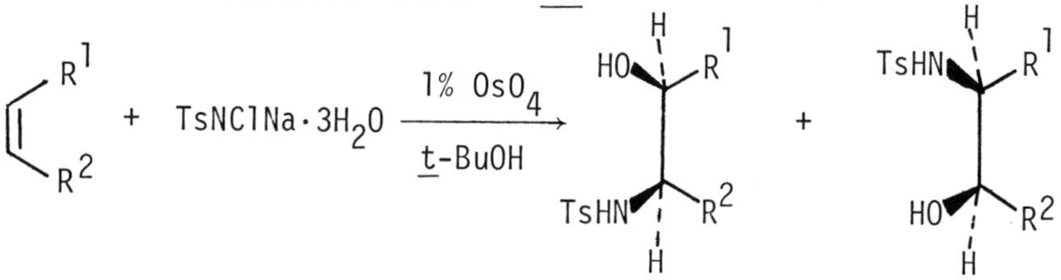

R^1R^2C=C + TsNClNa·3H$_2$O $\xrightarrow[\text{t-BuOH}]{1\% \text{ OsO}_4}$ HO-CHR^1-CH(NHTs)R^2 + TsHN-CHR^1-CH(OH)R^2

R^1, R^2 = alkyl, aryl, cyclic   Total yields 34-87%

II.F.3-6 J. Kalvoda and H. Kaufmann, *J. Chem. Soc., Chem. Commun.*, 209, 210 (1976).

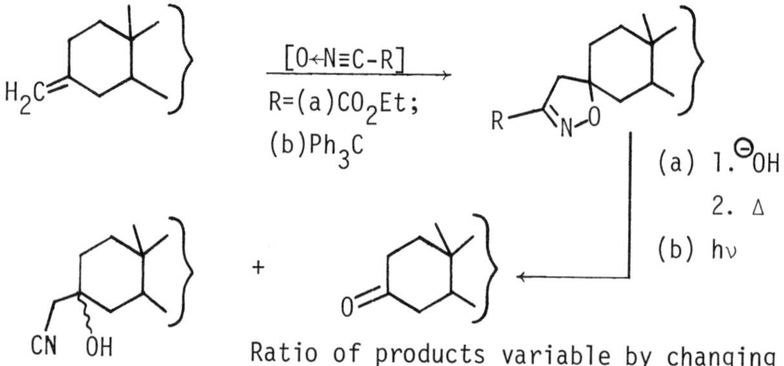

H$_2$C=(cyclohexyl) + [O←N≡C-R] R=(a)CO$_2$Et; (b)Ph$_3$C → isoxazoline spiro product

(a) 1. $^{\ominus}$OH 2. Δ
(b) hν

↓

CN-C(OH)-(cyclohexyl) + O=(cyclohexyl ketone)

Ratio of products variable by changing decarboxylation conditions.

II.F.3-7 G. M. Rubottom, J. M. Gruber, and G. M. Mong, J. Org. Chem., 41, 1673 (1976).

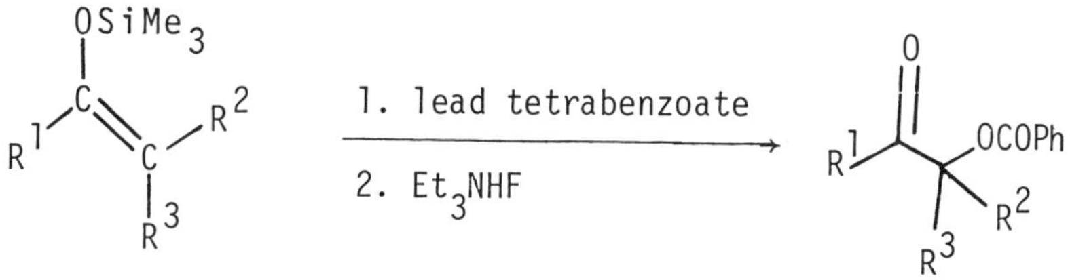

$R^1R^2C=C(OSiMe_3)R^3$ $\xrightarrow{\text{1. lead tetrabenzoate} \atop \text{2. Et}_3\text{NHF}}$ $R^1C(O)C(OCOPh)(R^2)(R^3)$

R's = H, alkyl, aryl

77-97%
(isolated)

II.G-1 P. Jacob III et al., J. Org. Chem., 41, 3627 (1976).

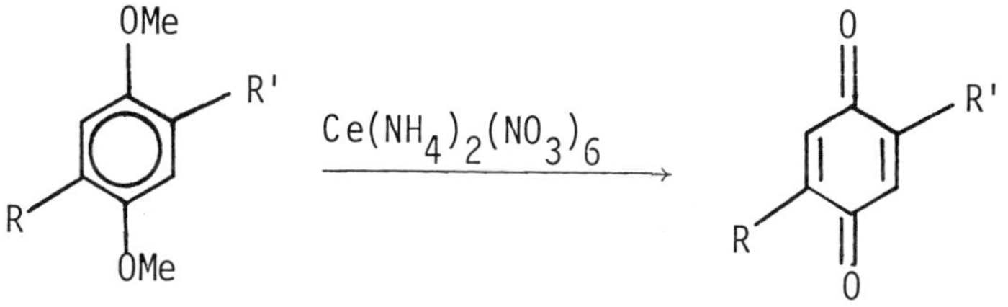

[2,5-dimethoxybenzene with R' and R substituents] $\xrightarrow{\text{Ce(NH}_4)_2(\text{NO}_3)_6}$ [1,4-benzoquinone with R' and R substituents]

R = H, Me, OMe

R' = H, 1° alkyl, vinyl, CH_2OH

~60-90%

II.G-2   A. McKillop, D. H. Perry, and M. Edwards, J. Org. Chem., 41, 282 (1976).

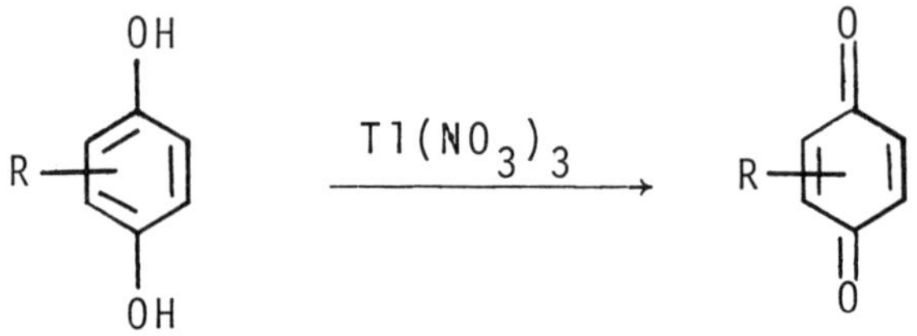

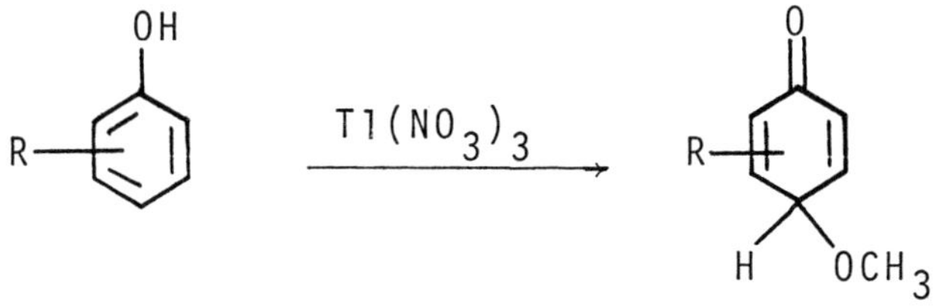

[hydroquinone] + Tl(NO$_3$)$_3$ → [1,4-benzoquinone]

[phenol] + Tl(NO$_3$)$_3$ → [4-methoxy-cyclohexa-2,5-dienone]

yields ~70-98%

II.H-1   H.-J. Liu, Can. J. Chem., 54, 3113 (1976).

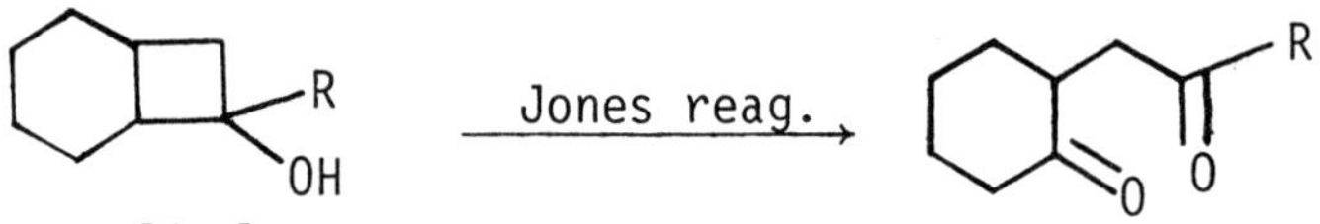

[bicyclic cyclobutanol, R = alkyl] —Jones reag.→ [2-(acyl-methyl)cyclohexanone]

R = alkyl

50-85%

II.H-2   B. Danieli and G. Palmisano, Chem. and Ind., 565 (1976).

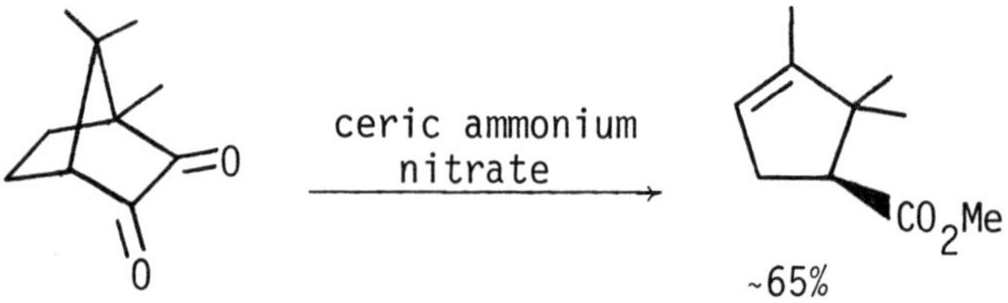

[camphor-2,3-dione] —ceric ammonium nitrate→ [cyclopentene-CO$_2$Me]

~65%

II.H-3/II.I-1   I. Saito et al., Synthesis, 255 (1976).

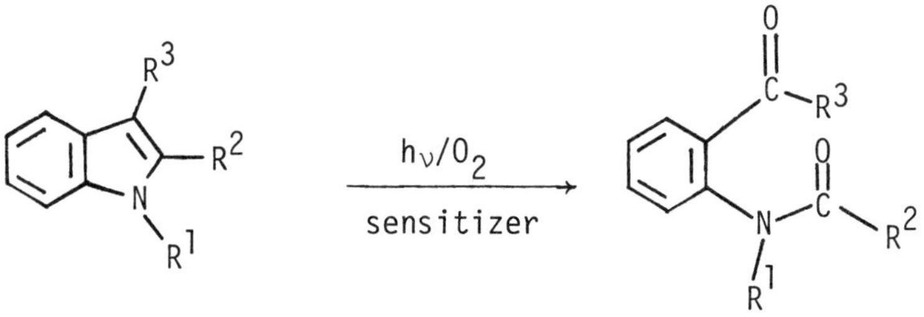

R^1 = Me, H
R^2 = H, Me, Ph
R^3 = Me, -COR, -CH$_2$COOH

generally 70-90%

II.H-4/II.I-2   R. W. Binkley, Synth. Commun., 6, 281 (1976).

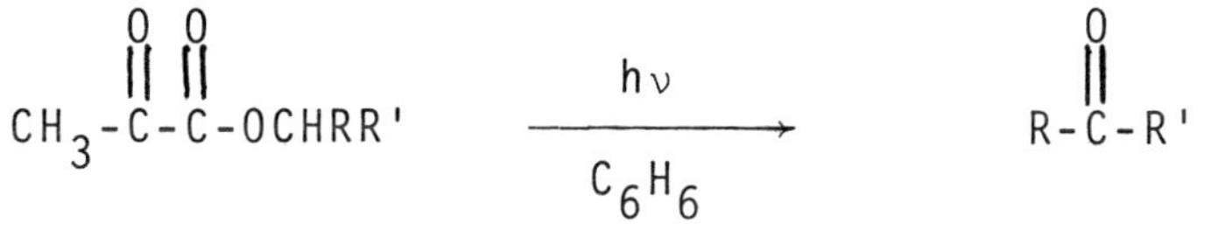

$$CH_3-\underset{\underset{O}{\|}}{C}-\underset{\underset{O}{\|}}{C}-OCHRR' \xrightarrow[C_6H_6]{h\nu} R-\underset{\underset{O}{\|}}{C}-R'$$

R,R' = H, alkyl, aryl   76-100%

II.I-3   N. Shimizer and P. D. Bartlett, J. Am. Chem. Soc., 98, 4193 (1976).

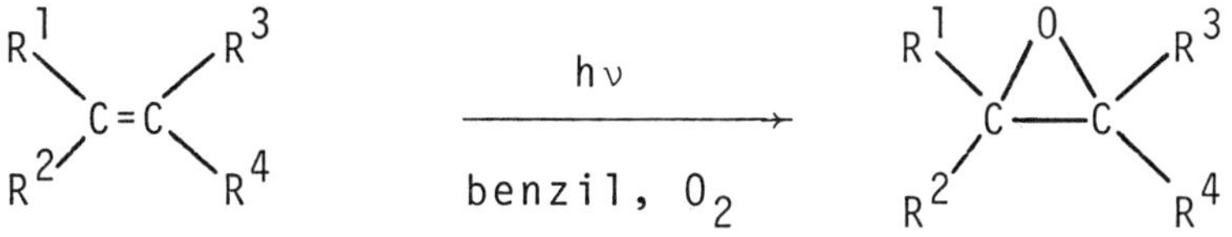

R's = alkyl, aryl, H

5-98%
trans product preferred

II.J-1 M. R. Czarny, Synth. Commun., 6, 285 (1976).

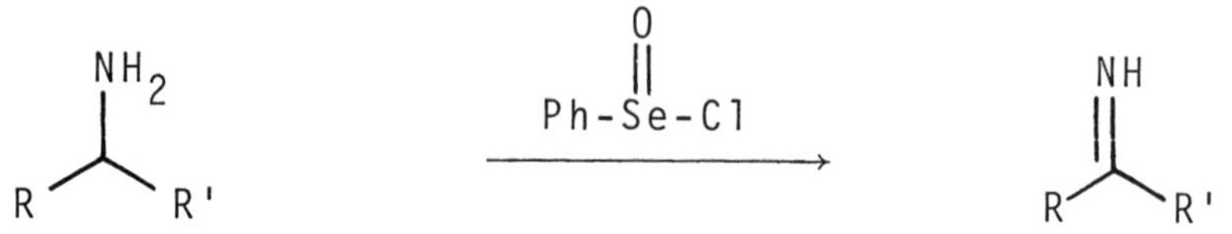

R(NH$_2$)R' + Ph-Se(=O)-Cl → R(=NH)R'   97%

R,R' = Ph; also works for 2-adamantylamine

Ph-CH$_2$-NH$_2$ + Ph-Se(=O)-Cl → PhCN   85%

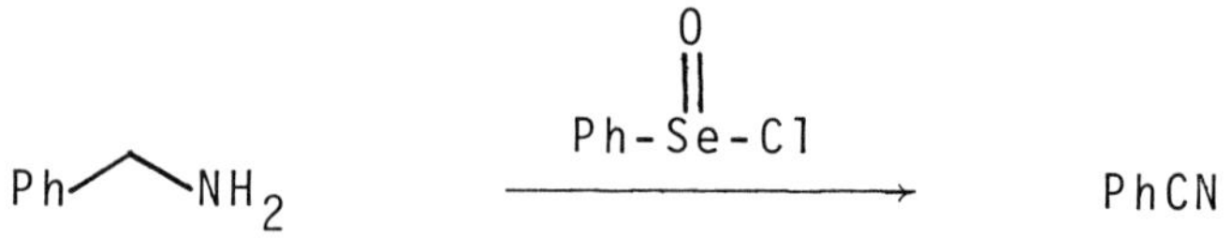

II.K-1 A. H. Haines, Chem. and Ind., 883 (1976).

Review: "Some Recent Developments in Oxidation in Organic Chemistry"

II.K-2 A. J. Fatiadi, Synthesis, 65, 133 (1976).

Review: "Active Manganese Dioxide Oxidation in Organic Chemistry"

II.K-3 R. N. Butler, Chem. and Ind., 499 (1976).

Review: "Lead Tetraacetate Acetoxylation of the Heteroallylic System ($>$C=X-YH)"

III.A-1   G. Brieger and T.-H. Fu, J. Chem. Soc., Chem. Commun., 757 (1976).

$$\text{Ar-}\overset{O}{\underset{\|}{C}}\text{-R} \xrightarrow[\text{limonene}]{\text{Pd/C} \atop \text{Cyclohexene or}} \text{Ar-CH}_2\text{-R}$$

R = H, Me   70-90%

Reaction fails when Ar is phenolic.

III.A-2   Y. Kikugawa, Chem. Pharm. Bull., 24, 1059 (1976).

$$\text{R-}\overset{O}{\underset{\|}{C}}\text{-OCH}_3 \xrightarrow[\text{CH}_2\text{Cl}_2, \text{ benzene,} \atop \text{etc.}]{\text{NaBH}_4 \cdot \text{PhNHCOCH}_3} \text{R-CH}_2\text{OH}$$

70-90%

R = alkyl, aryl, cyclic   Also reduces ketones, aldehydes.

III.A-3   T. Fujisawa, K. Sugimoto, and H. Ohta, J. Org. Chem., 41, 1667 (1976).

$$\underset{R \quad R'}{\overset{O}{\|}{C}} \xrightarrow[\text{FeCl}_2 \text{ or FeCl}_3]{\text{NaH}} \underset{R \quad R'}{\overset{H \quad OH}{C}}$$

R,R' = H, Ph, alkyl   75-82% (VPC)

Reaction fails with benzophenone.

III.A-4 R. Grandi et al., J. Org. Chem., **41**, 1755 (1976).

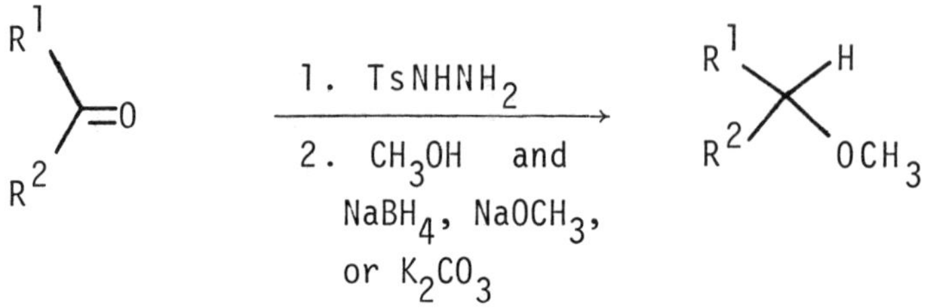

$$R^1R^2C=O \xrightarrow[\substack{2.\ CH_3OH\ \text{and} \\ NaBH_4,\ NaOCH_3, \\ \text{or}\ K_2CO_3}]{1.\ TsNHNH_2} R^1R^2C(H)(OCH_3)$$

R^1, R^2 = H, alkyl, aryl

Widely varying yields.

III.A-5 M. Sevrin, D. Van Ende and A. Krief, Tetrahedron Lett., 2643 (1976).

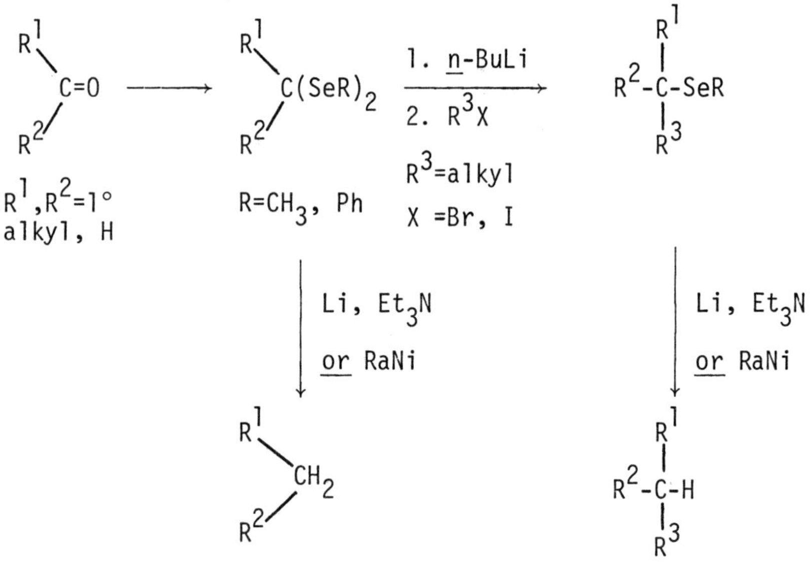

$$R^1R^2C=O \longrightarrow R^1R^2C(SeR)_2 \xrightarrow[2.\ R^3X]{1.\ \underline{n}\text{-BuLi}} R^2\text{-}\underset{R^3}{\underset{|}{\overset{R^1}{\overset{|}{C}}}}\text{-}SeR$$

$R^1, R^2 = 1°$ alkyl, H $R=CH_3$, Ph R^3=alkyl, X =Br, I

↓ Li, Et$_3$N or RaNi

$R^1R^2CH_2$

↓ Li, Et$_3$N or RaNi

$R^2\text{-}\underset{R^3}{\underset{|}{\overset{R^1}{\overset{|}{C}}}}\text{-}H$

85-92%

III.A-6 T. Fujisawa, K. Sugimoto, and H. Ohta, Chem. Lett., 581 (1976).

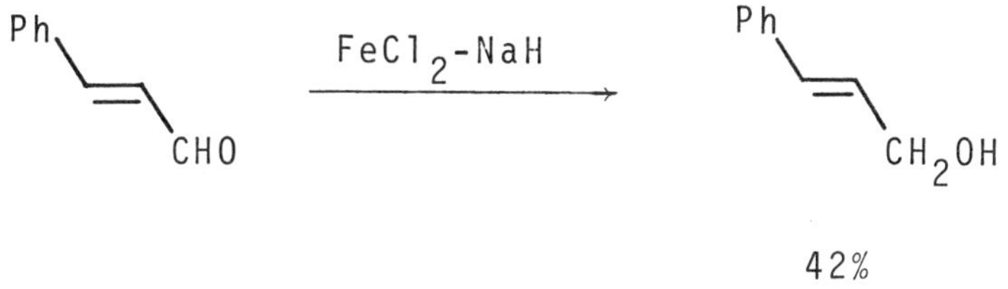

Ph–CH=CH–CHO →[FeCl$_2$-NaH] Ph–CH=CH–CH$_2$OH

42%

III.A-7 P. S. Gradeff and G. Formica, Tetrahedron Lett., 4681 (1976).

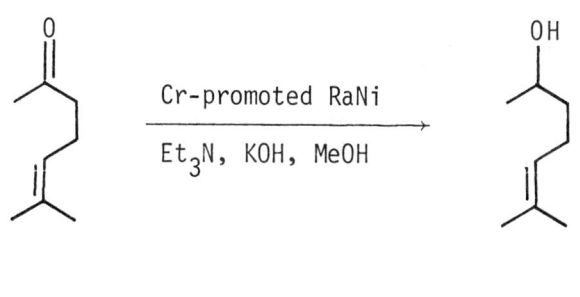

88%

III.A-8 S. Krishnamurthy and H. C. Brown, J. Am. Chem. Soc., 98, 3383 (1976).

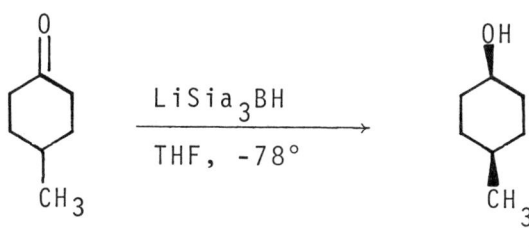

99%

Gives the less stable isomer, generally in about 99% yield. Nine examples cited.

III.A-9   Y. Yamamoto et al., J. Am. Chem. Soc., 98, 1965 (1976).

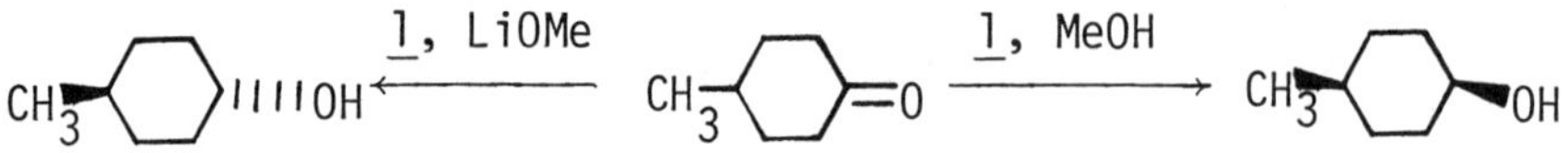

CH₃—⟨ ⟩····OH ←—— 1, LiOMe —— CH₃—⟨ ⟩=O ——— 1, MeOH ———→ CH₃—⟨ ⟩—OH

90%   84%

Several additional examples cited.

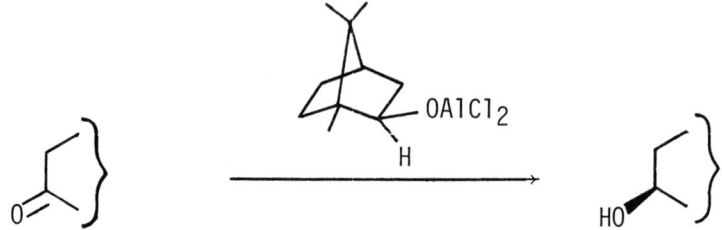

1 =

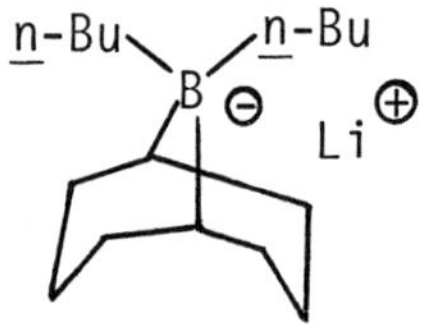

III.A-10   D. Nasipuri et al., J. Chem. Soc. Perkin I, 321 (1976).

various 3-oxo triterpenoids   ~70% chemical yield

75-100% axial alcohol

III.A-11 I. Ojima et al., J. Organometal. Chem., 122, 83 (1976).

$$\underset{R-C-R'}{\overset{O}{\|}} \xrightarrow[\text{Chiral Rh-phosphine complex}]{\text{1. Hydrosilane}} \xrightarrow{\text{2. MeOH, MeO}^{\ominus}} \underset{R-CH-R'}{\overset{OH}{|}}$$

R = alkyl

R' = alkyl, aryl

>90%

4-58% ee

III.A-12 T. Hayashi, T. Mise, and M. Kumada, Tetrahedron Lett., 4351 (1976).

$$\underset{CH_3}{\overset{R}{\diagdown}}C=O \xrightarrow[\text{Ligand-RH}^{\oplus}/\text{MeOH}]{H_2 (50 \text{ atm})} R-\overset{*}{C}H-CH_3 \atop \underset{OH}{|}$$

R = Ph, t-Bu, COOH

generally >90%
up to 83% ee

Ligand = [ferrocene with C(H)(Me)-NMe$_2$ and PPh$_2$ substituents, PPh$_2$ on lower ring]

III.A-13 E. C. Ashby and J. R. Boone, J. Org. Chem., 41, 2890 (1976).

A study of the stereochemistry of reduction of ketones by a variety of simple and complex metal hydrides of the main group elements.

III.A-14 S. Yamada, M. Kitamoto, and S. Terashima, Tetrahedron Lett., 3165 (1976).

[cyclopentane-1,3-dione with R substituent] →(LAH, (−)N-methyl-ephedrine)→ [hydroxy ketone with R, OH]

R = alkyl, ester

yields ~40%, 55% ee

III.A-15 B. S. Deol, D. D. Ridley and G. W. Simpson, Aust. J. Chem., 29, 2459 (1976).

$$R-C(=O)-CHR'-C(=O)-OEt \xrightarrow{yeast} R-CH(OH)-CHR'-C(=O)-OEt$$
                                                    *

R,R' = alkyl, aryl, cyclic

~70-80% high optical purity

Also works with $R-C(=O)-C(=O)-OMe$

and

$R-C(=O)-C(=O)-NH_2$

III.A-16 P. L. Russ and E. A. Caress, J. Org. Chem., 41, 149 (1976).

TsNHCH$_2$CH$_2$-N(-CH$_2$CO$_2$C$_6$Cl$_5$)(C(=O)CH$_3$) →(1. BH$_3$·THF; 2. H$_3$O$^{\oplus}$)→ TsNHCH$_2$CH$_2$-N(-CH$_2$COOH)(CH$_2$CH$_3$)

III.A-17 M. F. Grundon, D. G. McCleery, and J. W. Wilson, Tetrahedron Lett., 295 (1976).

$$\underset{R-C-R'}{\overset{O}{\|}} \quad \xrightarrow{(CH_3)_2CHCH_2\overset{\overset{BH_3}{\uparrow}{NH_2}}{\underset{*}{C}}HCOOCH_3} \quad \underset{\underset{H}{|}}{\overset{OH}{\underset{|}{R-\overset{*}{C}-R'}}}$$

15-22% ee

III.A-18 T.-L. Ho and G. A. Olah, Synthesis, 815 (1976).

$$\underset{R-Ph-\overset{O}{\overset{\|}{C}}-\overset{O}{\overset{\|}{C}}-Ph-R}{} \quad \xrightarrow[H_2O/THF]{VCl_2} \quad R-Ph-\overset{O}{\overset{\|}{C}}-\overset{OH}{\underset{|}{C}H}-Ph-R$$

R = H, Me, Ph, OMe

82-92%

$$\underset{R}{\overset{O}{\bigodot}}\overset{}{\underset{O}{}} \quad \xrightarrow[H_2O/THF]{VCl_2} \quad \underset{R}{\overset{OH}{\bigodot}}\overset{}{\underset{OH}{}}$$

R = Me, Cl, benzoquinone, etc.

~90%

III.A-19 K. Nishiyama et al., J. Chem. Soc., Chem. Commun., 101 (1976).

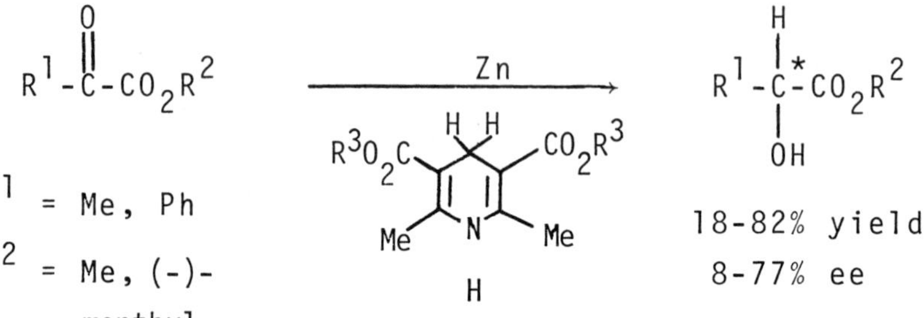

$$R^1-\underset{O}{\overset{\|}{C}}-CO_2R^2 \quad \xrightarrow{\text{Zn}} \quad R^1-\underset{OH}{\overset{H}{\underset{|}{C^*}}}-CO_2R^2$$

with dihydropyridine reagent:

R^3O_2C, CO_2R^3, Me, Me, N-H (1,4-dihydropyridine)

R^1 = Me, Ph
R^2 = Me, (-)-menthyl
R^3 = Et, (-)-menthyl

18-82% yield
8-77% ee

III.A-20 R. Kanazawa and T. Tokoroyama, Synthesis, 526 (1976).

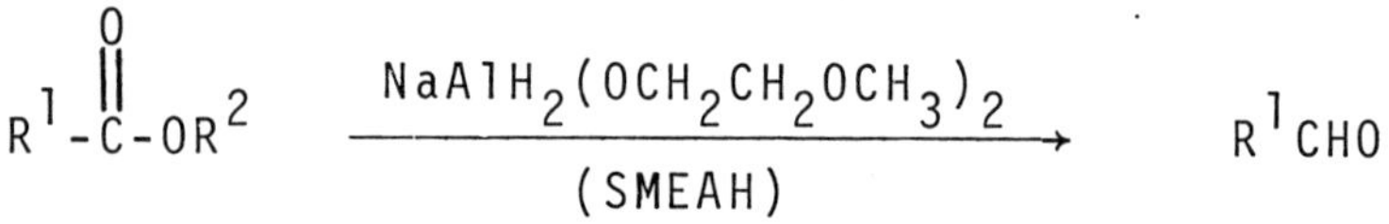

$$R^1-\underset{O}{\overset{\|}{C}}-OR^2 \quad \xrightarrow[\text{(SMEAH)}]{NaAlH_2(OCH_2CH_2OCH_3)_2} \quad R^1CHO$$

R^1 = alkyl, aryl, cinnamyl

[cyclic acetal with C=O, O] → [cyclic acetal with C(H)(OH), O]   SMEAH

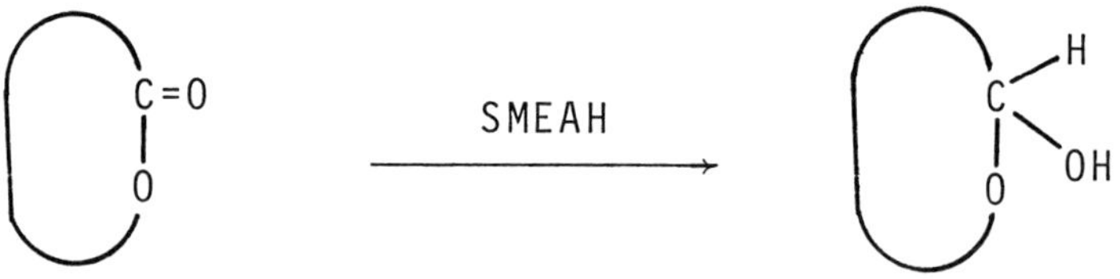

5 or 6-membered ring

60-90%

III.A-21 N. Chatterjie, J. G. Umans, and C. E. Inturrisi, J. Org. Chem., 41, 3624 (1976). See also J. Med. Chem., 18, 490 (1975).

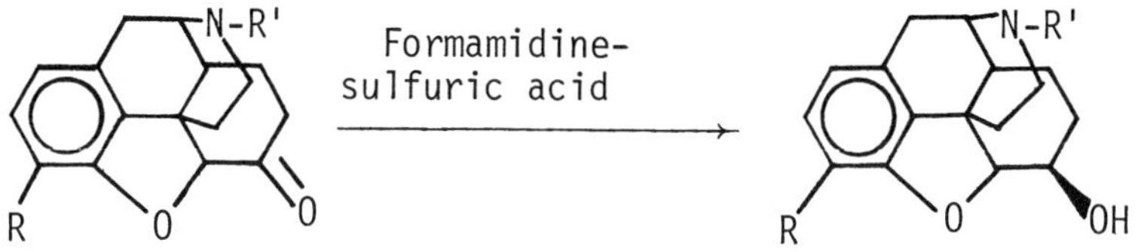

R = OH, OCH$_3$
R' = 1° alkyl

Formamidine-sulfuric acid

63%

III.A-22 Y. Maki et al., Chem. and Ind., 322 (1976).

$$R-\overset{O}{\underset{\|}{C}}-NH_2 \xrightarrow[HSCH_2CH_2SH]{NaBH_4, \text{ THF}} RCH_2NH_2$$

R = pentyl, phenyl   ~90%

III.A-23 N. Umino, T. Iwakuma, and N. Itoh, Tetrahedron Lett., 763 (1976).

$$R-\overset{O}{\underset{\|}{C}}-NHR' \xrightarrow[R''-COOH]{NaBH_4} R-CH_2NHR'$$

R = alkyl, aryl
R' = H, alkyl
R" = CH$_3$, CF$_3$, Ph

yields generally 60-90%

III.A-24 A. Rahman et al., Tetrahedron Lett., 219 (1976).

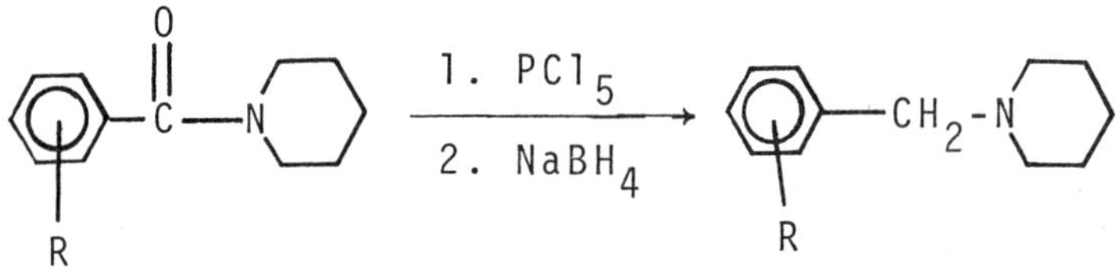

Ar(R)-C(=O)-N(piperidine) $\xrightarrow{\text{1. PCl}_5 \quad \text{2. NaBH}_4}$ Ar(R)-CH$_2$-N(piperidine)

75-90%

III.B-1 N. Umino, T. Iwakuma, and N. Itoh, Tetrahedron Lett., 2875 (1976).

$$R-CN \xrightarrow[\text{THF}]{\text{NaBH}_3(\text{OCOCF}_3)} RCH_2NH_2$$

R = alkyl, aryl   70-89%

III.C-1 T. Fujisawa and H. Ohta, Bull. Chem. Soc. Japan, 49, 2341 (1976).

$$RSSR + PhI \xrightarrow{h\nu} RSPh$$

R = alkyl, aryl   20-100%

III.C-2 J. Drabowicz and M. Mikolajczyk, Synthesis, 527 (1976).

$$R^1-\overset{O}{\underset{\|}{S}}-R^2 \xrightarrow{\text{LiAlH}_4, \text{TiCl}_4, \text{ether}} R^1-S-R^2$$

R's = aryl, 1°, 2°, 3° alkyl

REDUCTIONS

III.C-3 G. A. Olah, G.K. SuryaPrakash and T.-L. Ho, Synthesis, 810 (1976).

$$R_2S=O \xrightarrow[VCl_2, H_2O/THF]{MoOCl_3, Zn, THF} R_2S$$

R = Ph, Bz, n-Bu, $-(CH_2)_4-$ 74-91%

III.C-4 D. W. Chasar and T. M. Pratt, Synthesis, 262 (1976).

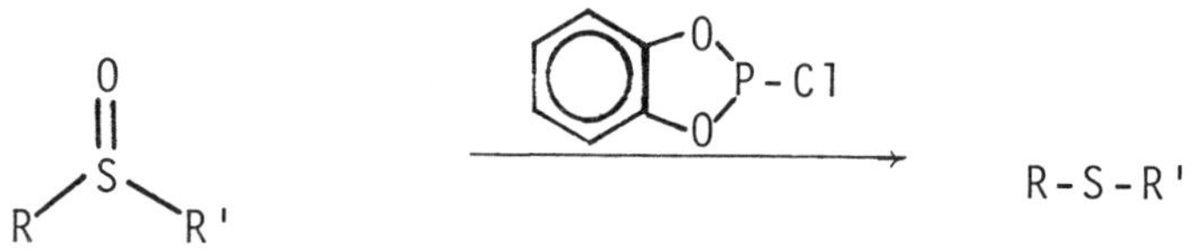

$$\underset{R}{\overset{O}{\underset{\|}{S}}}\diagdown R' \xrightarrow{\text{catechyl-P-Cl}} R-S-R'$$

R,R' = alkyl, aryl 72-100%

III.D-1 J. O. Morley, Synthesis, 528 (1976).

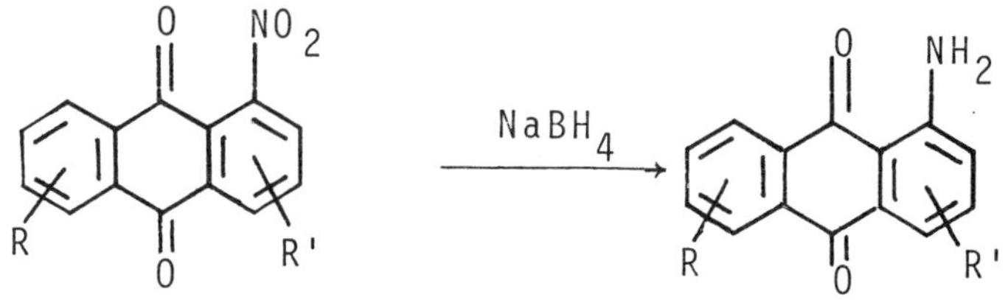

[anthraquinone with NO$_2$ group] $\xrightarrow{NaBH_4}$ [anthraquinone with NH$_2$ group]

R,R' = alkyl, COOH, SO$_3$Na, NH$_2$ ~80-90%

III.E.1-1   T. Fujisawa, K. Sugimoto, and H. Ohta, Chem. Lett., 581 (1976).

$$\text{R-CH=CH}_2 \xrightarrow{\text{FeCl}_2\text{-NaH}} \text{RCH}_2\text{CH}_3$$

$$\text{R-C≡CH} \nearrow \quad \sim 60\text{-}90\%$$

R = alkyl, Ph

III.E.1-2   N. J. Cusack et al., Tetrahedron, 32, 2157 (1976).

$$\text{R-CH=CH-R'} \xrightarrow[\text{CHMe}_2\text{-ArSO}_2\text{NHNH}_2]{\text{MeOH, Et}_3\text{N}} \text{R-CH}_2\text{-CH}_2\text{-R'}$$

R,R' = alkyl, aryl, alcohol, ester

61-99%

(Ar = 2,4,6-triisopropylphenyl: Me$_2$CH-, CHMe$_2$, CHMe$_2$ substituents with SO$_2$NHNH$_2$)

A convenient precursor for diimide reductions.

III.E.1-3   Y. Nakamura and H. Hirai, Chem. Lett., 165 (1976).

(P)—⟨C$_6$H$_4$⟩—iminodiacetate + PdCl$_2$

Selective catalyst for reduction of conjugated diolefins to monoolefins at 30° and 1 atm. H$_2$.

III.E.1-4 E. C. Ashby, J. J. Lin, and R. Kovar, J. Org. Chem., 41, 1939 (1976).

$$\underset{H}{\overset{R}{>}}C=C\underset{C-R}{\overset{H}{<}} \quad \xrightarrow{\text{LiAlH}_4\text{-CuI}} \quad RCH_2CH_2\overset{O}{\underset{\|}{C}}-R$$

R = CH_3, t-Bu

Yields up to 101% (VPC)

III.E.1-5 G. Descotes and D. Sinou, Tetrahedron Lett., 4083 (1976).

$$>C=C<\overset{O}{\underset{\|}{C}}- \quad \xrightarrow[\text{glucoside}]{\text{RuCl}_2(\text{PPh}_3)_3/H_2} \quad -\underset{H}{\overset{|}{C}}-\underset{H}{\overset{|}{C}}-\overset{O}{\underset{\|}{C}}-$$

Widely varying yields, up to 34% ee.

III.E.1-6 G. P. Boldrini and A. Umani-Ronchi, Synthesis, 596 (1976).

$$R^1-CH=CH-\overset{O}{\underset{\|}{C}}-R^2 + NaHCr_2(CO)_{10} \longrightarrow R^1CH_2CH_2-\overset{O}{\underset{\|}{C}}-R^2$$

R^1 = alkyl, aryl

R^2 = H, alkyl, aryl

40-80%

III.E.1-7 J. M. Fortunato and B. Ganem, J. Org. Chem., 41, 2194 (1976).

A detailed survey of the use of lithium and potassium tri-sec-butylborohydrides (L- and K-Selectrides) for the conjugate reduction and reductive alkylation of unsaturated ketones and esters.

III.E.1-8 F. Toda and M. Kanno, Bull. Chem. Soc. Japan, 49, 2643 (1976).

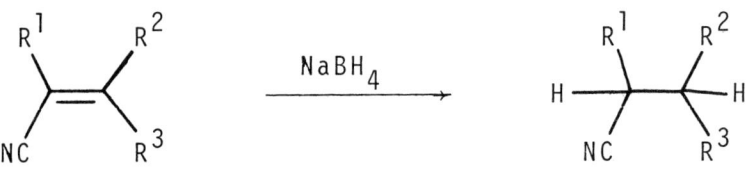

R^1 = aryl, n-alkyl   56-100%
R^2 = H, Ph, CN, NO_2, OMe
R^3 = H, Ph

III.E.1-9 R. O. Hutchins, et al., J. Org. Chem., 41, 3328 (1976).

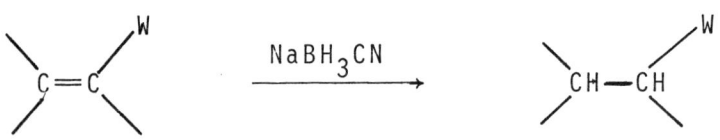

W = COOR, CN, NO_2 Yields generally
                              60-90%

III.E.1-10 F. Toda and K. Iida, Chem. Lett., 695 (1976).

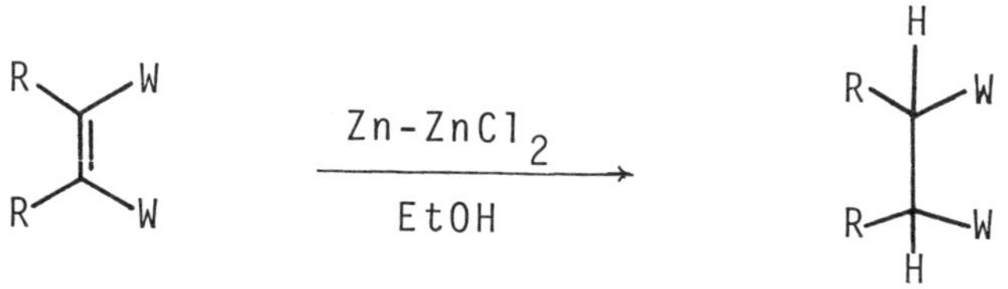

$$R_2C=CR_2(W,W) \xrightarrow{Zn-ZnCl_2, EtOH} R_2CH-CHR_2(W,W)$$

R = alkyl, aryl, vinyl
W = COR, COOR, CN, etc.

~70-90%

III.E.1-11 K. Junghaus, Chem. Ber., 109, 395 (1976).

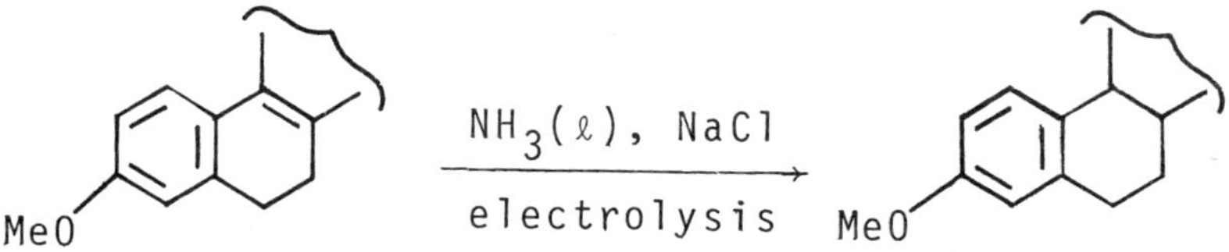

[methoxy-substituted dihydronaphthalene] $\xrightarrow{NH_3(\ell), NaCl, electrolysis}$ [reduced product]

Widely varying yields.

III.E.1-12 T. Hayashi et al., Tetrahedron Lett., 1133 (1976).

Use of **I** as a chiral hydrogenation catalyst:

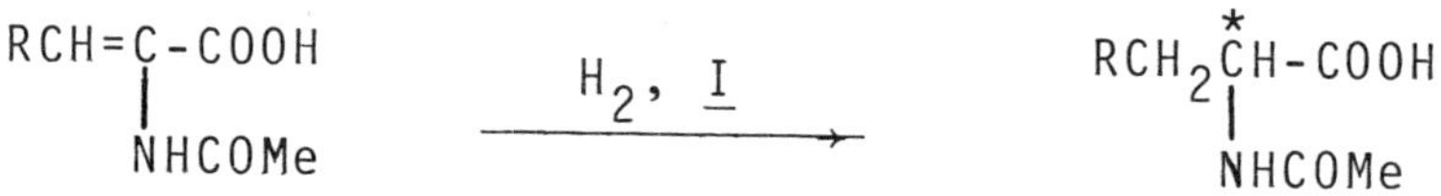

$$\text{RCH=C(NHCOMe)-COOH} \xrightarrow{H_2, \text{ I}} \text{RCH}_2\overset{*}{\text{C}}\text{H(NHCOMe)-COOH}$$

R = aryl

I = 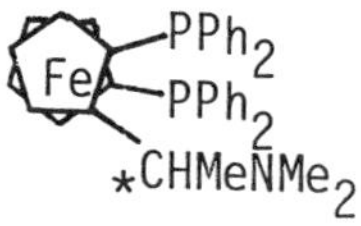 Fe(-PPh$_2$)(-PPh$_2$)(*CHMeNMe$_2$)

predominantly (S)
86-94% chemical yield
52-89% ee

III.E.1-13   K. M. Minachev, Y. S. Khodakov, and V. S. Nakhshunov, Russ. Chem. Rev., 45, 142 (1976).

Review: "Hydrogenation of Alkenes on Oxide Catalysts"

III.E.1-14/III.E.2-1   A. Sisak and F. Ungvary, Chem. Ber., 109, 531 (1976).

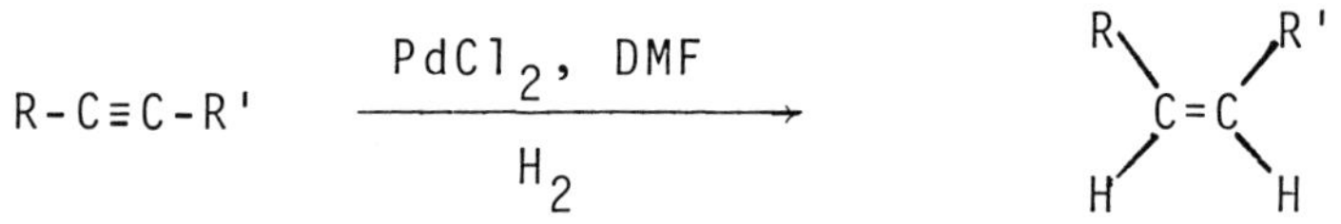

$$R-C\equiv C-R' \xrightarrow[H_2]{PdCl_2, DMF} \begin{array}{c} R \\ C=C \\ H H \end{array} R'$$

Nearly quantitative for 1-heptyne, 2-pentyne

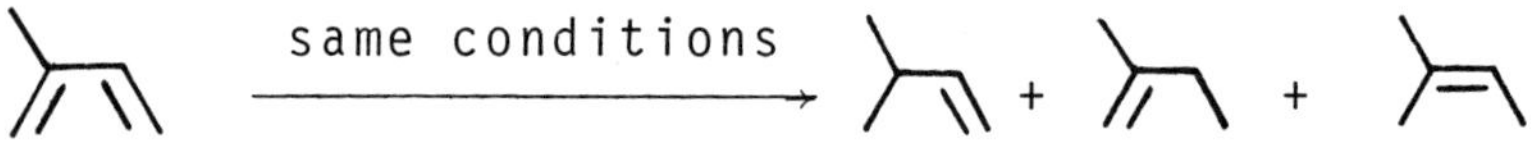

$$\text{(2-methyl-1,3-butadiene)} \xrightarrow{\text{same conditions}} \text{29\%} + \text{25\%} + \text{46\%}$$

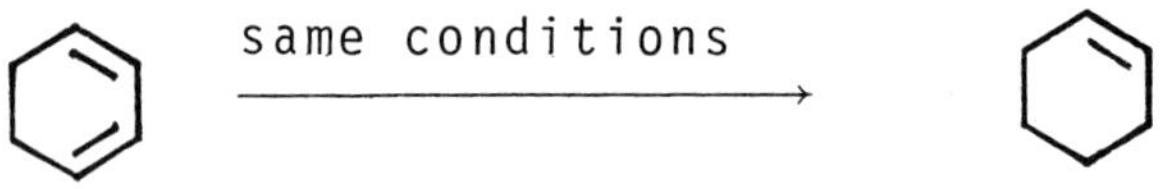

$$\text{(benzene)} \xrightarrow{\text{same conditions}} \text{(cyclohexene)} \quad 100\%$$

III.E.1-15/III.E.2-2 P. W. Chum and S. E. Wilson, Tetrahedron Lett., 15 (1976).

$$R-C\equiv C-R' \xrightarrow{LiAlH_4-TiCl_4} \underset{H}{\overset{R}{>}}C=C\underset{H}{\overset{R'}{<}}$$

1-octyne or 4-octyne

1-octene, 81%
cis-4-octene, 73%

$$\underset{H}{\overset{H}{>}}C=C\underset{\underline{n}-C_6H_{13}}{\overset{H}{<}} \xrightarrow{LiAlH_4-TiCl_4} \underline{n}\text{-octane}$$

92%

III.E.2-3 J. K. Crandall and F. Collonges, J. Org. Chem., 41, 4089 (1976).

$$R-C\equiv C-R' \xrightarrow{CuI + 2R''MgX} \underset{H}{\overset{R}{>}}C=C\underset{H}{\overset{R'}{<}}$$

R,R' = Me, $\underline{n}$-C$_5$H$_{11}$, Ph

R" = $\underline{n}$-Bu, Et

65-100% for relatively unhindered alkynes

III.E.2-4 R. R. Schrock and J. A. Osborn, J. Am. Chem. Soc., 98, 2143 (1976).

$$R^1-C\equiv C-R^2 \xrightarrow[\text{acetone, } H_2]{[Rh(NBD)(PPhMe_2)_3]^+ClO_4^-} \underset{R^1}{\overset{H}{>}}C=C\underset{R^2}{\overset{H}{<}}$$

R^1, R^2 = alkyl

>95%

III.E.2-5 E. L. Muetterties, Bull. Soc. Chim. Belges, 85, 451 (1976).

Review: "Metal Clusters in Catalysis-Reduction of Triple Bonds"

III.F.1-1 F. Bohlmann, J. Staffeldt, and W. Skuballa, Chem. Ber., 109, 1586 (1976).

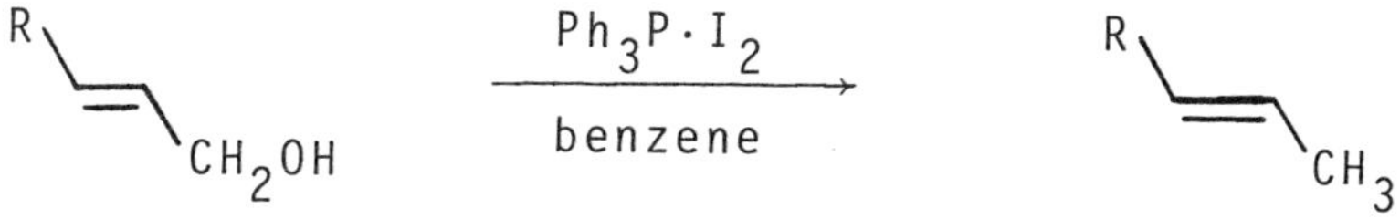

$$\text{R}-\text{CH}=\text{CH}-\text{CH}_2\text{OH} \xrightarrow[\text{benzene}]{Ph_3P \cdot I_2} \text{R}-\text{CH}=\text{CH}-\text{CH}_3$$

R = alkyl, aryl, acetylenic   24-60%

III.F.1-2 M. G. Adlington, M. Orfanopoulas, and J. L. Fry, Tetrahedron Lett., 2955 (1976).

$$\text{R-OH} \xrightarrow[\text{Et}_3\text{SiH}]{BF_3} \text{R-H}$$

R = 2°, 3°   poor to fair yields

REDUCTIONS

III.F.1-3   Y. Fujimoto and N. Ikekawa, <u>Chem. Pharm. Bull.</u>, <u>24</u>, 825 (1976).

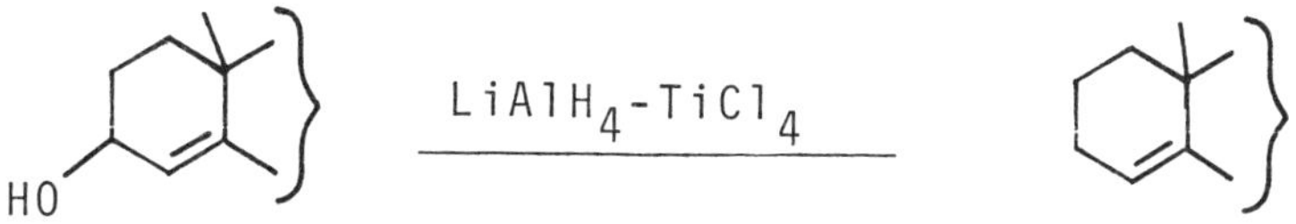

[Structure: HO-substituted bicyclic alkene] →(LiAlH$_4$-TiCl$_4$)→ [Structure: bicyclic alkene]   65%

Several additional examples.

III.F.1-4   H. C. Beyerman <u>et al.</u>, <u>Rec. Trav. Chim. Pays-Bas</u>, <u>95</u> 43 (1976).

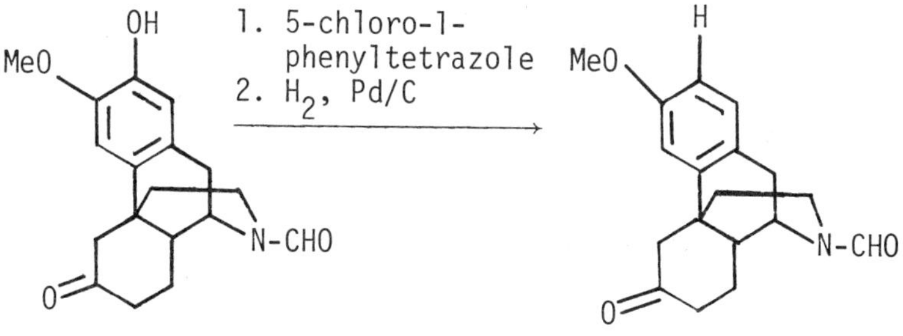

[Morphinan structure with MeO, OH, N-CHO, ketone] →(1. 5-chloro-1-phenyltetrazole 2. H$_2$, Pd/C)→ [Reduced morphinan structure with MeO, H, N-CHO, ketone]

~50%

III.F.1-5   S. Krishnamurthy and H. C. Brown, <u>J. Org. Chem.</u>, <u>41</u>, 3064 (1976).

R-OTs →(LiEt$_3$BH / THF)→ R-H   81-100%

R = 1°,2° alkyl, cyclic

e.g. [cyclohexyl-OTs] →(LiEt$_3$BH, THF, 25°, 12 hr)→ [cyclohexane]   80%

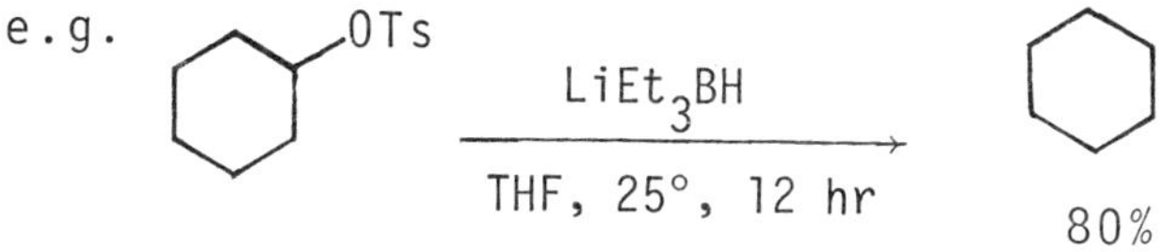

III.F.1-6 Y. Fujimoto and T. Tatsuno, *Tetrahedron Lett*., 3325 (1976).

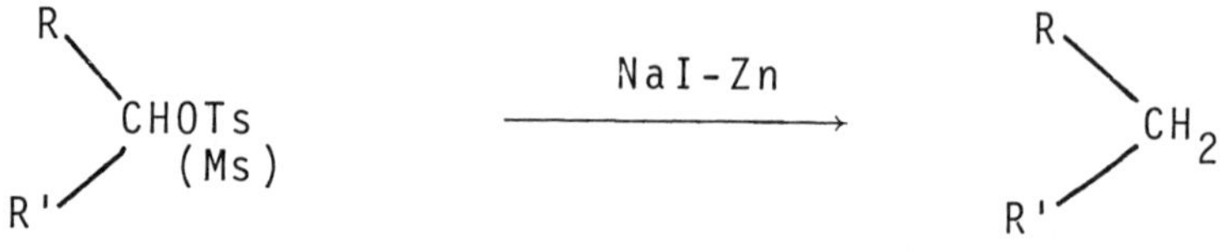

$$R_2R'C\text{-CHOTs(Ms)} \xrightarrow{NaI-Zn} R_2R'C\text{-CH}_2$$

R = alkyl
R' = H, alkyl

yields generally 70-90%

III.F.2-1 M. P. Doyle, C. C. McOsker, and C. T. West, *J. Org. Chem*., **41**, 1393 (1976).

$$RX + Et_3SiH \xrightarrow{AlCl_3} RH + Et_3SiX$$

R = 1°,2°,3° alkyl
X = Br, Cl

26-100% (VPC)

III.F.2-2 W. E. Parham, L. D. Jones and Y. A. Sayed, *J. Org. Chem*., **41**, 1184, 1187 (1976).

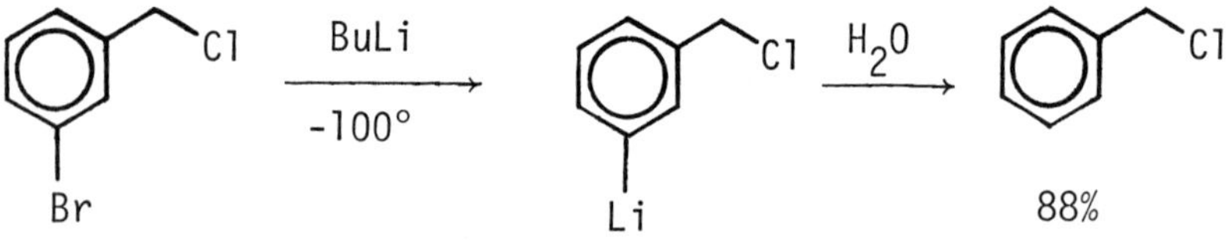

$$\text{3-Br-C}_6\text{H}_4\text{-CH}_2\text{Cl} \xrightarrow[-100°]{BuLi} \text{3-Li-C}_6\text{H}_4\text{-CH}_2\text{Cl} \xrightarrow{H_2O} \text{C}_6\text{H}_5\text{-CH}_2\text{Cl}$$

88%

Other electrophiles may also be added.

REDUCTIONS

III.F.2-3   T.-L. Ho and G. Olah, *Synthesis*, 807 (1976).

$$\underset{X}{\underset{|}{R-\overset{O}{\overset{\|}{C}}-CH-R'}} \xrightarrow[H_2O/THF]{2VCl_2} R-\overset{O}{\overset{\|}{C}}-CH_2-R'$$

X = Br, Cl
R = alkyl, aryl
R' = H, alkyl

80-96%

III.F.4-1   A. J. Boulton, *et al.*, *Tetrahedron Lett.*, 2689 (1976).

$$RCH_2NH_2 \;+\; \underset{ClO_4^{\ominus}}{\overset{Ph}{\underset{Ph\;\;\;O^{\oplus}\;\;\;Ph}{\bigcirc}}} \longrightarrow \underset{\underset{CH_2R}{ClO_4^{\ominus}}}{\overset{Ph}{\underset{Ph\;\;\;N^{\oplus}\;\;\;Ph}{\bigcirc}}} \xrightarrow[2.\;200°]{1.\;NaBH_4} R-CH_3$$

R = vinyl, phenyl, furyl, pyridyl

30-60% overall

III.F.4-2   A. Rahman and A. Basha, *J. Chem. Soc., Chem. Commun.*, 594 (1976).

$$R-\bigcirc-\overset{O}{\overset{\|}{C}}-N\bigcirc \xrightarrow{POCl_3} Ar-\underset{\oplus}{\overset{Cl}{\overset{|}{C}}=N}\bigcirc \xrightarrow[2.\;H_2O]{1.\;Zn} ArCHO$$

R = H, Cl, Br

90-95%

III.G-1 P. B. Dervan and M. A. Shippey, *J. Am. Chem. Soc.*, **98**, 1265 (1976).

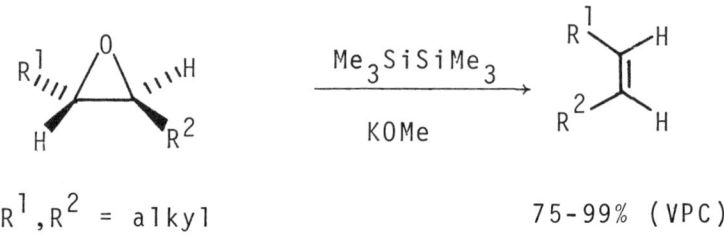

R^1, R^2 = alkyl

75-99% (VPC)

III.G-2 M. T. Reetz and M. Plachky, *Synthesis*, 199 (1976).

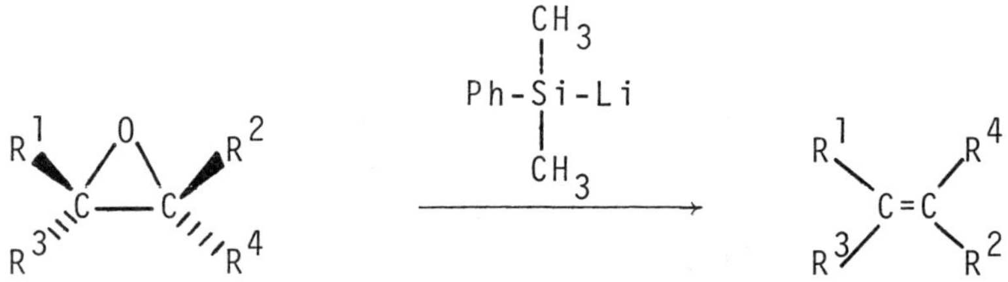

Epoxidized Substrate	Yield
trans-stilbene	75% (97% *cis*)
cis-stilbene	83% (99% *trans*)
Styrene	polymer
1-Pentene	60%
1-Octene	64%

III.G-3 T. Kempe, T. Norin, and R. Caputo, *Acta Chem. Scand. B*, **30**, 366 (1976).

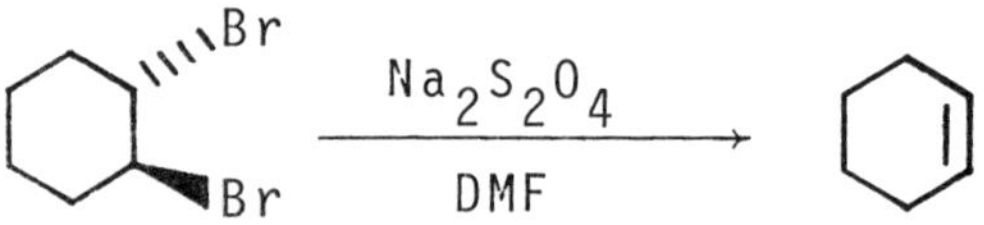

60%

III.G-4   J. E. McMurray and M. P. Fleming, J. Org. Chem., 41, 896 (1976).

$$\begin{array}{c} \diagdown C-OH \\ \diagup | \\ \diagdown C-OH \\ \diagup \end{array} \quad \xrightarrow{\text{active Ti}^0} \quad \begin{array}{c} \diagdown C \\ \diagup \| \\ \diagdown C \\ \diagup \end{array}$$

55-85%

Reaction fails with glycols which cannot form a five-membered ring:

$$\underset{\diagup \diagdown}{O \overset{Ti}{\diagdown} O}$$

III.G-5   V. Calo et al., Synthesis, 200 (1976).

$$\underset{R^2 \diagdown \quad \diagup R^4}{R^1 \diagup X \diagdown R^3} \quad \xrightarrow{\underset{S}{\overset{CH_3}{\underset{|}{N}}} =Se} \quad \underset{R^2}{\overset{R^1}{\diagdown}} C=C \underset{R^4}{\overset{R^3}{\diagup}}$$

R's = alkyl, aryl, cyclic, Cl   87-100% (VPC)

X = O, S

III.G-6   P. Dowd and L. K. Marwaha, J. Org. Chem., 41, 4035 (1976).

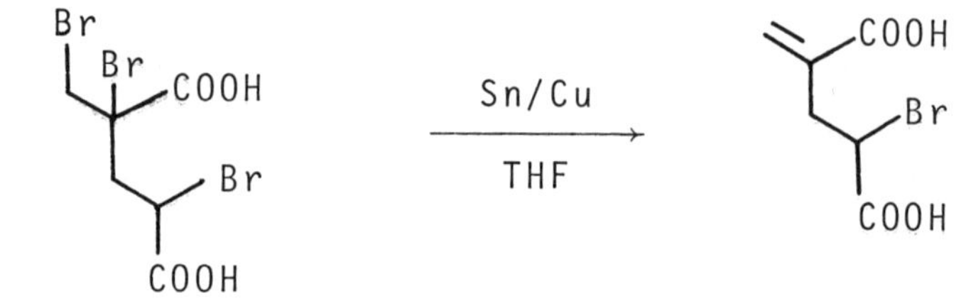

[Reaction: tribromo diacid + Sn/Cu in THF → bis-methylene dibromo diacid product, 81%]

[Reaction: tetrabromide (Br-CH2-C(=)-C(=)-CH2-Br type) + Sn/Cu in THF → bis-methylene dibromide, 53%]

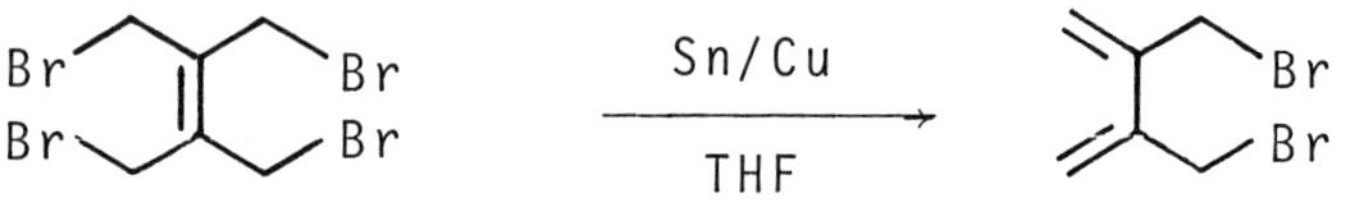

III.G-7   S. Mageswaran and M.U.S. Sultanbawa, J. Chem. Soc. Perkin I, 884 (1976).

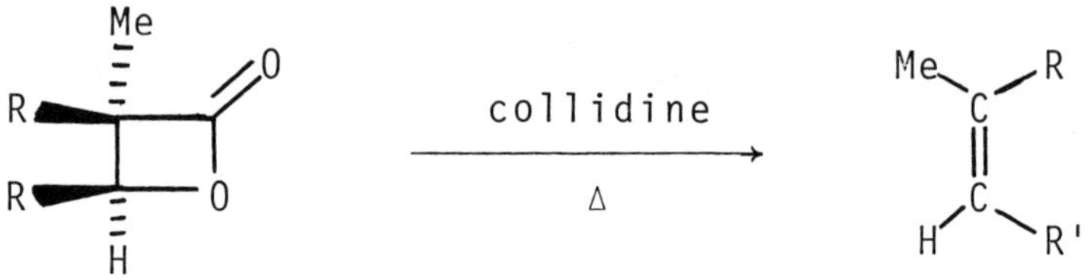

[β-lactone with Me, H, R, R substituents] —collidine, Δ→ [alkene Me, R, H, R']

R,R' = alkyl   ~60%

REDUCTIONS

III.H-1 A. E. Greene, A. Cruz, and P. Crabbé, Tetrahedron Lett., 2707 (1976).

[cyclopentanone-CO$_2$R] $\xrightarrow[\text{K}_2\text{CO}_3,\ \text{Butanone reflux}]{\text{Al}_2\text{O}_3\ \ \ \text{I(CH}_2)_6\text{CO}_2\text{Et}}$ [cyclopentanone]

R = H, Me, Et   70-96%

III.H-2 M. Mori and Y. Ban, Chem. Pharm. Bull., 24, 1992 (1976).

[CH$_2$=CH-CH$_2$-N(Ac)-R] $\xrightarrow{\text{Pd-CuCl}_2\text{-AcONa}}$ [CH$_3$-CH$_2$-CH$_2$-N(HAc)-R...]

HNAc—R

59%, R = Bz
61%, R = PhCH$_2$CH$_2$

III.H-3 M. T. Thomas, E. G. Breitholle, and A. G. Fallis, Synth. Commun., 6, 113 (1976).

[cyclohexanone=CHPh] $\xrightarrow[\text{KOH, crown ether}]{\text{4-aminobutyric acid}}$ [cyclohexanone]

40-56%

III.I-1 H. C. Brown and N. Ravindran, *J. Am. Chem. Soc.*, **98**, 1785, 1798 (1976).

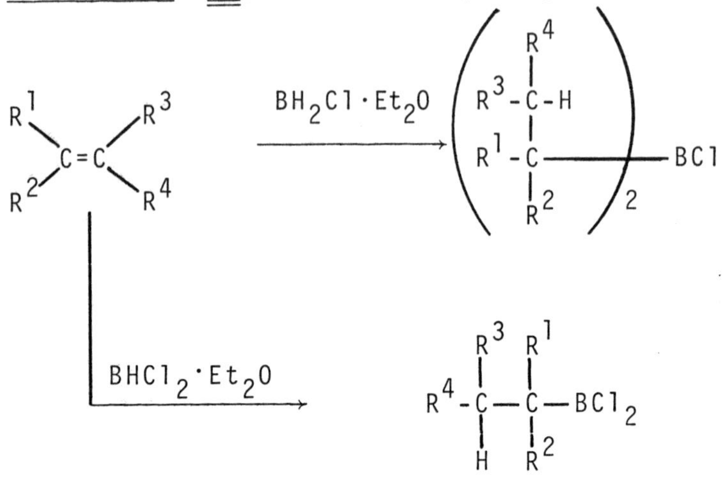

$$R^1R^2C=CR^3R^4 \xrightarrow{BH_2Cl \cdot Et_2O} \left(R^3\text{-}C(R^4)\text{-}H, R^1\text{-}C(R^2)\text{-} \right)_2 BCl$$

$$\xrightarrow{BHCl_2 \cdot Et_2O} R^4\text{-}C(H)(R^3)\text{-}C(R^1)(R^2)\text{-}BCl_2$$

Usual sense of addition, yields generally >80%.

III.I-2 G. W. Kabalka, D.T.C. Yang, and J. D. Baker, Jr., *J. Org. Chem.*, **41**, 574 (1976).

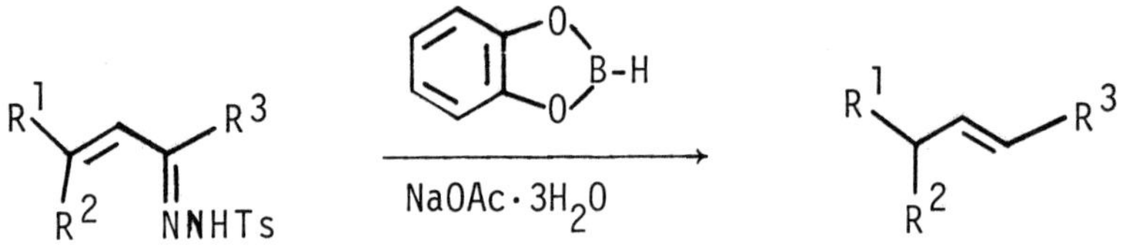

$$R^1R^2C=C(R^3)\text{-}NNHTs \xrightarrow[NaOAc \cdot 3H_2O]{catecholborane} R^1R^2CH\text{-}CH=CHR^3$$

$R^1, R^2 = $ H, alkyl, aryl   53-77%
$R^3 = $ H, alkyl

III.I-3 A. Pelter, *Chem. and Ind.*, 888 (1976).

Review: "Reductions Involving Diborane and Its Derivatives"

III.I-4 C. F. Lane and G. W. Kabalka, Tetrahedron, 32, 981 (1976).

Review: "Catecholborane. A new hydroboration reagent."

III.I-5 H. C. Brown, S. Krishnamurthy, and N. M. Yoon, J. Org. Chem., 41, 1778 (1976).

An exhaustive review of the uses of 9-BBN as a reducing agent in organic chemistry.

III.I-6 C. F. Lane, Chem. Rev., 76, 773 (1976).

Review: "Reduction of Organic Compounds with Diborane"

III.J-1 H. Ishikawa and T. Mukaiyama, Chem. Lett., 737 (1976).

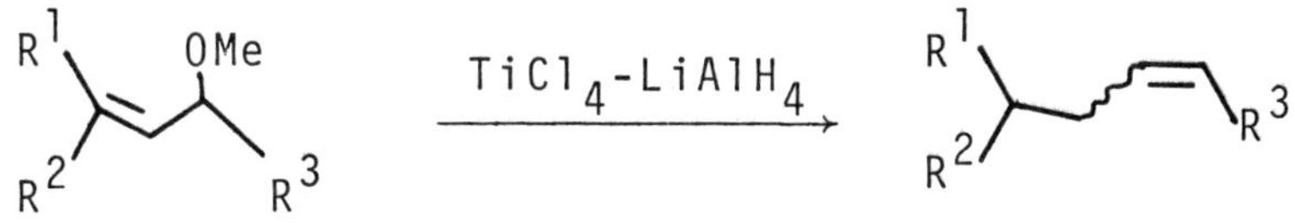

R^1, R^2 = H, Me, Et
R^3 = alkyl, Bz, cyclohexyl

~80%

III.J-2   D. J. Raber and W. C. Guida, <u>J. Org. Chem.</u>, <u>41</u>, 690 (1976).

  Use of $Bu_4\overset{\oplus}{N}\ \overset{\ominus}{B}H_4$ as a reducing agent. Very similar to $NaBH_4$, but may be used in CH_2Cl_2 solution.

III.J-3   E.R.H. Walker, <u>Chem. Soc. Rev.</u>, <u>5</u>, 23 (1976).

  Review: "The Functional Group Selectivity of Complex Hydride Reducing Agents"

III.J-4   O. N. Chupakhin and I. Ya. Postovskii, <u>Russ. Chem. Rev.</u>, <u>45</u>, 454 (1976).

  Review: "Nucleophilic Substitution of Hydrogen in Aromatic Systems"

III.J-5   H. B. Kagan, <u>Pure and Appl. Chem.</u>, <u>43</u>, 401 (1975).

  Review: "Asymmetric catalysis by chiral rhodium complexes in hydrogenation and hydrosilylation reactions"

IV.A-1 Y. Girault, M. Decouzon, and M. Azzaro, Tetrahedron Lett., 1175 (1976).

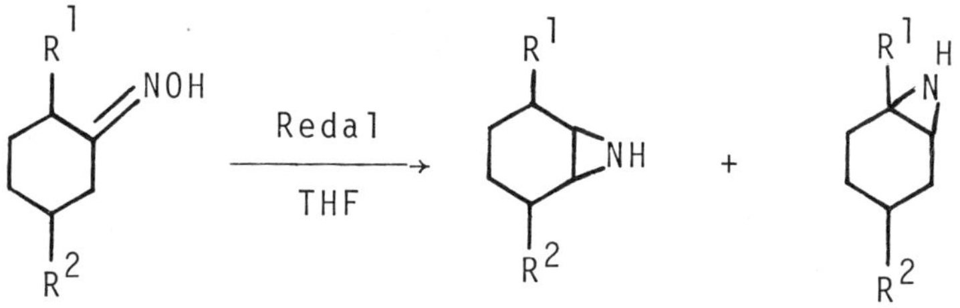

R^1, R^2 = alkyl

IV.A-2 K. Imai. Y, Kawazoe, and T. Taguchi, Chem. Pharm. Bull., 24, 1083 (1976).

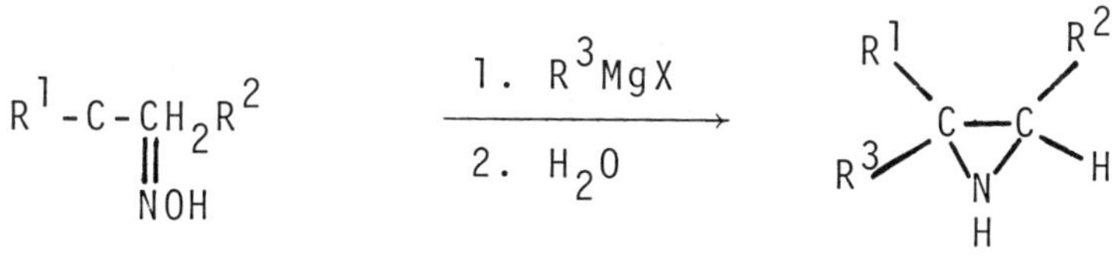

R's = Ph, Bz, alkyl, etc.   Widely varying yields.

IV.A-3 Y. Tamura et al., Synthesis, 35 (1976).

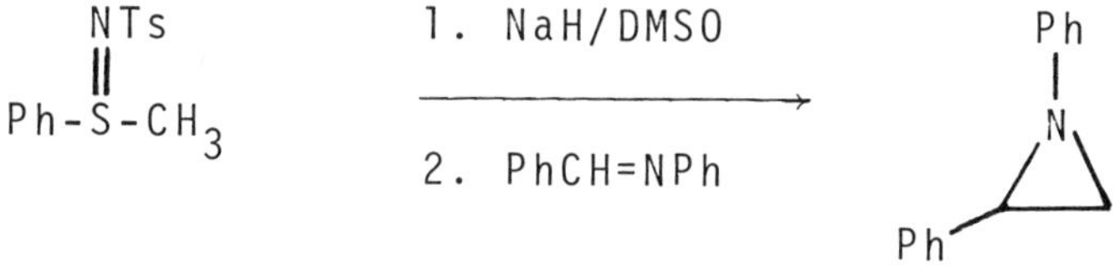

47%

IV.A-4   I. Shahak, Y. Ittah, and J. Blum, <u>Tetrahedron Lett.</u>, 4003 (1976).

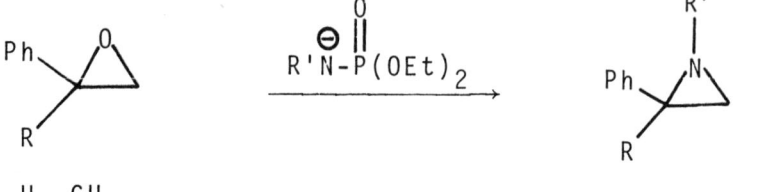

R = H, CH$_3$
R' = Ph, Bz, <u>t</u>-Bu

generally low yields (10-58%)

IV.A-5   A. Zwierzak and K. Osowska, <u>Angew. Chem. Int. Ed.</u>, <u>15</u>, 302 (1976).

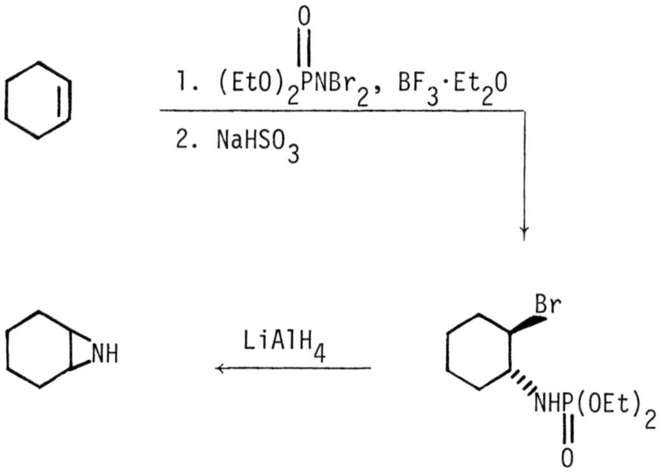

IV.A-6   Y. Nomura, N. Hatanaka, and Y. Takeuchi, Chem. Lett., 901 (1976).

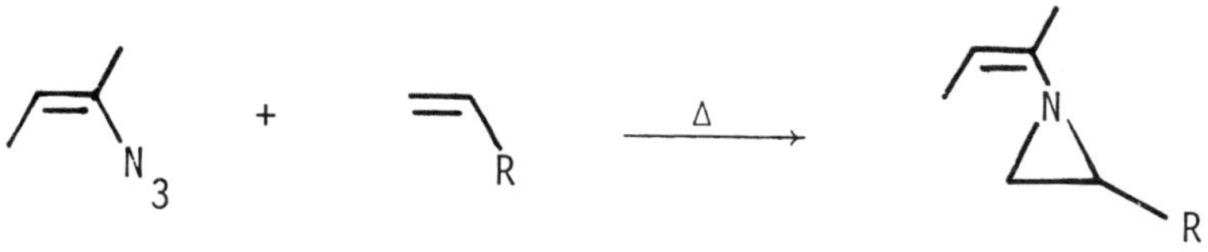

73% (R=CN)

83% (R=CO_2Me)

IV.B-1   H.-D. Scharf and E. Wolters, Angew. Chem. Int. Ed., 15, 682 (1976).

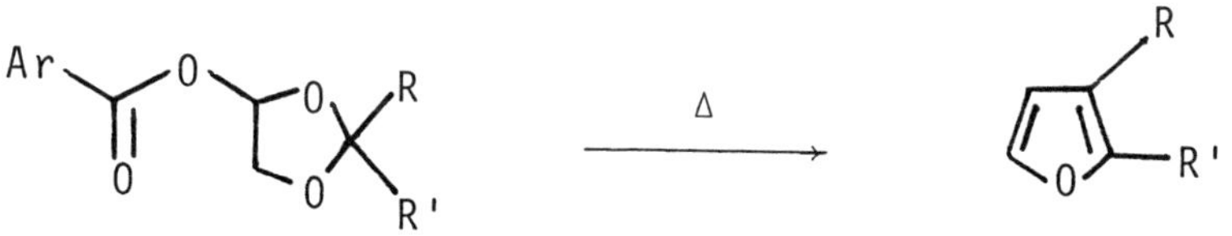

R,R' = H, Me, Et,   30-63%
       -(CH_2)$_3$-

IV.B-2   G. M. El-Naggar and M. A. Eldawy, Indian J. Chem., 14B, 47 (1976).

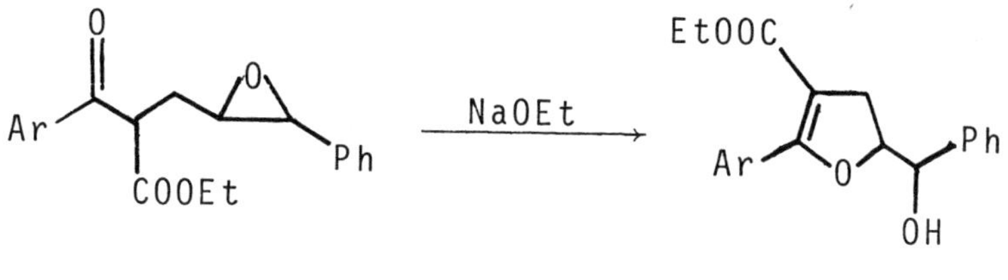

~60-80%

IV.B-3 P. Cagniant and G. Kirsch, <u>Comptes Rendus (C)</u>, <u>282</u>, 465 (1976).

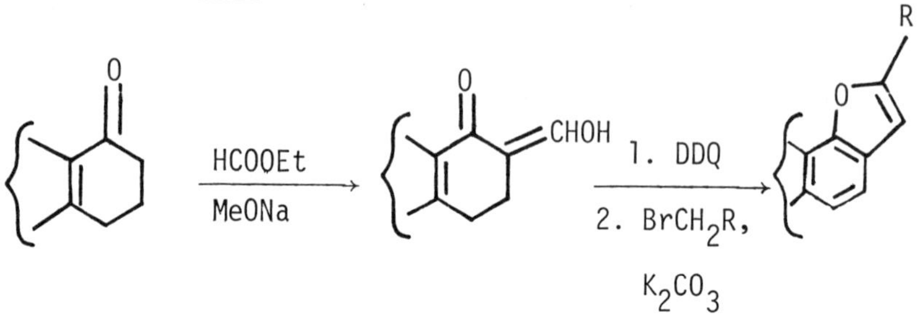

R = COR, COOR, CN, NO$_2$, etc.

~35-55% overall

IV.B-4 M. Yamamoto, <u>J. Chem. Soc. Perkin I</u>, 1688 (1976).

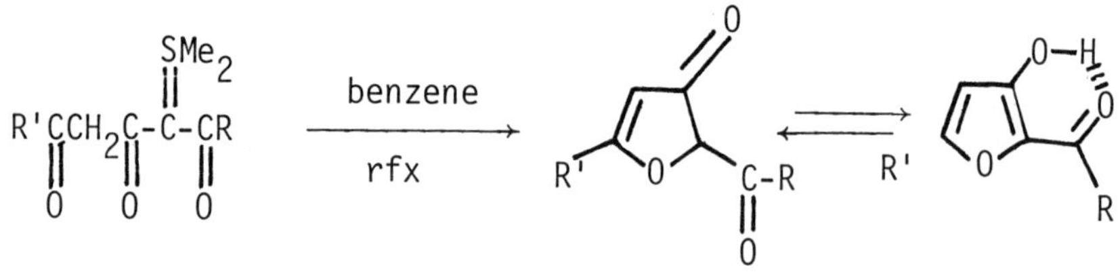

R = Me, Ph, OEt

R' = Me, Ph, OEt, NHPh

80-95%

IV.B-5 W. K. Anderson, E. J. LaVoie, and J. C. Bottaro, <u>J. Chem. Soc. Perkin I</u>, 1 (1976).

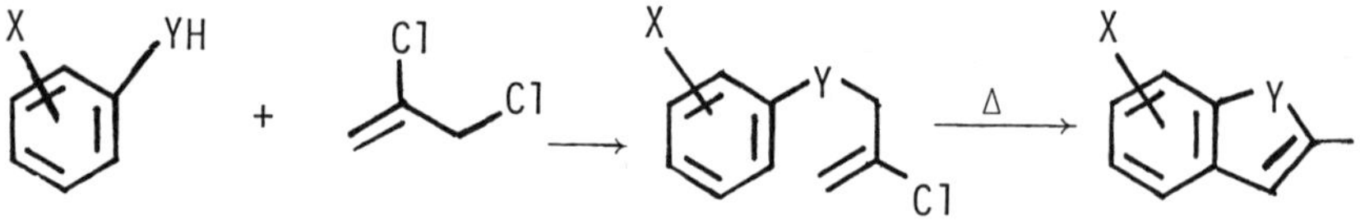

X = H, Cl, Me, OMe

Y = O, S

Widely varying yields.

IV.B-6 L. Cadona and P. D. Croce, <u>Synthesis</u>, 800 (1976).

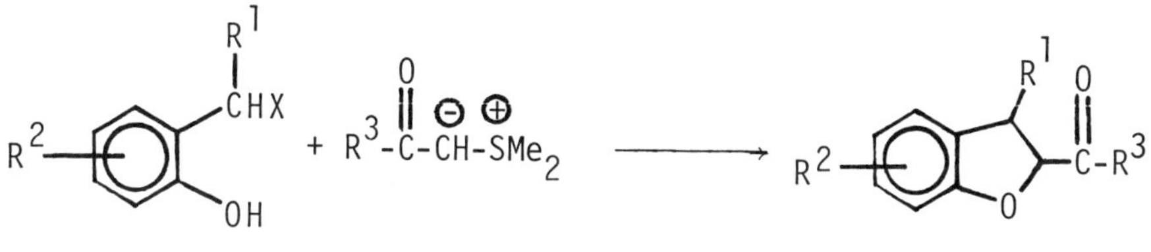

R^3 = OEt, Ar

R^1 = H, Me, Ph
R^2 = H, Cl, NO_2, Me
X = $^+NMe_3I^-$, OH

50-85%

IV.C-1 L. S. Hegedus, G. F. Allen, and E. L. Waterman, <u>J. Am. Chem. Soc.</u>, <u>98</u>, 2674 (1976).

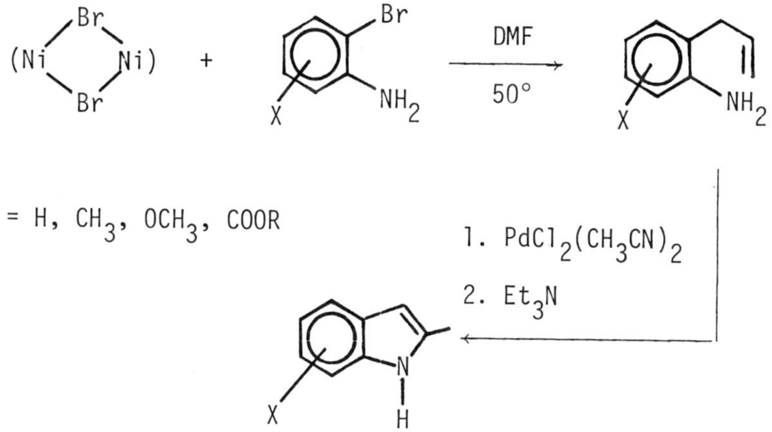
DMF
50°
1. $PdCl_2(CH_3CN)_2$
2. Et_3N

X = H, CH_3, OCH_3, COOR

74-84%

IV.C-2 C. Germain and J. Bourdais, J. Het. Chem., 13, 1209 (1976).

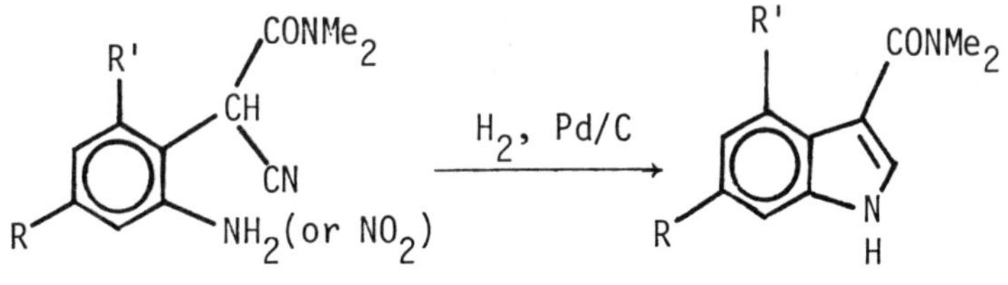
CONNMe2
R'
CH
CN
R
NH2(or NO2)
H2, Pd/C
R'
CONMe2
R
N
H

R = H, Me, OMe, SO_2NMe_2

R' = H, OMe, NH_2

~40-67%

IV.C-3 H. Bartsch, Monat. für Chem., 107, 663 (1976).

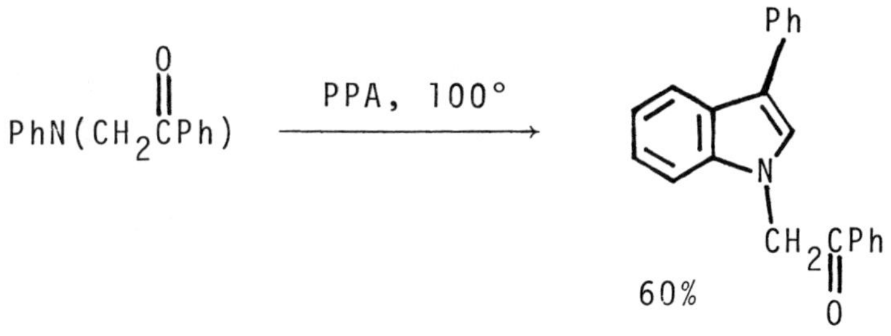
O
PhN(CH2CPh)
PPA, 100°
Ph
N
CH2CPh
O

60%

Further reaction to the pyrroloindole can be effected.

IV.C-4 Y. Makisumi and S. Takada, Chem. Pharm. Bull., 24, 770 (1976).

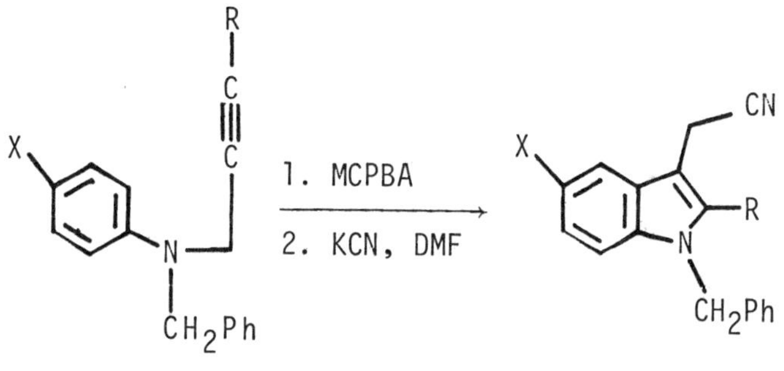

R
C
C
X
N
CH2Ph
1. MCPBA
2. KCN, DMF
CN
X
R
N
CH2Ph

R = H, CH_3

X = H, Cl, OCH_3

~30-50%

IV.C-5 D. E. Ames and O. Ribeiro, J. Chem. Soc. Perkin I, 1073 (1976).

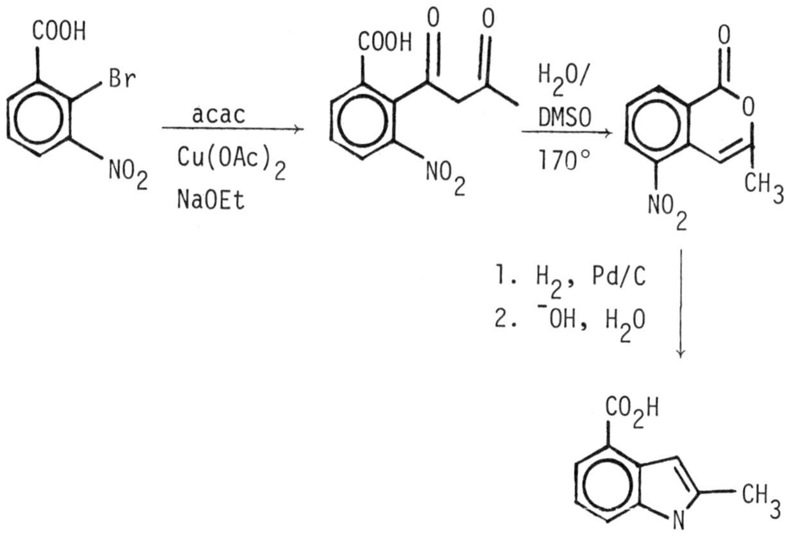

COOH
Br
NO2
acac
Cu(OAc)2
NaOEt
COOH
O
O
NO2
H2O/
DMSO
170°
O
O
CH3
NO2
1. H2, Pd/C
2. ⁻OH, H2O
CO2H
CH3
N

individual yields ~68-70%

IV.C-6 K. Freter, F. Hess, and K. Grozinger, <u>Liebigs Ann. Chem.</u>, 241 (1976).

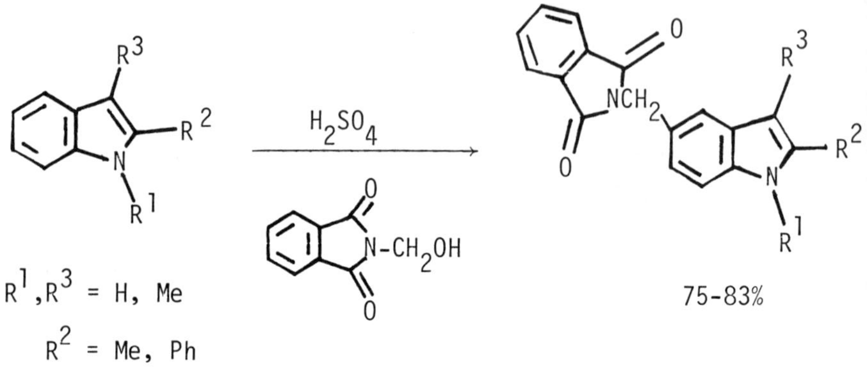
H_2SO_4
N-CH_2OH
NCH_2
R^1,R^3 = H, Me
R^2 = Me, Ph
75-83%

IV.C-7 M. K. Eberle and L. Brzechffa, <u>J. Org. Chem.</u>, <u>41</u>, 3775 (1976).

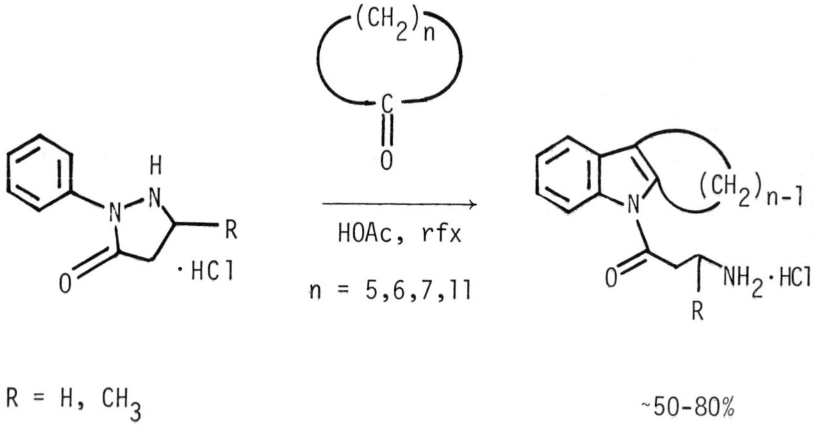
$(CH_2)_n$
HOAc, rfx
n = 5,6,7,11
·HCl
$(CH_2)_{n-1}$
NH_2·HCl
R = H, CH_3
~50-80%

IV.C-8/IV.D-1   H. Aoyama et al., J. Chem. Soc. Perkin I, 1556 (1976).

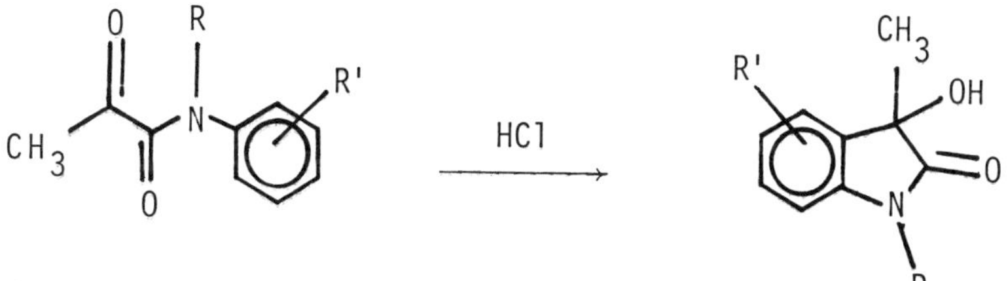

R = Me, Et
R' = Me, OMe, Cl

73-96%

IV.D-2   T. Hasegawa, H. Aoyama, and Y. Omote, J. Chem. Soc. Perkin I, 2054 (1976).

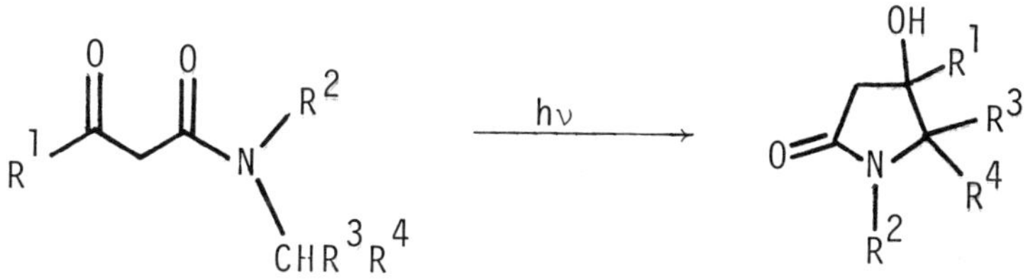

R^1 = Me, Ph
R^2, R^3, R^4 = Me_2, Et_2, $\underline{i}$-Pr_2, Bz_2, -$(CH_2)_5$-, etc.

~70-90%

IV.D-3   A. Kasahara, Chem. and Ind., 1032 (1976).

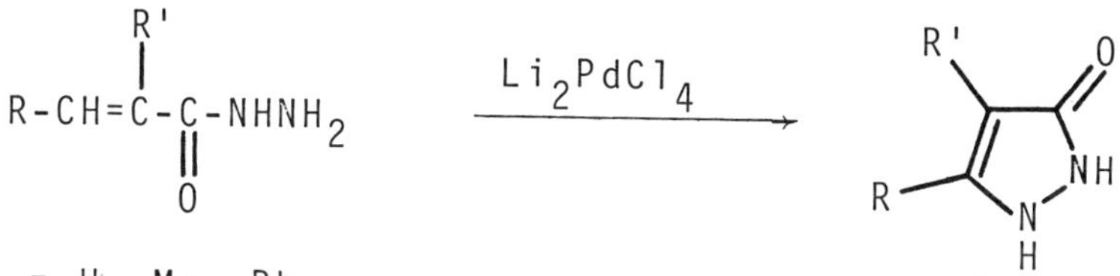

$R-CH=C-C-NHNH_2$ with R' substituent, $\xrightarrow{Li_2PdCl_4}$

R's = H, Me, Ph

33-42%

IV.D-4   A. Baba, Y. Oshiro, and T. Agawa, Chem. Lett., 11 (1976).

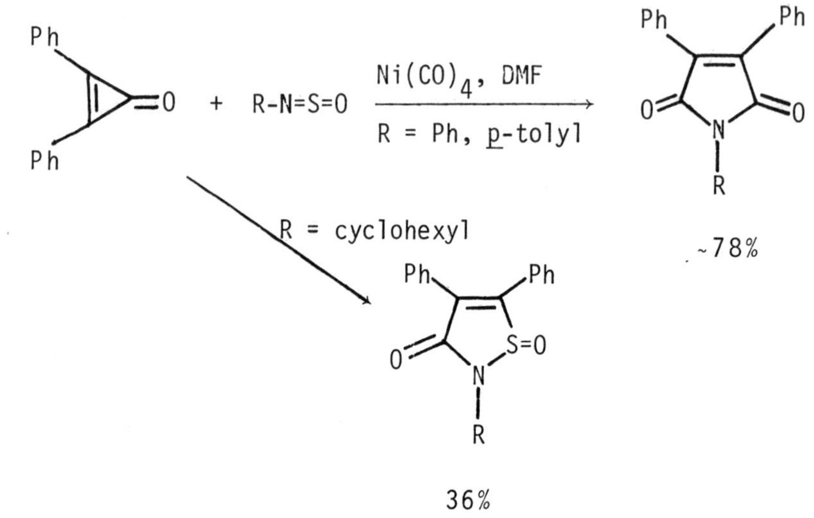

Ph₂C=C=O (diphenylcyclopropenone) + R-N=S=O → (Ni(CO)$_4$, DMF; R = Ph, p-tolyl) → N-substituted 3,4-diphenylmaleimide ~78%

For R = cyclohexyl → 3,4-diphenyl-isothiazol-3(2H)-one S-oxide 36%

IV.D-5   M. A. Ponomareva-Stepnaya et al., J. Gen. Chem. (USSR), 45, 2451 (1975).

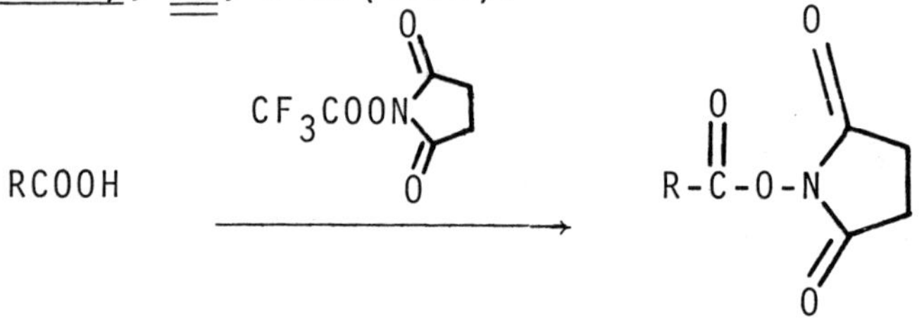

RCOOH + N-hydroxysuccinimide trifluoroacetate (CF$_3$COON-succinimide) → R-C(O)-O-N(succinimide)   52-99%

RCOOH = amino acid

racemization not observed

IV.D-6 G. Bettoni et al., J. Org. Chem., 41, 2780 (1976).

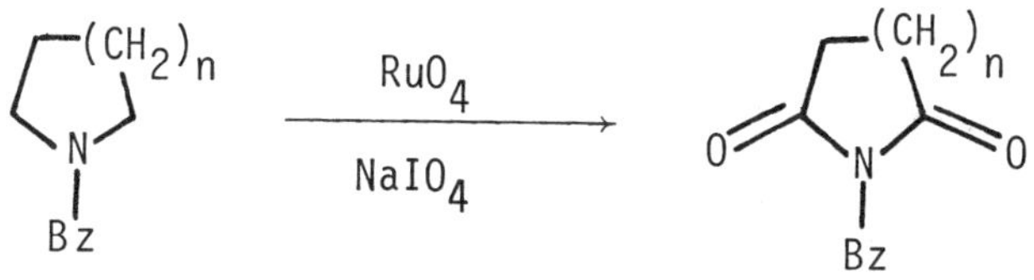

n = 1, 2

Several examples
yields ~30-40% (isolated)

IV.D-7 J.-C. Gramain, R. Remuson, and Y. Troin, J. Chem. Soc., Chem. Commun., 194 (1976).

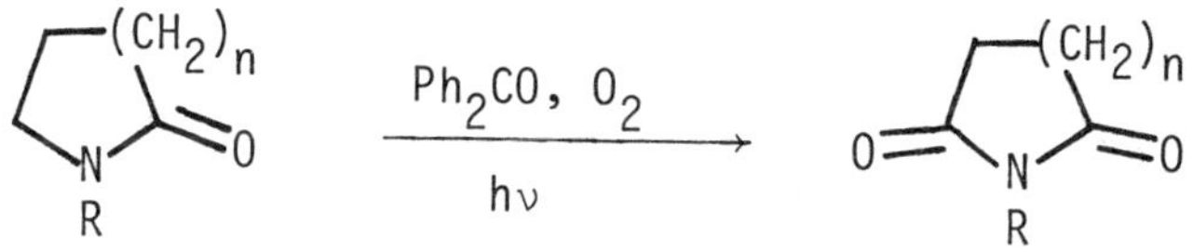

n = 1, 2

~60% (isolated)

IV.D-8 M. Ogata and H. Matsumoto, Chem. and Ind., 1067 (1976).

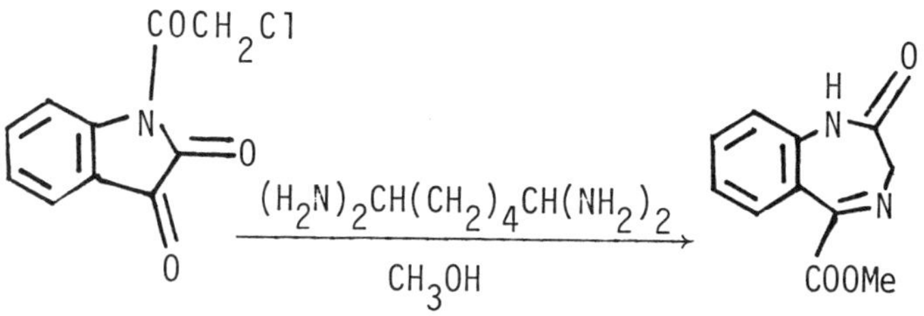

$(H_2N)_2CH(CH_2)_4CH(NH_2)_2$
CH_3OH

51%

IV.D-9   M. S. Manhas et al., Synthesis, 689 (1976).

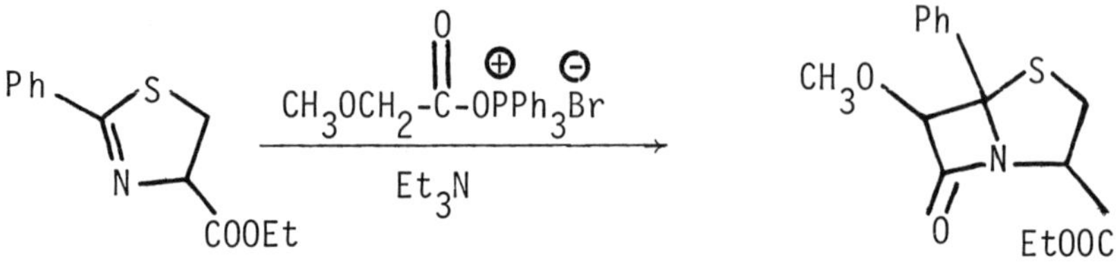

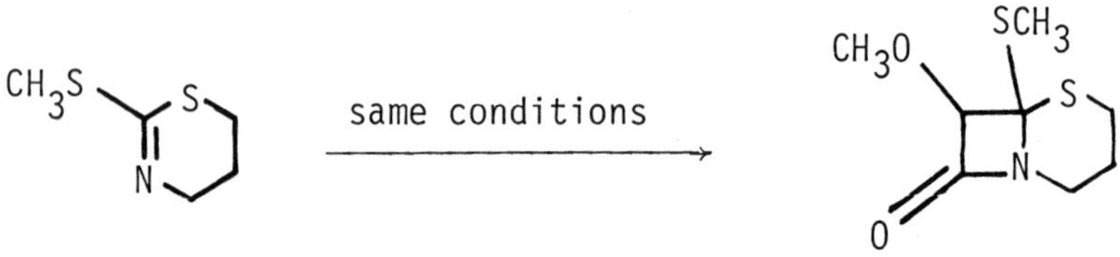

Ph–C(S–CH₂–CH(COOEt)–N=) + CH₃OCH₂-C(=O)-OPPh₃⁺Br⁻ / Et₃N → bicyclic β-lactam with CH₃O, Ph, S, N, EtOOC substituents

CH₃S–C(S–CH₂–CH₂–N=) —same conditions→ bicyclic β-lactam with CH₃O, SCH₃, S, N substituents

---

IV.D-10   D. F. Sullivan et al., J. Org. Chem., 41, 1112 (1976).

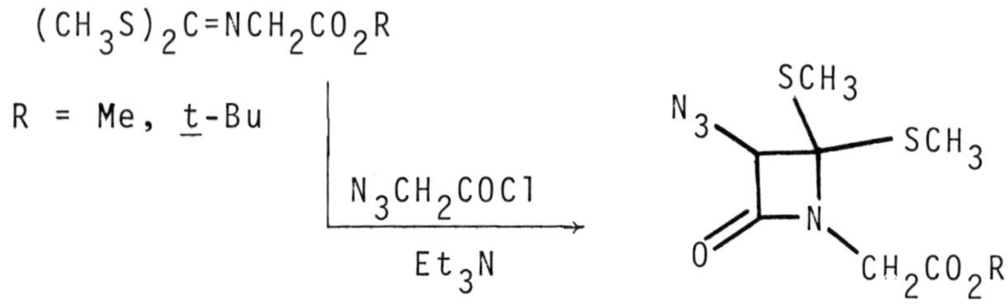

$(CH_3S)_2C=NCH_2CO_2R$

R = Me, t-Bu

N_3CH_2COCl / Et₃N

→ β-lactam with N₃, SCH₃, SCH₃, CH₂CO₂R substituents

~80%

IV.D-11 H. H. Wasserman and B. H. Lipshutz, Tetrahedron Lett., 4613 (1976).

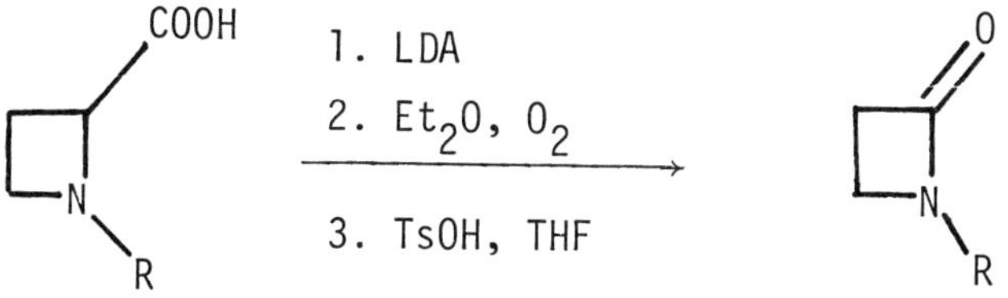

[azetidine-2-COOH with N-R] →
1. LDA
2. Et$_2$O, O$_2$
3. TsOH, THF
→ [azetidin-2-one with N-R]

R = 1° alkyl, t-butyl, acetal, cyclic   50-61%

IV.D-12 M. S. Manhas et al., Synth. Commun., 6, 435 (1976).

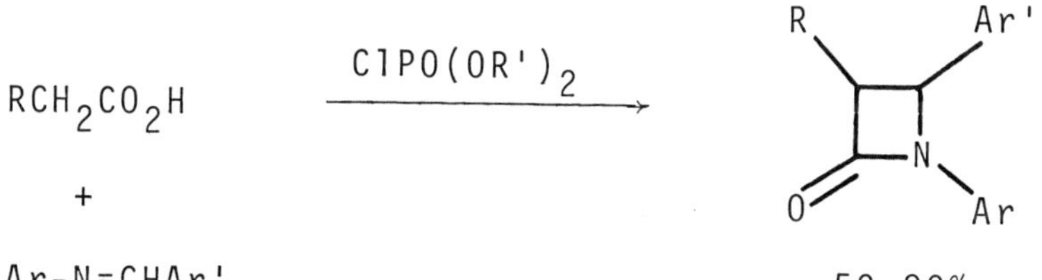

RCH$_2$CO$_2$H

+

Ar-N=CHAr'

$\xrightarrow{\text{ClPO(OR')}_2}$

[β-lactam with R, Ar' on C and N-Ar]

50-80%

IV.D-13 N. S. Isaacs, Chem. Soc. Rev., 5, 181 (1976).

Review: "Synthetic Routes to β-Lactams"

IV.D-14 P. G. Sammes, Chem. Rev., 76, 113 (1976).

Review: "Recent Chemistry of the β-Lactam Antibiotics"

IV.D-15 Y. Kishi, <u>Pure and Appl. Chem.</u>, <u>43</u>, 423 (1975).

    Review: "Synthetic Studies in the Fields of Natural Products"

    (penicillins and cephalosporins)

IV.D-16 H. Bhagwatheeswaran, S. P. Gaur, and P. C. Jain, <u>Synthesis</u>, 615 (1976).

$$\text{Ar-CHO} + O_2NCH_2CH_2CH_2COOEt \xrightarrow{NH_4OAc} \underset{\underset{H}{\bigg|}}{\underset{Ar}{\diagdown}}\!\!\!\underbrace{}_{}\!\!\!\overset{NO_2}{\diagup}\!\!=\!O$$

widely varying yields

IV.D-17 W. F. Keir and H.C.S. Wood, <u>J. Chem. Soc., Perkin I</u>, 1847 (1976).

$$H_2N-\overset{O}{\overset{\|}{C}}-\underset{\underset{HN=NHR^1}{|}}{CR^2H} \xrightarrow[\text{EtOH, NaOEt}]{\text{formamide}} \text{pyrimidinone}$$

R^1 = H, alkyl, aryl

R^2 = H, PhN_2, NH_2, NHCHO

yields ~70-90%

IV.D-18   J. Becher and E. G. Frandsen, Tetrahedron Lett., 3347 (1976).

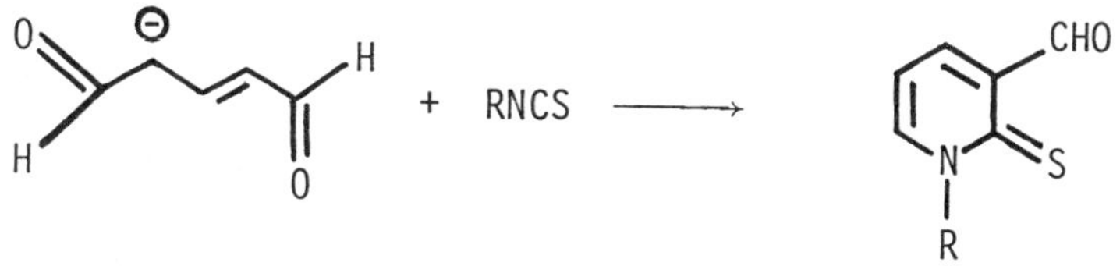

R = alkyl, aryl

yields generally 60-90%

IV.D-19   C. Bischoff and H. Herma, J. Prakt. Chem., 318, 891 (1976).

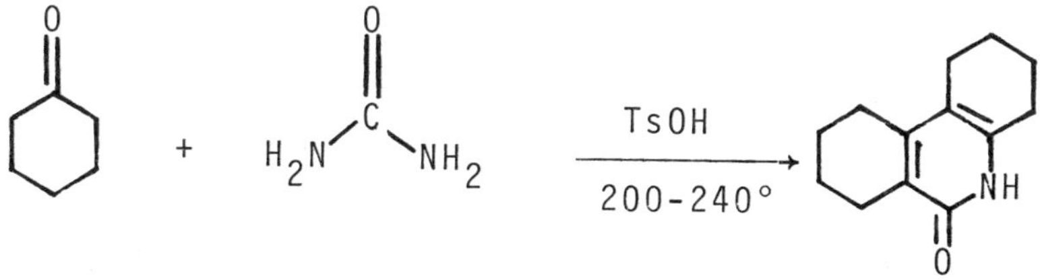

TsOH, 200-240°

64%

IV.D-20   H. L. Wehrmeister, J. Het. Chem., 13, 61 (1976).

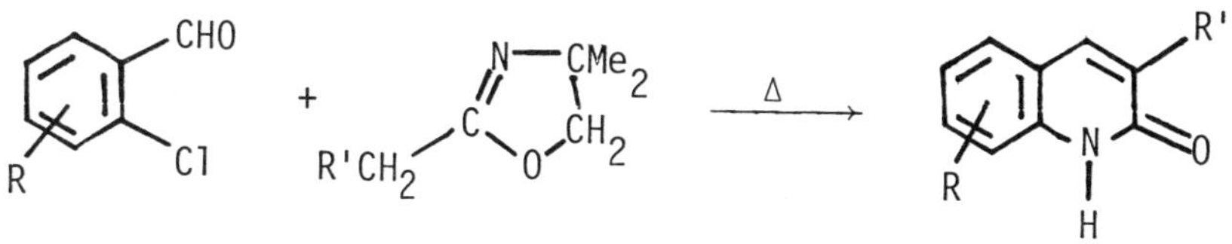

R = Cl, NO_2 R' = Me, Ph

Yields vary widely

IV.D-21   M. Y. Shandala and N. H. Al-Jobour, <u>Aust. J. Chem.</u>, <u>29</u>, 1583 (1976).

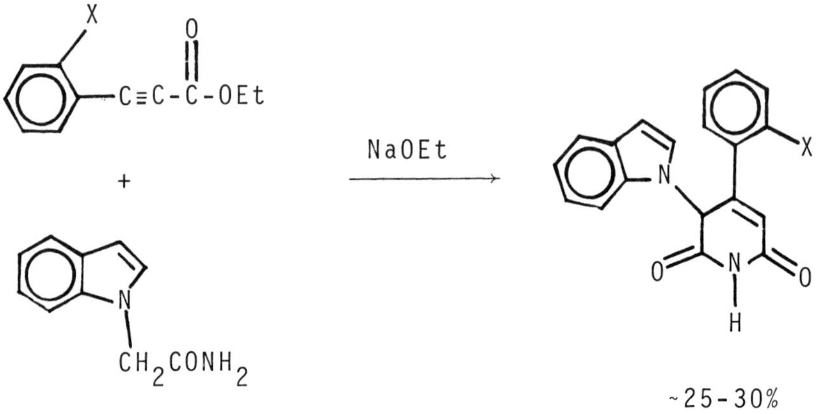

[Reaction: o-X-C₆H₄-C≡C-C(=O)-OEt + indole-N-CH₂CONH₂ → (NaOEt) → 3-(indol-1-yl)-4-(o-X-phenyl)-2,6-dioxopiperidine product, ~25-30%]

IV.D-22/IV.E-1   H. J. Bestmann, G. Schmid, and D. Sandmeier, <u>Angew. Chem. Int. Ed.</u>, <u>15</u>, 115 (1976).

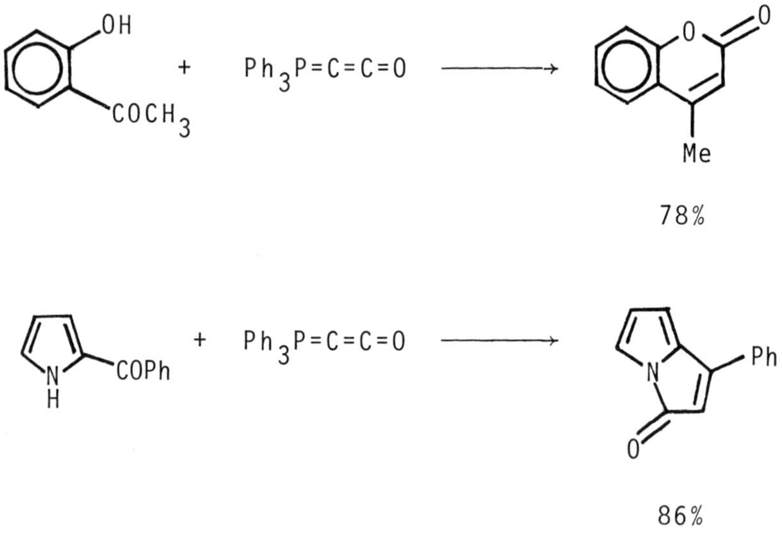

[Reaction: o-hydroxyacetophenone + Ph₃P=C=C=O → 4-methylcoumarin, 78%]

[Reaction: 2-benzoylpyrrole + Ph₃P=C=C=O → 1-phenyl-3-oxo-pyrrolizine (Ph-substituted), 86%]

IV.D-23/IV.E-2 W. Ott, G. Kollenz, and E. Ziegler, Synthesis, 546 (1976).

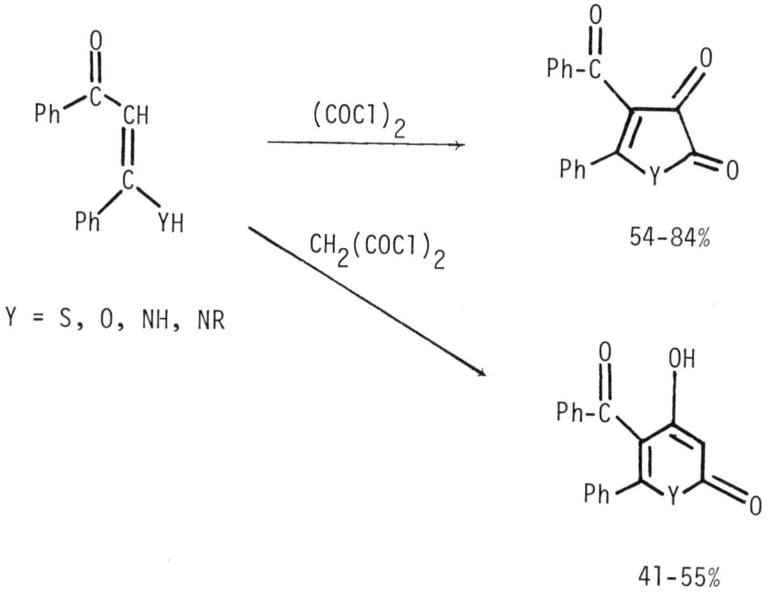

Y = S, O, NH, NR

54-84%

41-55%

IV.D-24/IV.E-3 F. A. L'Eplattenier et al., Synthesis, 543 (1976).

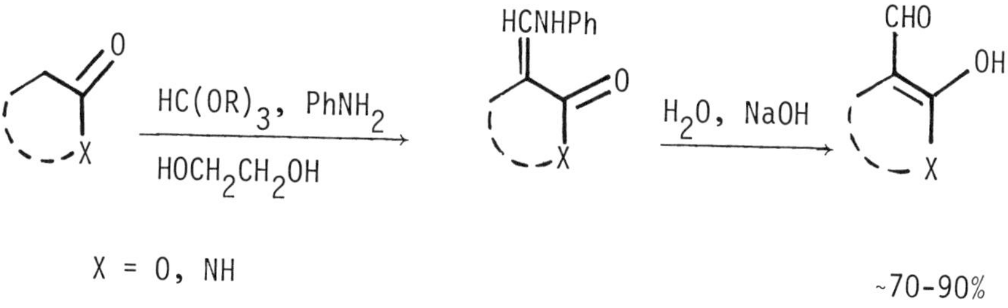

X = O, NH

~70-90%

IV.E-4   S. Danishefsky, T. Kitahara, M. Tsai and
J. Dynak, J. Org. Chem., **41**, 1669 (1976).

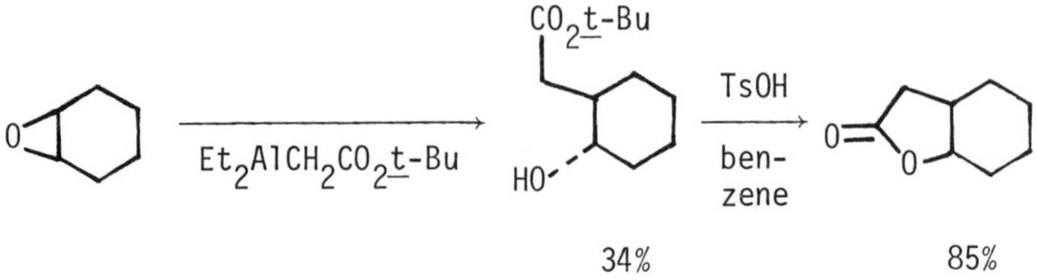

[Reaction: 7-oxabicyclo[4.1.0]heptane + Et$_2$AlCH$_2$CO$_2$t-Bu → 2-(CH$_2$CO$_2$t-Bu)cyclohexanol (34%) → (TsOH, benzene) → hexahydrobenzofuran-2(3H)-one (85%)]

IV.E-5   T. F. Murray, V. Varma, and J. R. Norton,
J. Chem. Soc., Chem. Commun., 907 (1976).

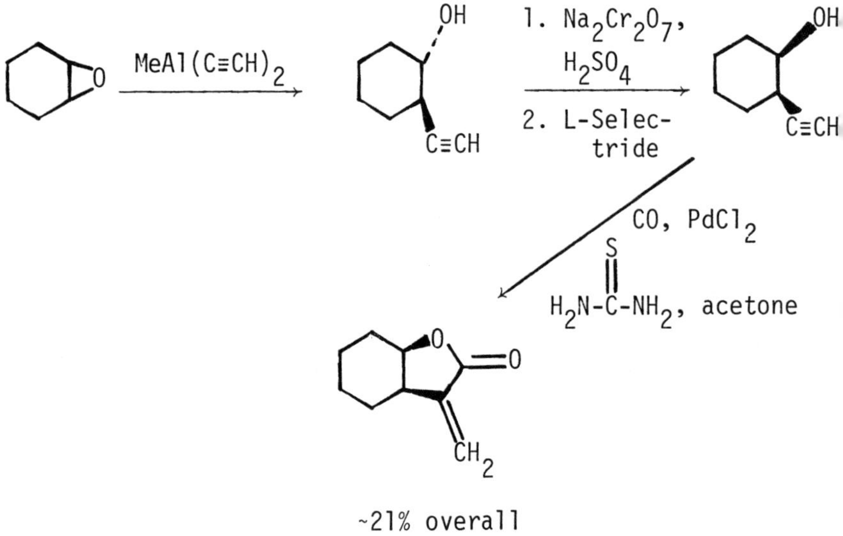

[Reaction: 7-oxabicyclo[4.1.0]heptane + MeAl(C≡CH)$_2$ → 2-ethynylcyclohexanol;
1. Na$_2$Cr$_2$O$_7$, H$_2$SO$_4$; 2. L-Selectride → trans-2-ethynylcyclohexanol;
CO, PdCl$_2$, H$_2$N-C(=S)-NH$_2$, acetone → α-methylene-hexahydrobenzofuran-2(3H)-one]

~21% overall

IV.E-6   M. Bertrand *et al.*, *Tetrahedron Lett.*, 3305 (1976).

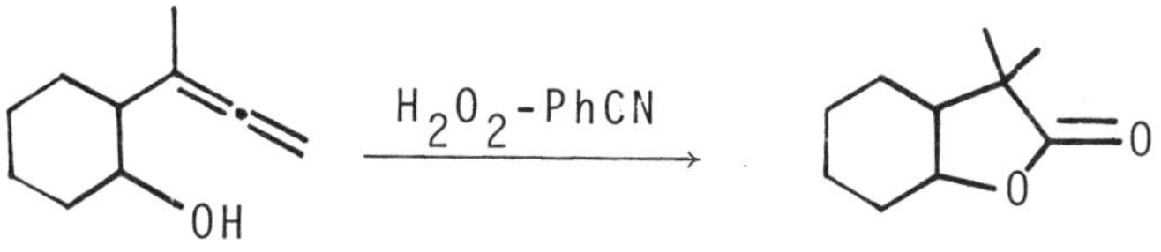

[structure: 2-(propa-1,2-dienyl with Me)cyclohexanol] → H_2O_2-PhCN → [bicyclic lactone]

est. 80% yield

IV.E-7   J. F. LeBorgne *et al.*, *Synthesis*, 238 (1976).

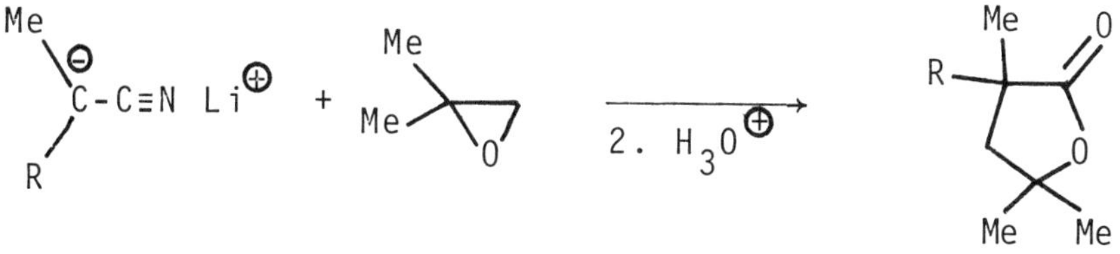

$$\text{Me}\underset{R}{\overset{\ominus}{C}}-C\equiv N \ Li^{\oplus} \ + \ \text{Me-Me epoxide} \ \xrightarrow{2. \ H_3O^{\oplus}} \ \text{γ-butyrolactone product}$$

72%, R=H
89%, R=Me

IV.E-8   G. Domschke, *Z. Chem.*, 16, 13 (1976).

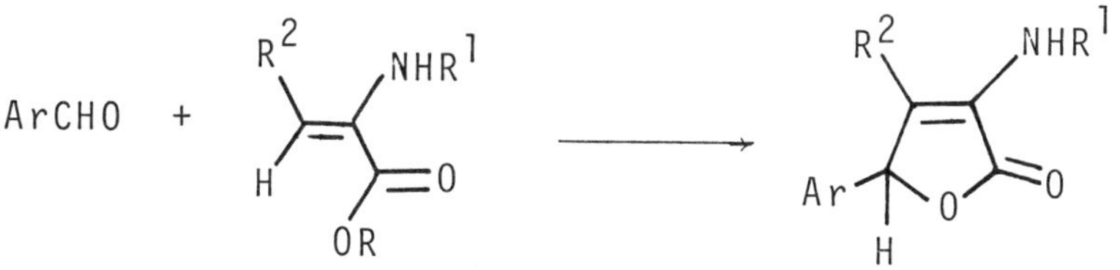

ArCHO + [enol ether with R^2, NHR^1, OR] ⟶ [furanone with Ar, R^2, NHR^1]

R^1 = H, alkyl, aryl
R^2 = COR, COOR

~60-90%

IV.E-9 R. C. Larock and B. Riefling, Tetrahedron Lett., 4661 (1976).

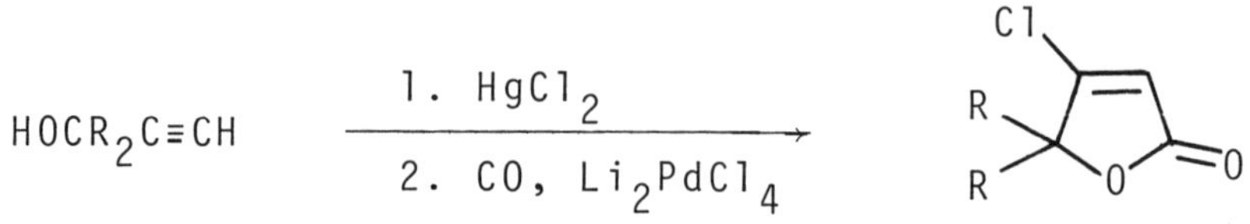

$HOCR_2C\equiv CH$ $\xrightarrow{\begin{array}{c}1.\ HgCl_2\\ 2.\ CO,\ Li_2PdCl_4\end{array}}$

$R_2 = H_2,\ Me_2,\ -(CH_2)_{4,5}$ ~90%

IV.E-10 S. F. Martin and D. R. Moore, Tetrahedron Lett., 4459 (1976).

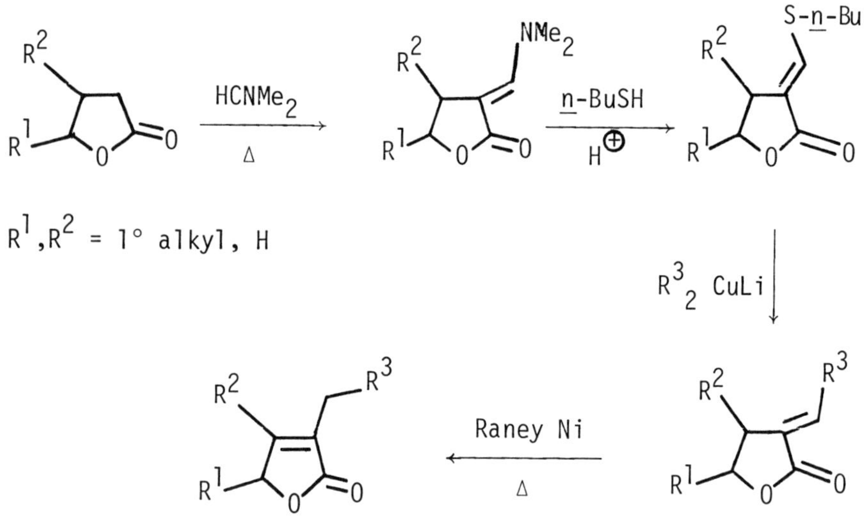

$R^1, R^2 = 1°$ alkyl, H

Overall yields ~50-60%

$R^3 = Me,\ \underline{n}-Bu$

IV.E-11   R. M. Carlson and A. R. Oyler, *J. Org. Chem.*, **41**, 4065 (1976).

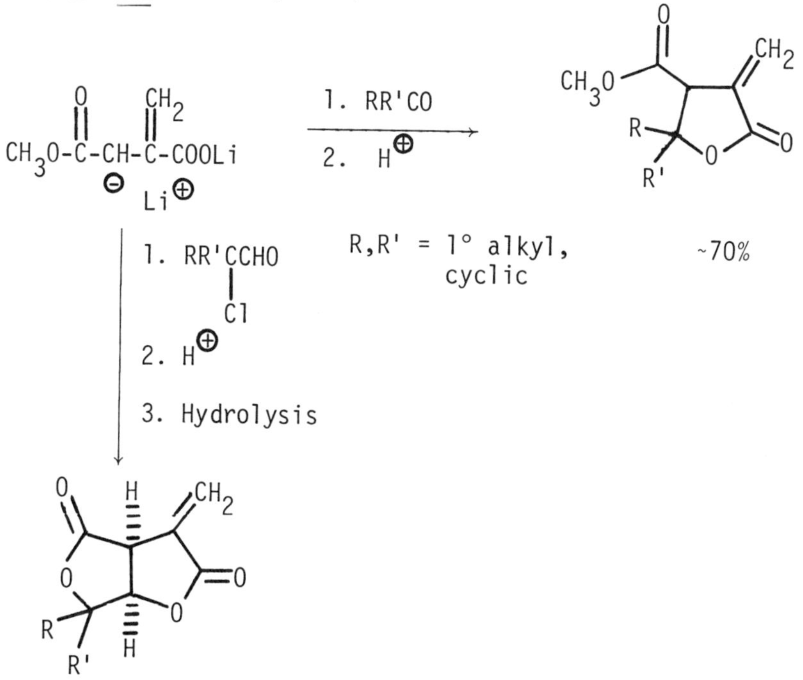

$$CH_3O-\overset{O}{\underset{}{C}}-CH-\overset{CH_2}{\underset{}{C}}-COOLi \quad \underset{Li^{\oplus}}{\overset{\ominus}{}}$$

1. RR'CO
2. $H^{\oplus}$

→ [lactone with CH_3O-C(=O), R, R', =CH_2, O, =O] ~70%

1. RR'CCHO
   |
   Cl
2. $H^{\oplus}$
3. Hydrolysis

↓

[bicyclic structure] ~70%

IV.E-12   K. Kurata, S. Tanaka, and K. Takahashi, *Chem. Pharm. Bull.*, **24**, 538 (1976).

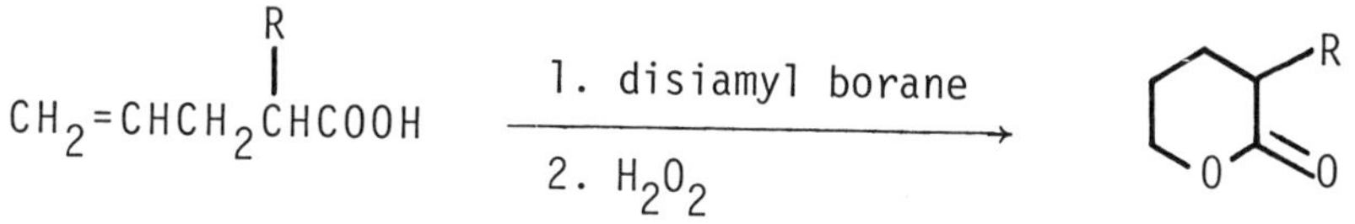

$$CH_2=CHCH_2\overset{R}{\underset{}{C}}HCOOH$$

1. disiamyl borane
2. H_2O_2

→ [δ-lactone with R substituent]   54-84%

(obtained from malonic ester)

R = H, 1°, 2°, 3° alkyl

IV.E-13   R. H. Carter et al., J. Chem. Soc. Perkin I, 1438 (1976).

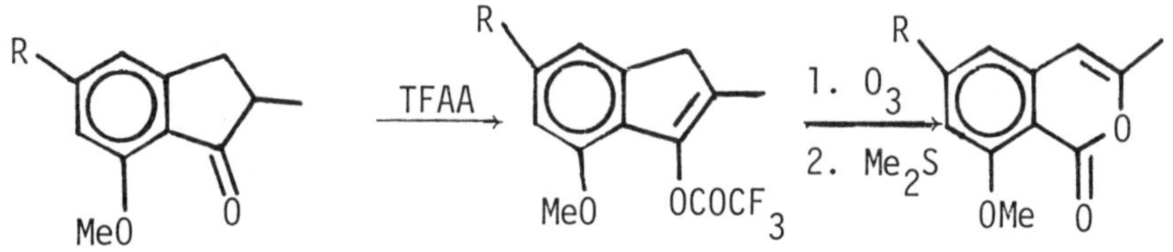

R = H, OMe

40% overall

IV.E-14   J. Ficini et al., Tetrahedron Lett., 683 (1976).

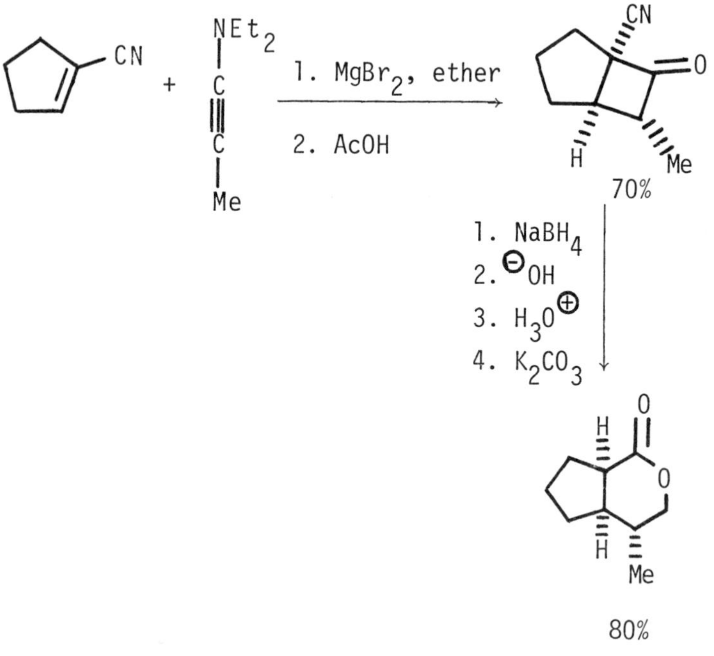

70%

80%

SYNTHESIS OF HETEROCYCLES

IV.E-15 A. I. Meyers and C. E. Whitten, <u>Tetrahedron Lett.</u>, 1947 (1976).

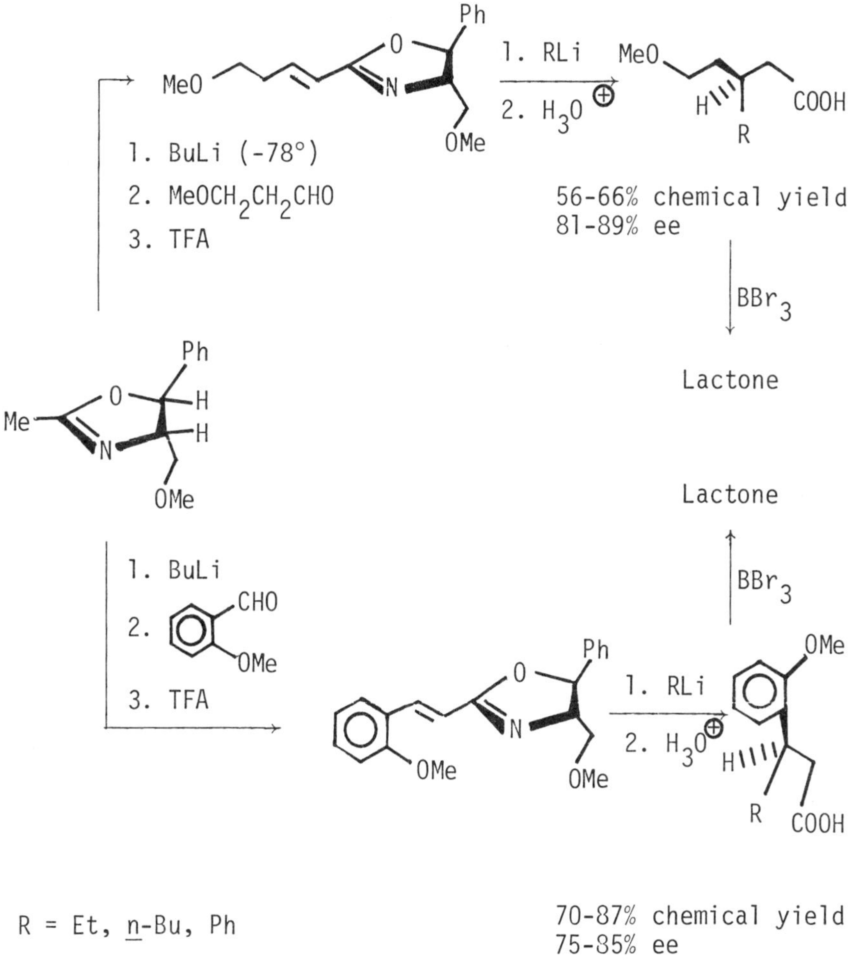

56-66% chemical yield
81-89% ee

R = Et, n-Bu, Ph

70-87% chemical yield
75-85% ee

IV.E-16   T. Sato et al., Bull. Chem. Soc. Japan, 49, 1055 (1976).

R-CHO + [dihydropyranone] $\xrightarrow{BF_3 \cdot Et_2O}$ [acetyl tetrahydropyranone product]

R = H, alkyl, aryl, $ClCH_2$, etc.   60-90%

IV.E-17   E. J. Corey and D. J. Brunelle, Tetrahedron Lett., 3409 (1976).

$HO(CH_2)_n COOH$ + (imidazole-S)$_2$ + Ph_3P → lactone $\overset{O}{\underset{O}{C}}-(CH_2)_n$

t-Bu

n ≈ 15   R = Me, i-Pr   ~80%

IV.E-18   T. Mukaiyama, M. Usui, and K. Saigo, Chem. Lett., 49 (1976).

$HO(CH_2)_n COOH$ $\xrightarrow{\text{Cl-pyridinium-Me } I^\ominus}$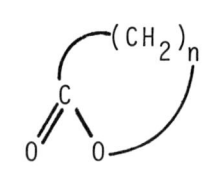

n = 5,7,10,11,14   Reaction fails
                                    for n = 6.

IV.E-19   T. Kurihara, Y. Nakajima, and O. Mitsunobu, Tetrahedron Lett., 2455 (1976).

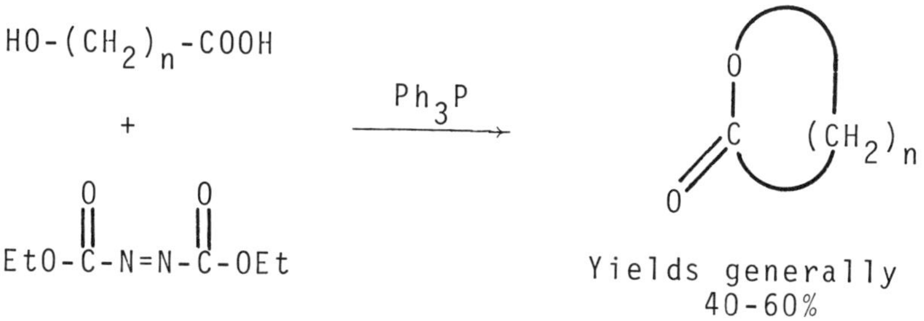

HO-$(CH_2)_n$-COOH
+
EtO-C(=O)-N=N-C(=O)-OEt

$Ph_3P \longrightarrow$

[cyclic lactone with $(CH_2)_n$]

n = 2-11

Yields generally 40-60%

IV.E-20   L. T. Scott and J. O. Naples, Synthesis, 738 (1976).

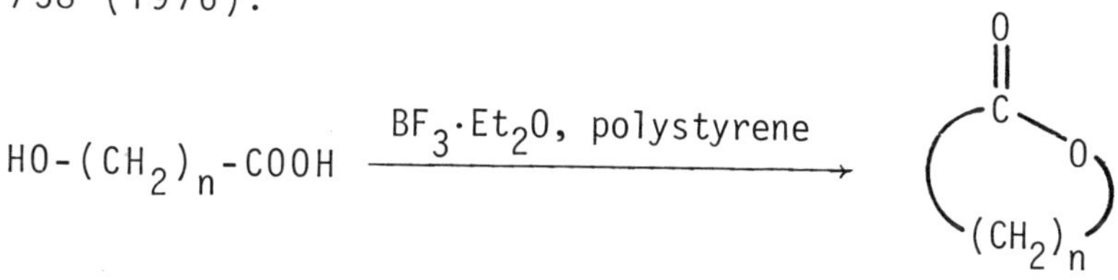

HO-$(CH_2)_n$-COOH $\xrightarrow{BF_3 \cdot Et_2O,\ polystyrene}$ [cyclic lactone with $(CH_2)_n$]

41% (n=11) - 78% (n=17)

IV.E-21   Y. S. Rao, Chem. Rev., 76, 625 (1976).

Review: "Recent Advances in the Chemistry of Unsaturated Lactones"

IV.E-22   J. R. Mahajan, Synthesis, 110 (1976).

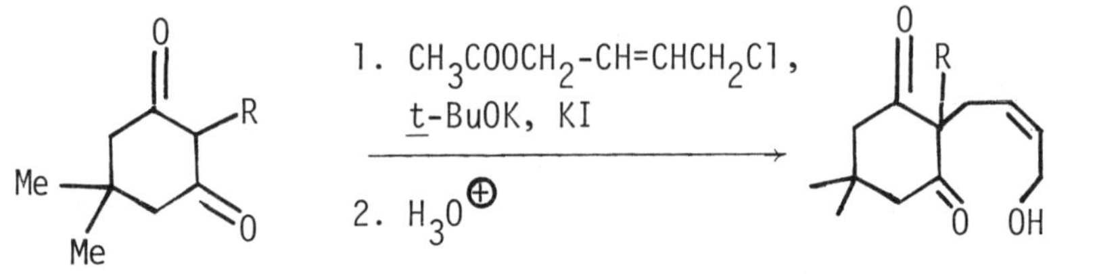

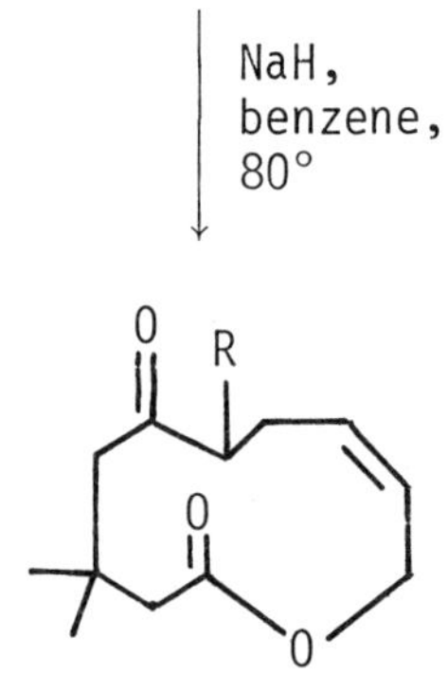

1. $CH_3COOCH_2-CH=CHCH_2Cl$, t-BuOK, KI
2. $H_3O^{\oplus}$

NaH, benzene, 80°

R	Yield (pure)
methyl	34%
allyl	48%
benzyl	50%

IV.F-1   Y. Wakatsuki and H. Yamazaki, Synthesis, 26 (1976).

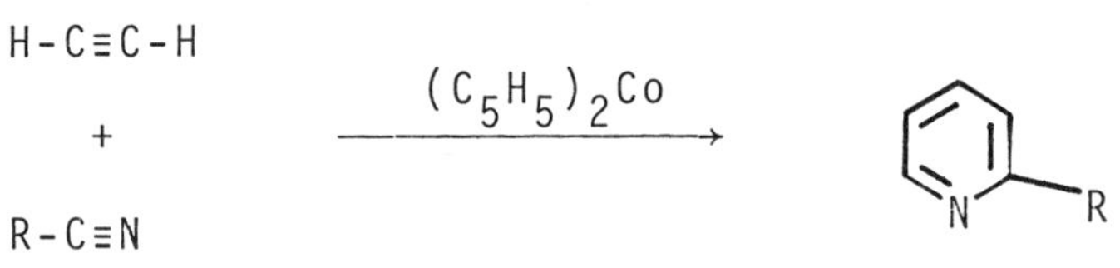

$H-C\equiv C-H$
+
$R-C\equiv N$

$(C_5H_5)_2Co$

R = alkyl, aryl, vinyl   20-73%

IV.F-2   A. L. Cossey et al., Aust. J. Chem., 29, 1039 (1976).

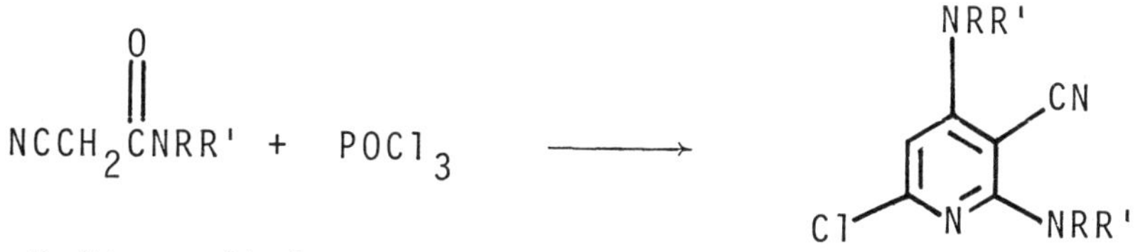

NCCH$_2$CNRR' + POCl$_3$ $\longrightarrow$

R,R' = alkyl

76-86%

NCCH$_2$CNHR + POCl$_3$ $\longrightarrow$

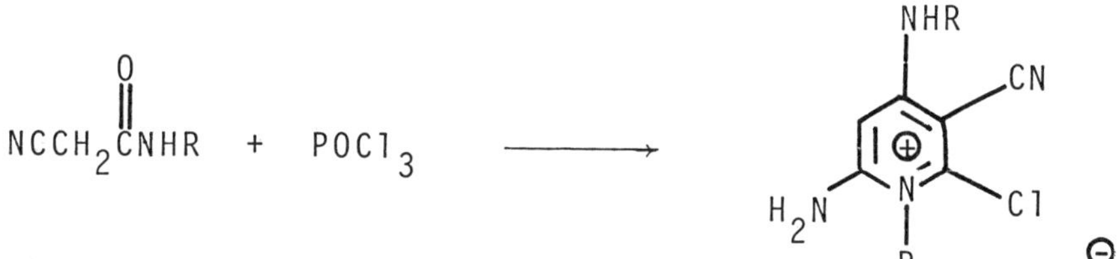

R = alkyl

39-86%

IV.F-3   J. N. Bonfiglio et al., J. Am. Chem. Soc., 98, 2344 (1976).

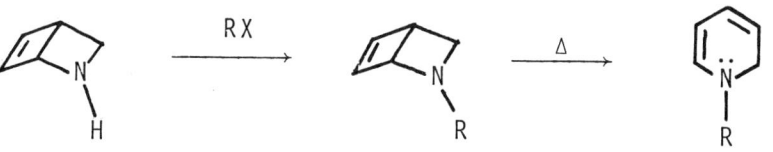

R = alkyl, benzyl, unsaturated, etc.

18-54% overall

IV.F-4   Y. Akita et al., Chem. Pharm. Bull., 24, 1839 (1976).

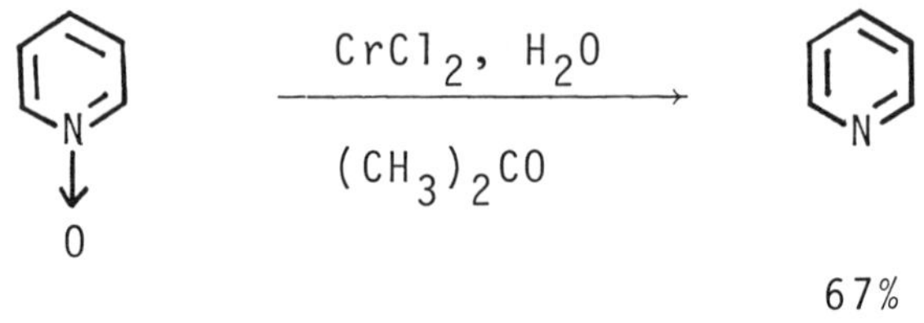

pyridine N-oxide →[CrCl$_2$, H$_2$O / (CH$_3$)$_2$CO]→ pyridine   67%

(Also works for chloropyridine, quinoline, and pyrazine derivative N-oxides, yields 41-95%.)

IV.F-5   G. Dauphin and D. Planat, Tetrahedron Lett., 4065 (1976).

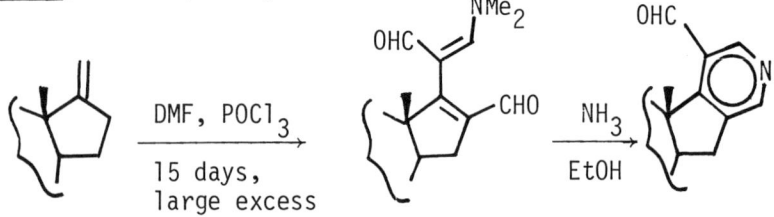

IV.F-6   E. B. Pedersen, Acta Chem. Scand. B, 30, 136 (1976).

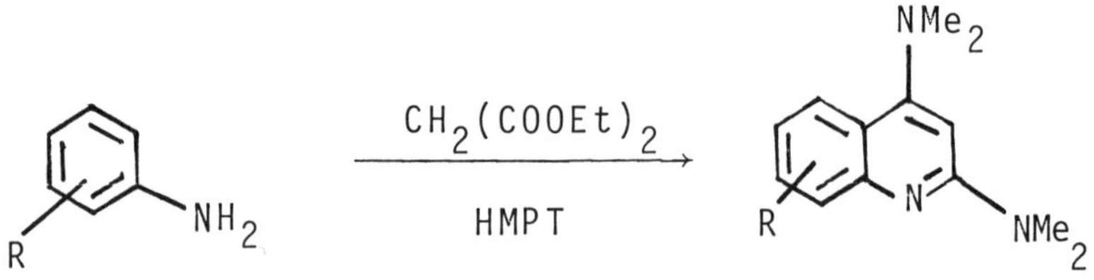

3-R-aniline →[CH$_2$(COOEt)$_2$ / HMPT]→ 2,4-bis(NMe$_2$)-quinoline

R = H, Me, OMe, Cl   25-30%

IV.F-7 A. Walser, G. Zenchoff, and R. I. Fryer, J. Het. Chem., 13, 131 (1976).

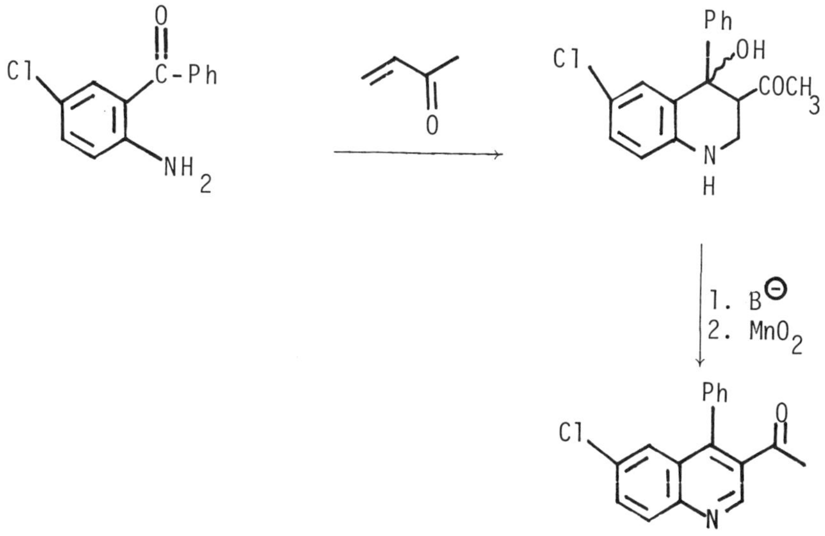

1. B⁻
2. MnO$_2$

overall ~40%

IV.F-8 F. X. Smith and G. G. Evans, J. Het. Chem., 13, 1025 (1976).

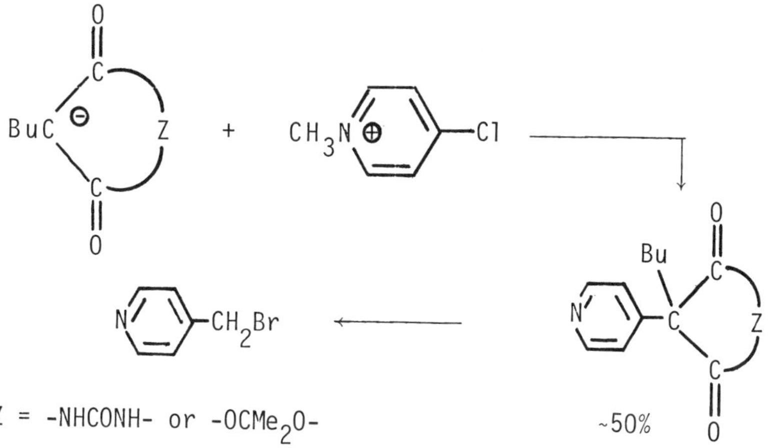

Z = -NHCONH- or -OCMe$_2$O-   ~50%

IV.F-9   S. Kwon et al., Synthesis, 249 (1976).

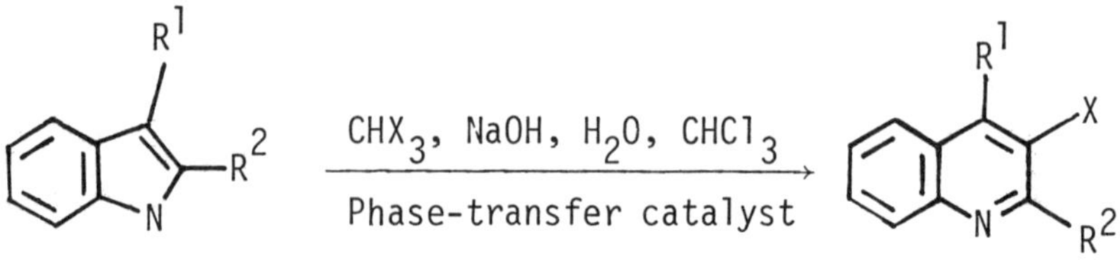

CHX_3, NaOH, H_2O, $CHCl_3$
Phase-transfer catalyst

X = Cl, Br

R^1 = H, CH_3
R^2 = H, CH_3, Ph

24-68%

IV.F-10   G. Coudert et al., Synthesis, 764 (1976).

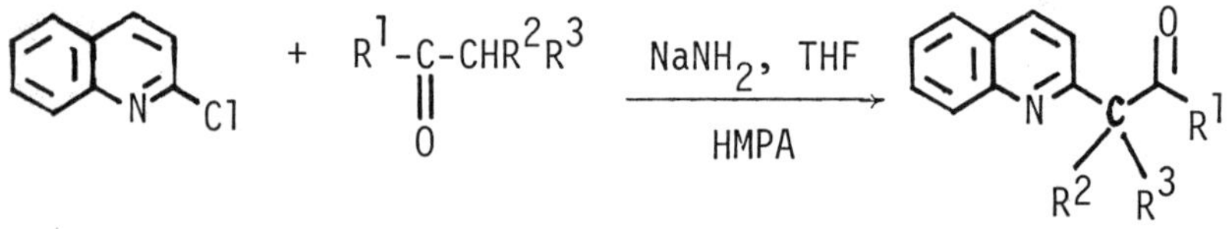

R^1-C(=O)-CHR^{2}R^3 $\xrightarrow{NaNH_2, THF}{HMPA}$

60-78%

same conditions

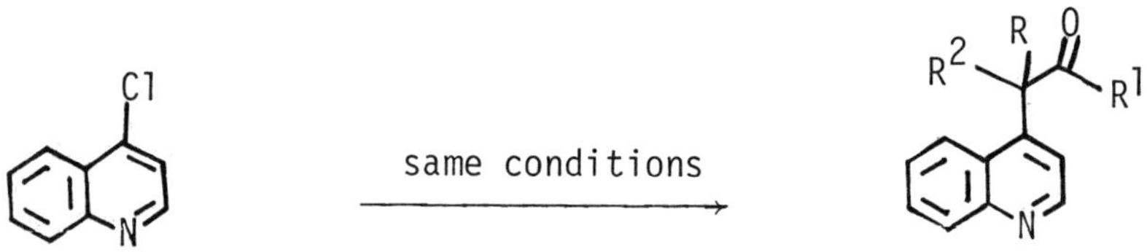

42-90%

R's = H, aryl, 1°,2° alkyl, cyclic

IV.F-11   F. Kröhnke, Synthesis, 1 (1976).

Review: "The Specific Synthesis of Pyridines and Oligopyridines"

IV.G-1 H. Wamhoff and B. Wehling, Synthesis, 51 (1976).

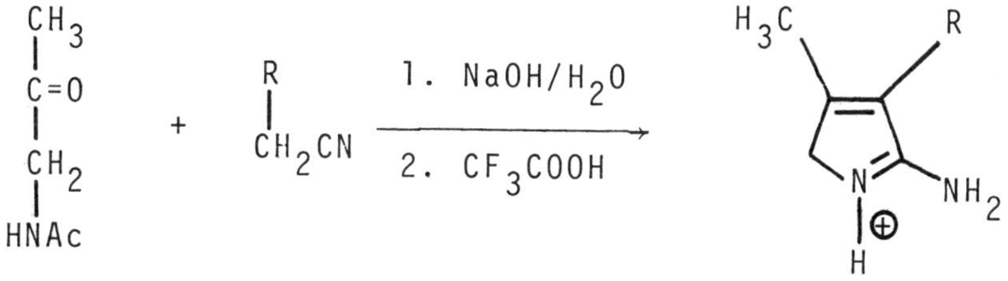

$$\begin{array}{c} CH_3 \\ | \\ C=O \\ | \\ CH_2 \\ | \\ HNAc \end{array} + \begin{array}{c} R \\ | \\ CH_2CN \end{array} \xrightarrow[2.\ CF_3COOH]{1.\ NaOH/H_2O}$$

[pyrrole product with H_3C, R substituents, NH_2, N-OH]

R = CN   73%
R = COO$\underline{t}$-Bu   41%

IV.G-2 Z. Yoshida, T. Harada, and Y. Tamaru, Tetrahedron Lett., 3823 (1976).

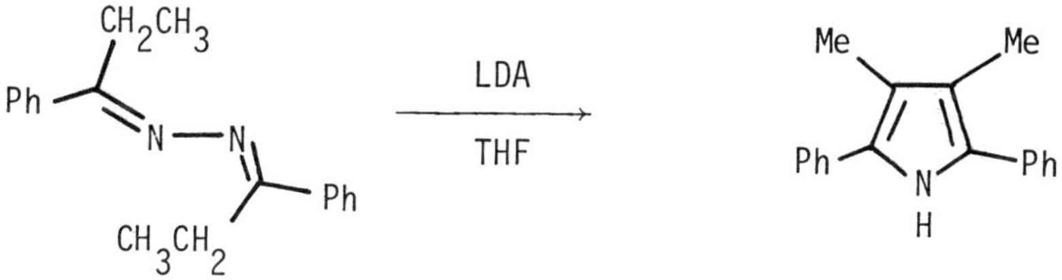

[Ph-C(CH$_2$CH$_3$)=N-N=C(CH$_2$CH$_3$)-Ph] $\xrightarrow[\text{THF}]{\text{LDA}}$ [3,4-dimethyl-2,5-diphenylpyrrole]

52%

IV.G-3 H. Sliwa and D. Blondeau, Tetrahedron Lett., 933 (1976).

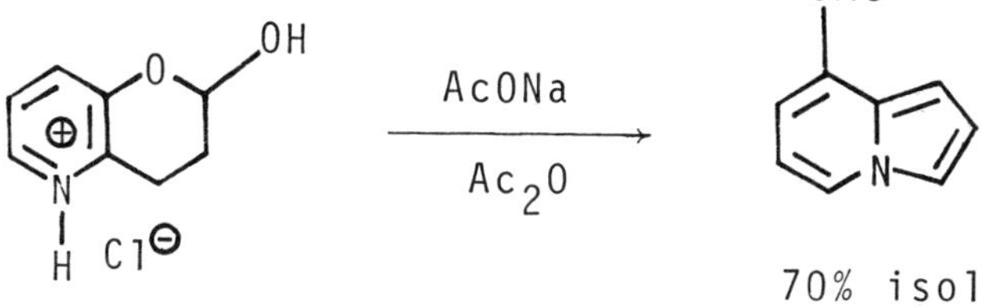

[pyridinium chromanol with Cl$^{\ominus}$] $\xrightarrow[Ac_2O]{AcONa}$ [8-OAc indolizine]

70% isolated

IV.G-4   R. R. Fraser and S. Passannanti, <u>Synthesis</u>, 540 (1976).

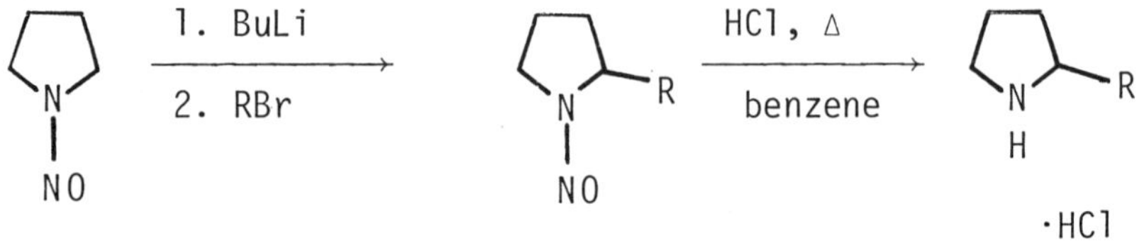

[pyrrolidine-N-O] →(1. BuLi; 2. RBr)→ [2-R-pyrrolidine-N-O] →(HCl, Δ / benzene)→ [2-R-pyrrolidine-N-H] · HCl

R = benzyl, 1° alkyl

(may be repeated)

IV.G-5   J. Leroy and C. Wakselman, <u>Can. J. Chem.</u>, <u>54</u>, 218 (1976).

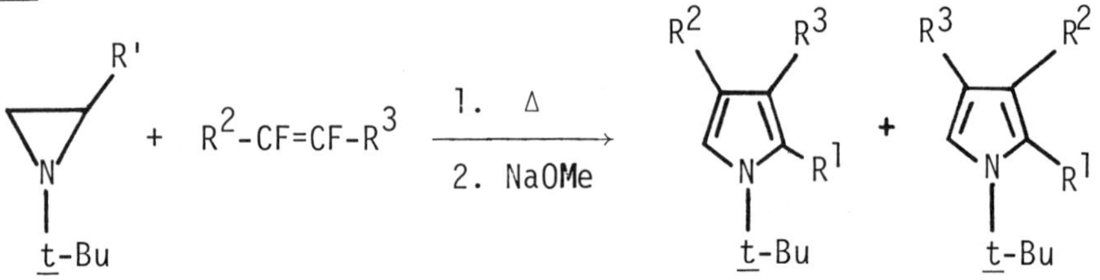

[N-t-Bu aziridine with R'] + R^2-CF=CF-R^3 →(1. Δ; 2. NaOMe)→ [pyrrole with R^2, R^3, R^1, N-t-Bu] + [pyrrole isomer with R^3, R^2, R^1, N-t-Bu]

R^1 = CN, COOR   Yields ~20-60%

R^2, R^3 = H, F, CF_3

IV.G-6   T. Uchida and K. Matsumoto, <u>Synthesis</u>, 209 (1976).

Review: "Methods for the Construction of the Indolizine Nucleus"

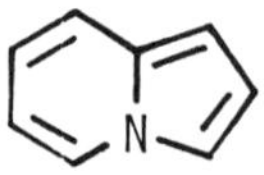

[indolizine structure]

SYNTHESIS OF HETEROCYCLES

IV.G-7   J. M. Patterson, Synthesis, 281 (1976).

Review: "Recent Synthetic Methods for Pyrroles and Pyrrolenines (2H- or 3H-Pyrroles)"

IV.H-1   A. A. Akhrem, F. A. Lakhvich, and V. A. Khripach, J. Gen. Chem. (USSR), 45, 2534 (1975).

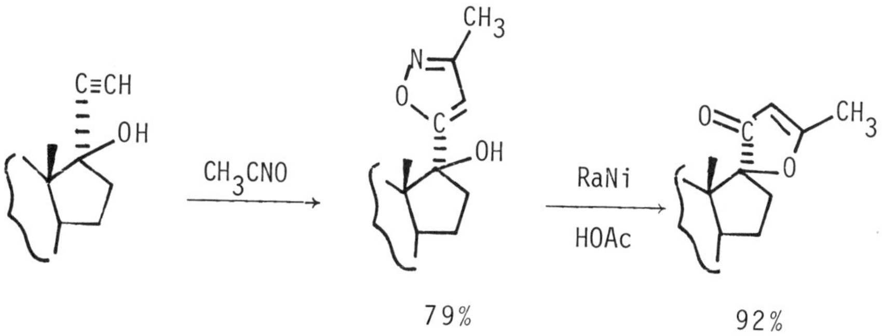

79%   92%

IV.H-2   R. J. Bass, J. Chem. Soc., Chem. Commun., 78 (1976).

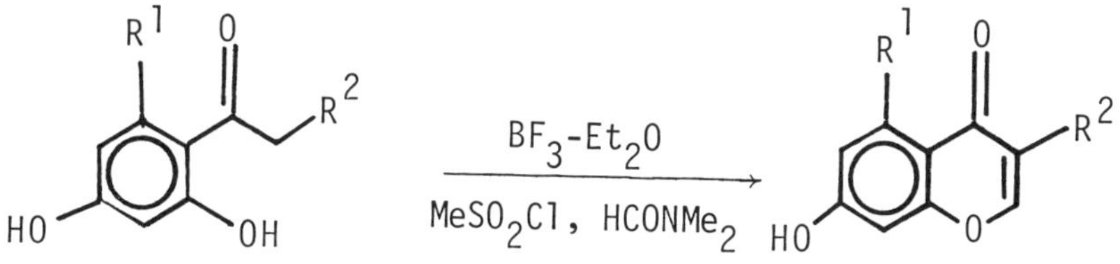

R^1 = H, OH

R^2 = aryl, Bz, OMe, cyclopentyl

60-98%

IV.H-3   J. Martens, K. Praefcke, and U. Schulze, Synthesis, 532 (1976).

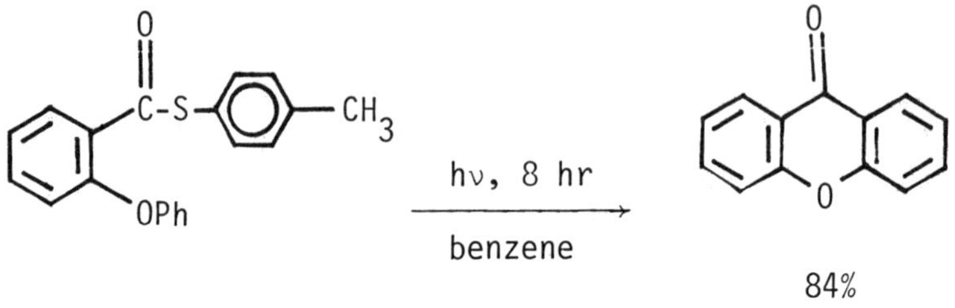

[reaction scheme: 2-phenoxy thioester (S-tolyl) → xanthone, hν, 8 hr, benzene, 84%]

IV.H-4   G.J.P. Becket and G. P. Ellis, Tetrahedron Lett., 719 (1976).

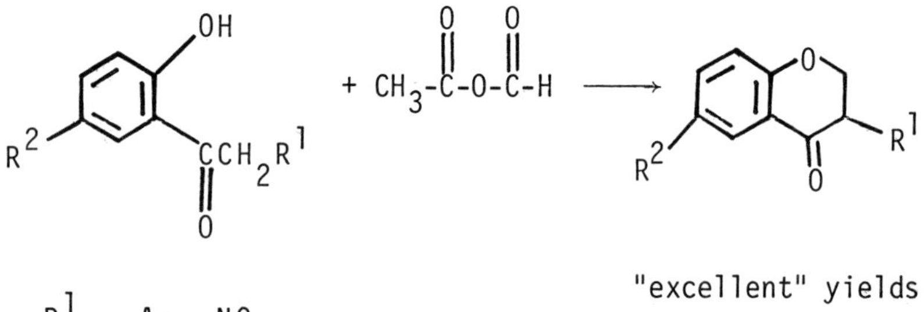

[reaction scheme: 2-hydroxyaryl ketone + CH₃-C(O)-O-C(O)-H → chromanone]

R¹ = Ac, NO₂
R² = H, NO₂

"excellent" yields

IV.H-5   G. Pfeiffer and H. Bauer, Liebigs Ann. Chem., 383 (1976).

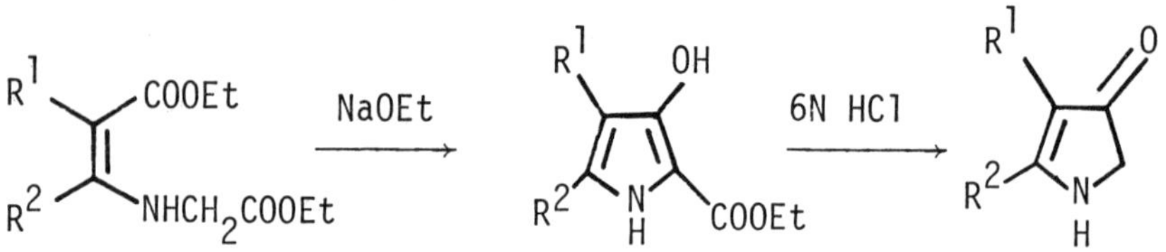

[reaction scheme: R¹-C(COOEt)=C(R²)-NHCH₂COOEt →(NaOEt) 3-hydroxypyrrole-2-carboxylate →(6N HCl) 3-oxo-pyrroline]

R¹,R² = Me, n-Bu, cyclic

>90%

IV.H-6   D. S. Black and A. B. Boscacci, <u>Aust. J. Chem.</u>, <u>29</u>, 2561 (1976).

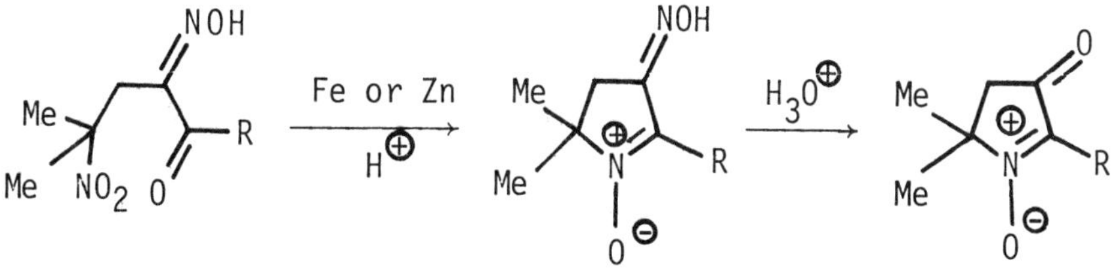

Me₂C(NO₂)-CH₂-C(=NOH)-R $\xrightarrow{\text{Fe or Zn} \atop H^{\oplus}}$ [2,2-dimethyl-5-R-1-oxido-3-hydroxyimino-Δ¹-pyrrolinium] $\xrightarrow{H_3O^{\oplus}}$ [2,2-dimethyl-5-R-1-oxido-3-oxo-Δ⁴-pyrrolinium]

~50%

IV.H-7   H. O. House and L. F. Lee, <u>J. Org. Chem.</u>, <u>41</u>, 863 (1976).

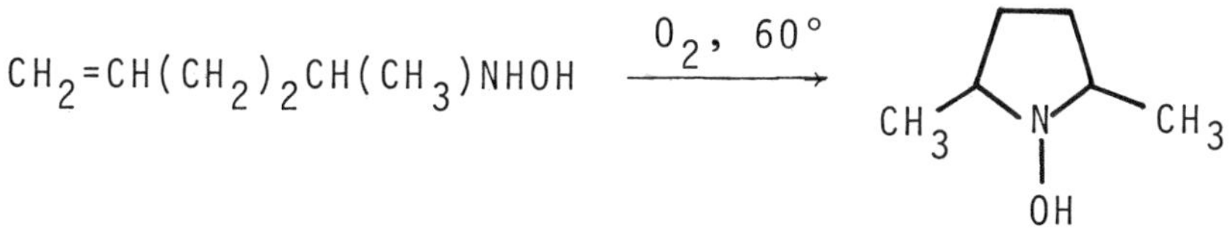

$CH_2=CH(CH_2)_2CH(CH_3)NHOH$ $\xrightarrow{O_2, \ 60°}$ 2,5-dimethyl-1-hydroxypyrrolidine

~80%

$CH_2=CH(CH_2)_3CH_2NHOH$ $\xrightarrow{145°, \ \text{xylene}}$ 2-methyl-1-hydroxypiperidine

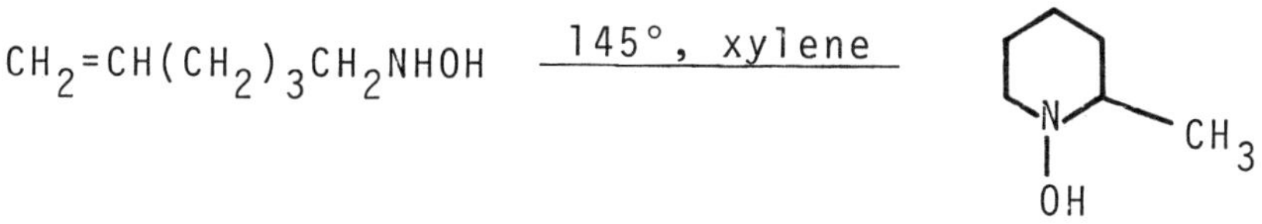

~70-80%

IV.H-8   P. K. Larsen and H. Hjeds, Acta Chem. Scand. (B), 30, 884 (1976).

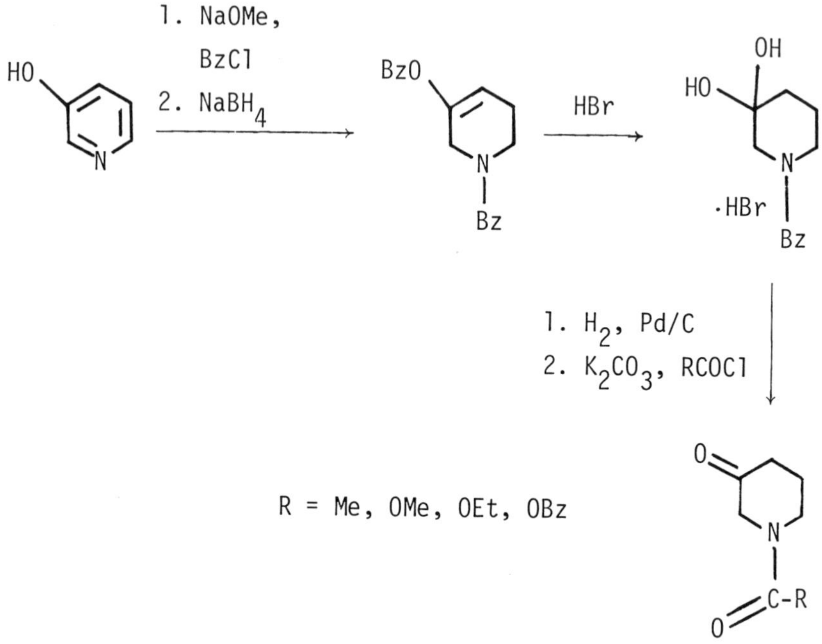

1. NaOMe, BzCl
2. NaBH$_4$

HBr

1. H$_2$, Pd/C
2. K$_2$CO$_3$, RCOCl

R = Me, OMe, OEt, OBz

~30% overall

IV.H-9   C. R. Johnson and K. Tanaka, Synthesis, 413 (1976).

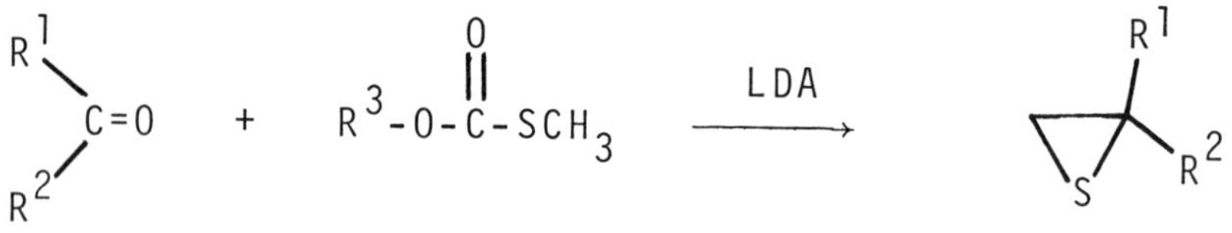

$R^1R^2C=O$ + $R^3-O-\overset{O}{\underset{\|}{C}}-SCH_3$ $\xrightarrow{LDA}$

R^1, R^2 = H, alkyl, aryl
R^3 = (-)-menthyl

55-77% yield
13-34% ee

VI.H-10 A. I. Meyers and M. E. Ford, J. Org. Chem., 41, 1735 (1976).

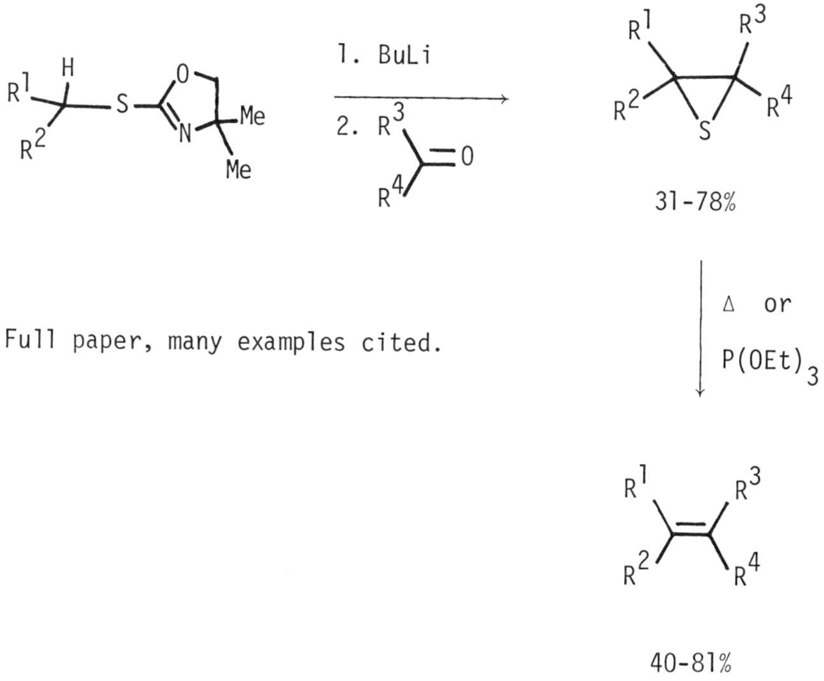

$R^1R^2CH-S-\text{(oxazoline with Me, Me)}$ $\xrightarrow[\text{2. } R^3R^4C=O]{\text{1. BuLi}}$ thiirane with R^1, R^2, R^3, R^4 substituents

31-78%

Full paper, many examples cited.

$\downarrow \Delta \text{ or } P(OEt)_3$

alkene $R^1R^2C=CR^3R^4$

40-81%

IV.H-11 I. L. Knunyants, Doklady Chem., 224, 521 (1975).

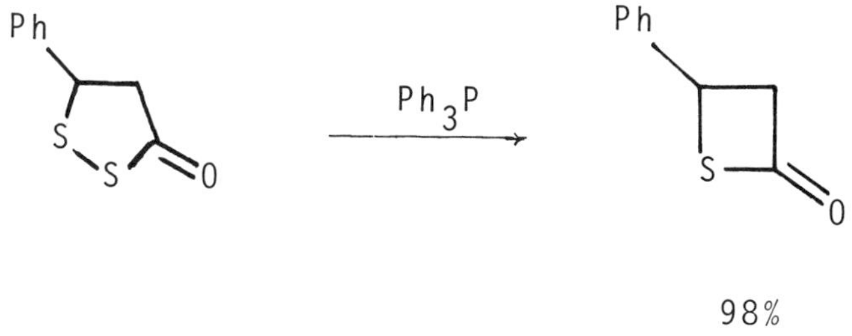

Ph-substituted 1,2-dithiolan-3-one $\xrightarrow{Ph_3P}$ Ph-substituted thietan-2-one

98%

IV.H-12   F. Duus, Tetrahedron, 32, 2817 (1976).

$$R^1-\underset{\underset{O}{\|}}{C}-\underset{\underset{}{R^2}}{CH}-\underset{\underset{}{R^3}}{CH}-\underset{\underset{O}{\|}}{C}-R^4 \xrightarrow[HCl]{H_2S}$$

[thiophene product with R^1, R^2, R^3, R^4 substituents]

R's = Me, Ph, t-Bu, COOR

Widely varying yields.

IV.H-13   W. Ried and L. Kaiser, Synthesis, 120 (1976).

$$Ph-C\equiv C-\underset{\underset{}{\overset{\overset{S}{\|}}{}}}{C}-NHPh$$

+

RCH_2Br

$\xrightarrow{Et_3N, \Delta}$

[thiophene product with Ph, R, NHPh substituents]

45-83%

R = NO_2, CN, p-NO_2Ph

IV.H-14   H. Gotthardt, M. C. Weisshuhn, and B. Christl, Chem. Ber., 109, 753 (1976).

[thiolate starting material with R, R', O⁻, ⊕] $\xrightarrow[\Delta]{MeO_2C-C\equiv C-CO_2Me}$ [thiophene product with R, R', MeOOC, CO_2Me]

R = Me, aryl
R' = aryl

~70-90%

IV.H-15   H. Blatt et al., Aust. J. Chem., 29, 883 (1976).

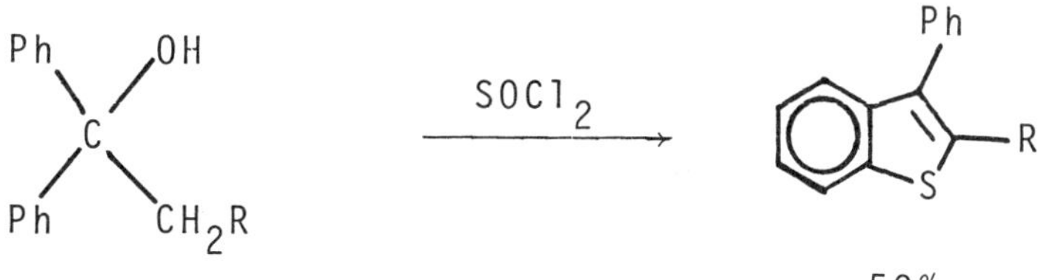

Ph₂C(OH)(CH₂R) →[SOCl₂] 3-Ph-2-R-benzothiophene

R = H, Me, Et, Ph

~50%

IV.H-16   T. Higa and A. J. Krubsack, J. Org. Chem., 41, 3399 (1976).

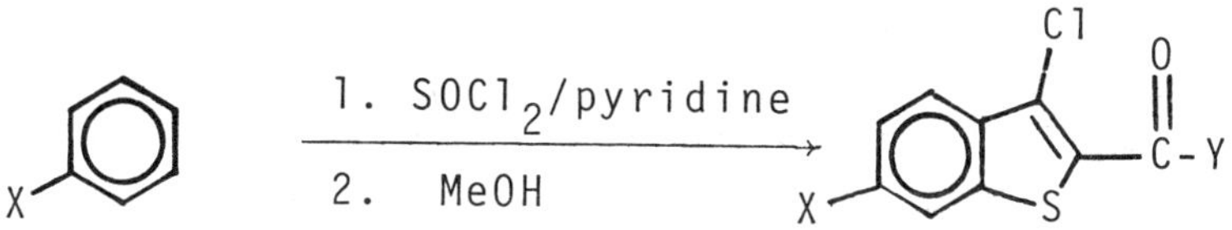

X-C₆H₅ →[1. SOCl₂/pyridine; 2. MeOH] 3-Cl-2-(C(=O)-Y)-6-X-benzothiophene

X = H, CH₃, OCH₃, NO₂   Y = Cl, OCH₃

46-69%

IV.H-17   J. M. McIntosh and H. Khalil, Can. J. Chem., 54, 1923 (1976).

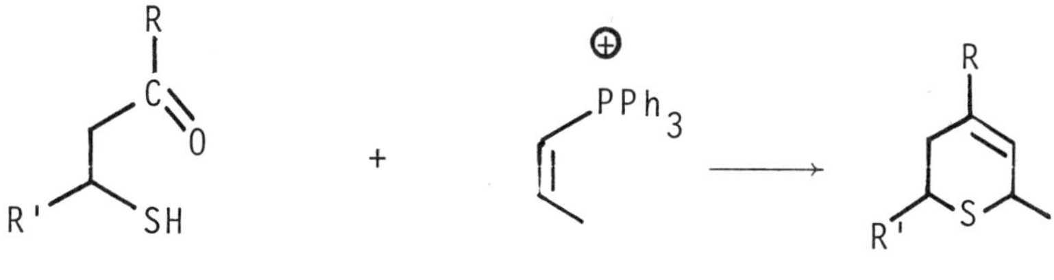

R'CH(SH)CH₂C(R)=O   +   CH₂=CH-CH(PPh₃)⁺   →   2-R'-4-R-6-methyl-dihydrothiopyran

R = H, Me
R' = Me, Ph

generally ~50-80%

IV.I.1.a-1   J. P. Ferris and R. W. Trimmer, J. Org. Chem., 41, 19 (1976).

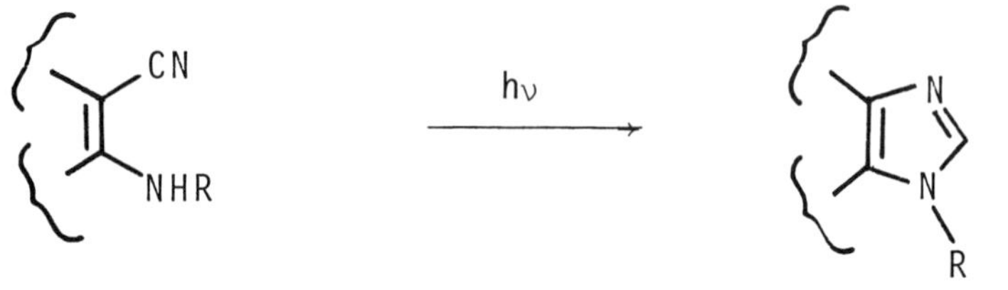

R = H, CH$_3$, t-Bu

Yields ~50-80%

IV.I.1.a-2   D. Pocar and R. Stradi, Tetrahedron Lett., 1839 (1976).

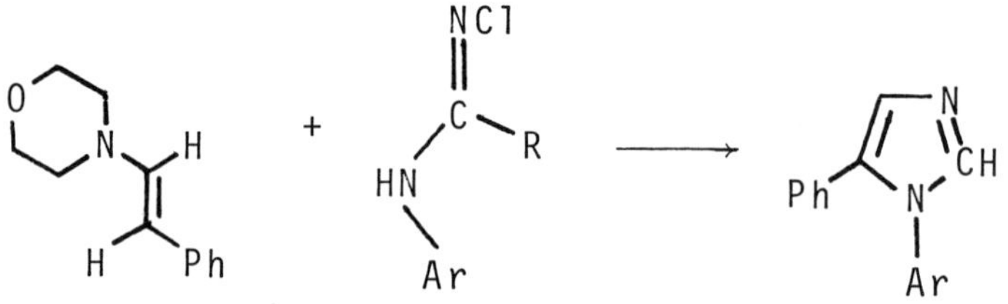

"good yields"

IV.I.1.a-3   G. M. Devasia, Tetrahedron Lett., 571 (1976).

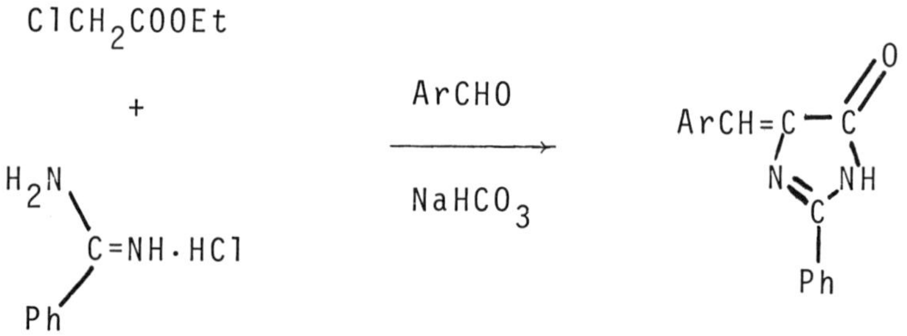

ClCH$_2$COOEt +

$\underset{Ph}{\overset{H_2N}{\diagdown}}$C=NH·HCl

ArCHO, NaHCO$_3$ →

Yields ~40-55%

IV.I.1.a-4   Y. Kikugawa and L. A. Cohen, <u>Chem. Pharm. Bull.</u>, 24, 3205 (1976).

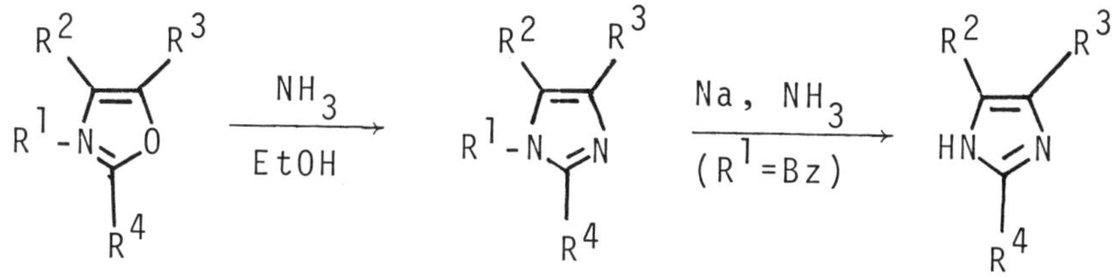

R^1 = Me, Bz   61-97%
R^{2-4} = H, Me, Et, Ph

IV.I.1.a-5   N. Latif, N. Mishriky, and N. S. Girgis, <u>Chem. and Ind.</u>, 28 (1976).

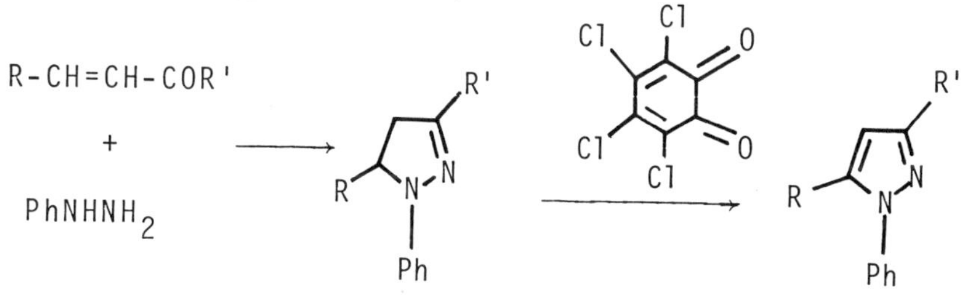

R-CH=CH-COR'
+
$PhNHNH_2$

No yields given.

IV.I.1.a-6   R. Baumes, R. Jacquier, and G. Tarrago, <u>Bull. Soc. Chim. France</u>, 260 (1976).

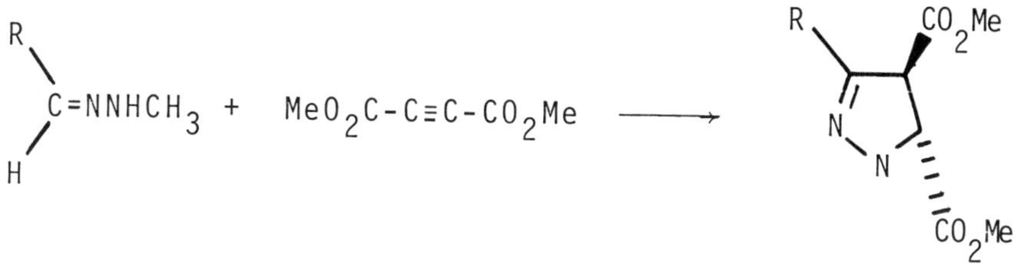

$\begin{array}{c}R\\ \diagdown\\ C=NNHCH_3\\ \diagup\\ H\end{array}$ + $MeO_2C-C\equiv C-CO_2Me \longrightarrow$

R = alkyl, Ph   45-60%

IV.I.1.a-7   J. F. Hansen and D. E. Vietti, J. Org. Chem., 41, 2871 (1976).

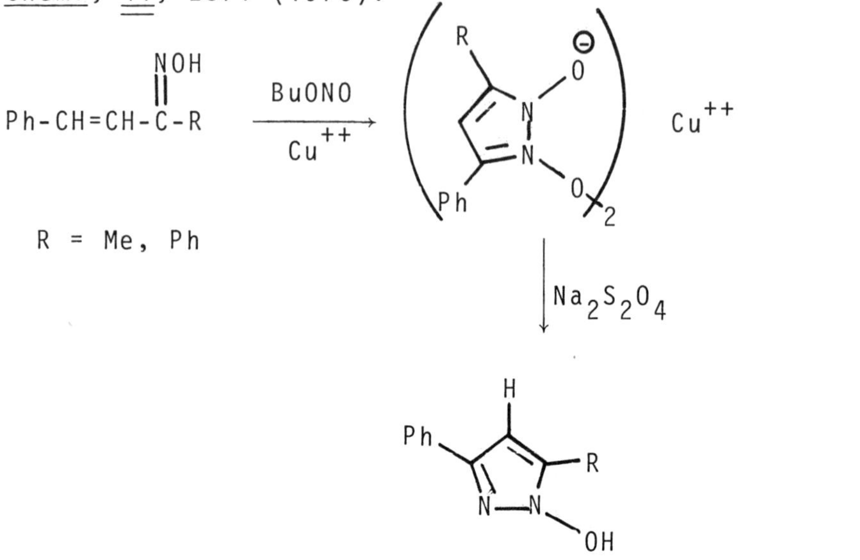

$$\text{Ph-CH=CH-C(=NOH)-R} \xrightarrow{\text{BuONO} \atop Cu^{++}} \left(\begin{array}{c} \text{pyrazole-N,N-dioxide complex} \end{array} \right) Cu^{++}$$

R = Me, Ph

$\downarrow Na_2S_2O_4$

Ph–(pyrazole)–R, N–OH

IV.I.1.a-8   P. Frêche, A. Gorgues, and E. Levas, Tetrahedron Lett., 1495 (1976).

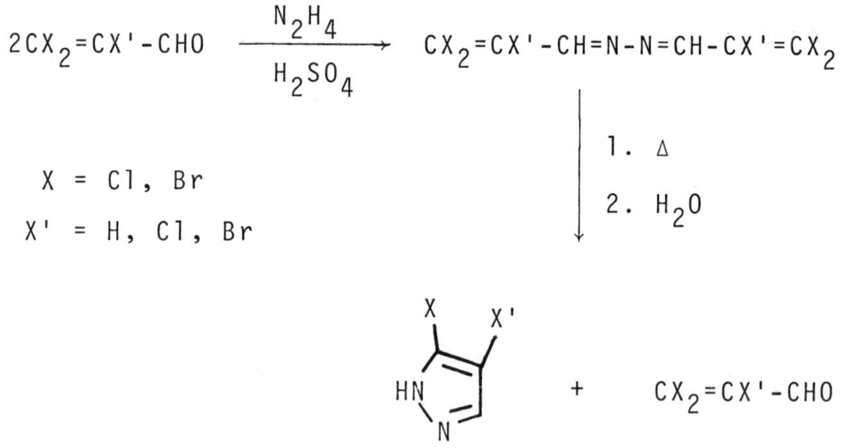

$$2\,CX_2=CX'-CHO \xrightarrow[H_2SO_4]{N_2H_4} CX_2=CX'-CH=N-N=CH-CX'=CX_2$$

X = Cl, Br
X' = H, Cl, Br

$\downarrow \begin{array}{l} 1.\ \Delta \\ 2.\ H_2O \end{array}$

X,X'-pyrazole (HN) + $CX_2=CX'-CHO$

Overall yields 30-70%

IV.I.1.a-9   C. F. Beam et al., Synth. Commun., 6, 5 (1976).

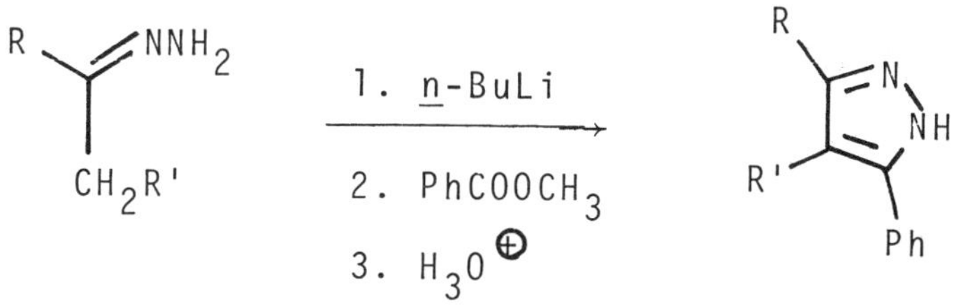

R−C(=NNH$_2$)−CH$_2$R'

1. n-BuLi
2. PhCOOCH$_3$
3. H$_3$O$^{\oplus}$

R,R' = 1° alkyl, aryl, cyclic

→ 3-R, 4-R', 5-Ph pyrazole (NH)

27-73%

IV.I.1.a-10   A. Kasahara, Chem. and Ind., 1032 (1976).

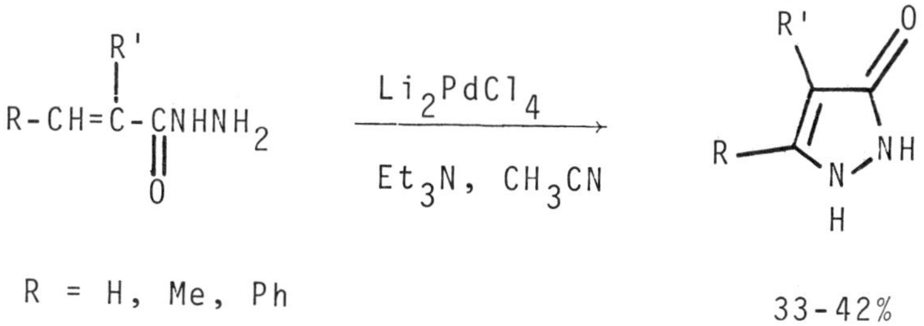

R−CH=C(R')−C(=O)NHNH$_2$

$\xrightarrow{\text{Li}_2\text{PdCl}_4}$
Et$_3$N, CH$_3$CN

→ 3-R, 4-R'-pyrazol-5(4H)-one (NH-NH)

R = H, Me, Ph
R' = H, Me

33-42%

IV.I.1.a-11   N. Suzuki, T. Yamabayashi, and Y. Izawa, Bull. Chem. Soc. Japan, 49, 353 (1976).

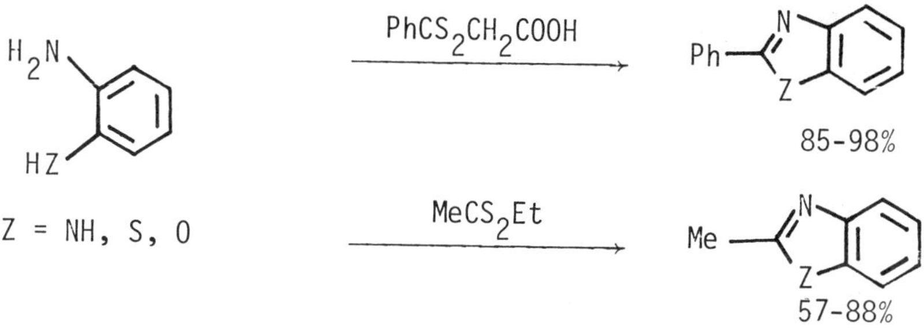

2-H$_2$N, 1-HZ benzene

Z = NH, S, O

$\xrightarrow{\text{PhCS}_2\text{CH}_2\text{COOH}}$ 2-Ph benzazole

85-98%

$\xrightarrow{\text{MeCS}_2\text{Et}}$ 2-Me benzazole

57-88%

IV.I.1.a-12   A. Botta, Liebigs Ann. Chem., 336 (1976).

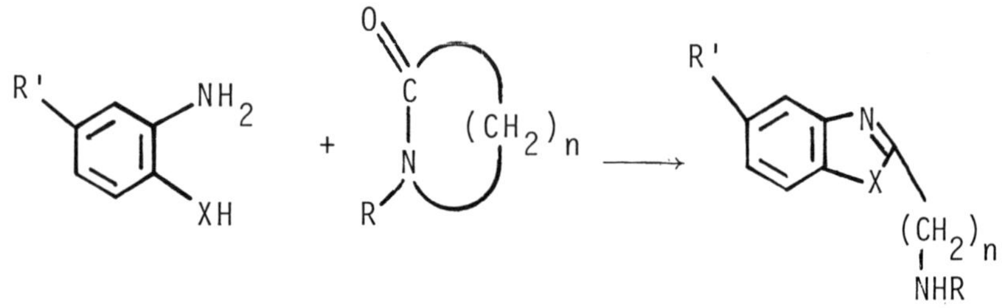

X = NH, O, S
R' = H, Me

R = H, Me, Ph
n = 3,4,5,11

~50-80%

IV.I.1.b-1   A. Kreutzberger and U.-H. Tesch, Chem. Ber., 109, 3255 (1976).

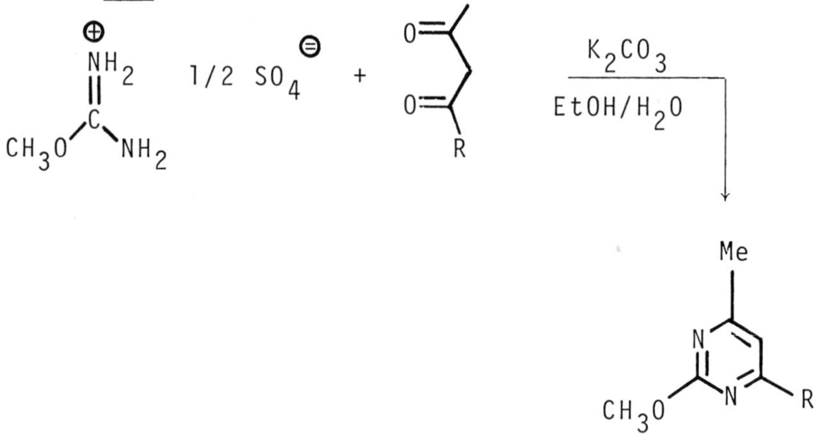

70%, R = Me
56%, R = Et

IV.I.1.b-2 J.-P. Gallemaers, D. Christophe, and
R. Promel, Tetrahedron Lett., 693 (1976).

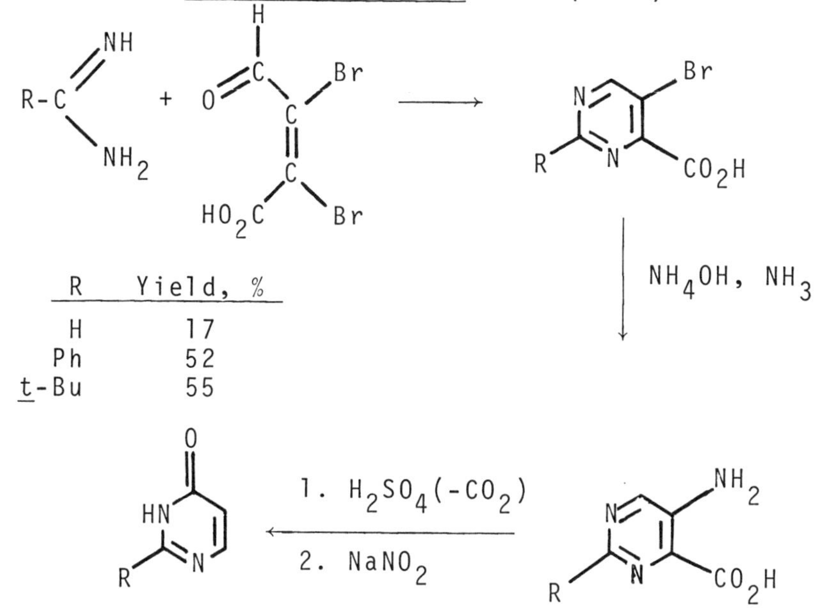

$$R-C(=NH)NH_2 + O=C(Br)-C(CO_2H)=C(Br)-CH \longrightarrow$$ 5-bromo-4-carboxy-2-R-pyrimidine

R	Yield, %
H	17
Ph	52
t-Bu	55

$\xrightarrow{NH_4OH,\ NH_3}$ 5-amino-4-carboxy-2-R-pyrimidine

$\xleftarrow[\text{2. NaNO}_2]{\text{1. H}_2\text{SO}_4(-CO_2)}$ 2-R-pyrimidin-4(3H)-one

---

IV.I.1.b-3 T. Hirayama et al., Chem. Pharm. Bull., 24, 26, 507 (1976).

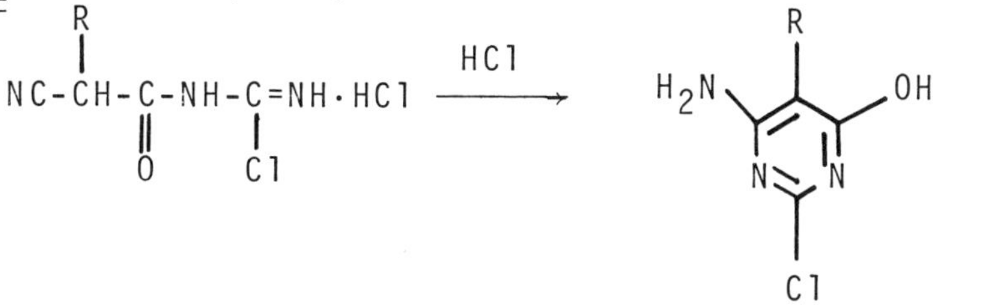

$$NC-CH(R)-C(=O)-NH-C(Cl)=NH \cdot HCl \xrightarrow{HCl} $$ 6-amino-2-chloro-5-R-pyrimidin-4-ol

R = H, Me, Ph   56-97%

Also several similar pyrimidine syntheses.

IV.I.1.b-4 S.M.S. Chauhan and H. Junjappa, Tetrahedron, 32, 1779, 1911 (1976).

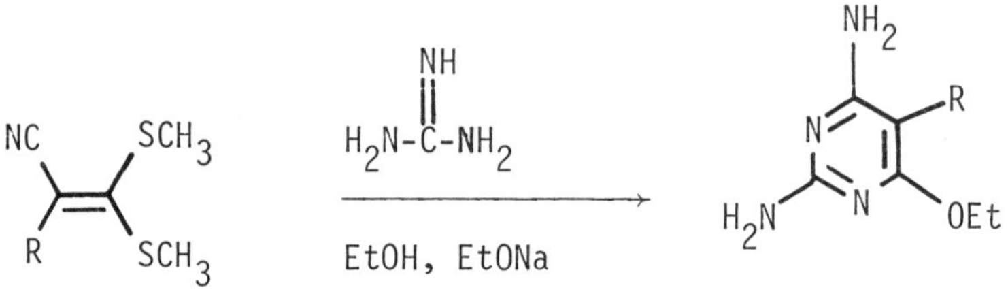

R = CN, Ar

50-60%

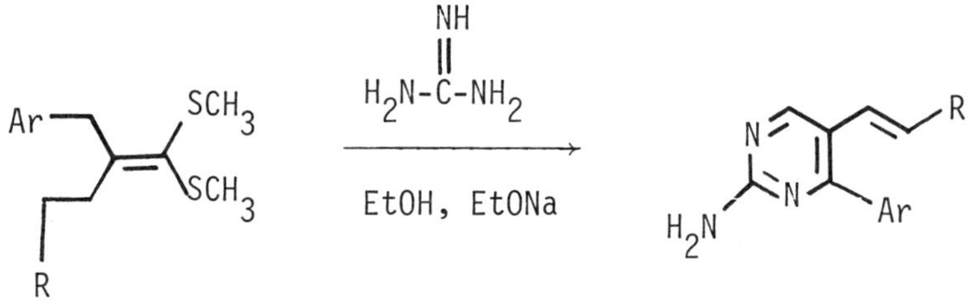

R = H, CH_3

"fair" yields

IV.I.1.b-5 C. H. Foster and E.U. Elam, J. Org. Chem., 41, 2646 (1976).

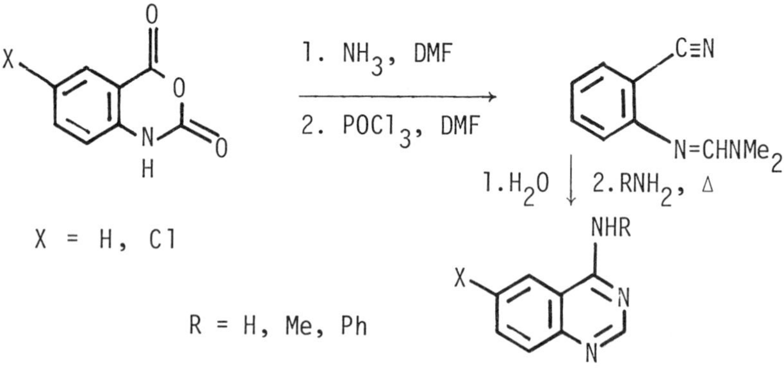

X = H, Cl

R = H, Me, Ph

44-79%

IV.I.1.b-6   D. G. McMinn, Synthesis, 824 (1976).

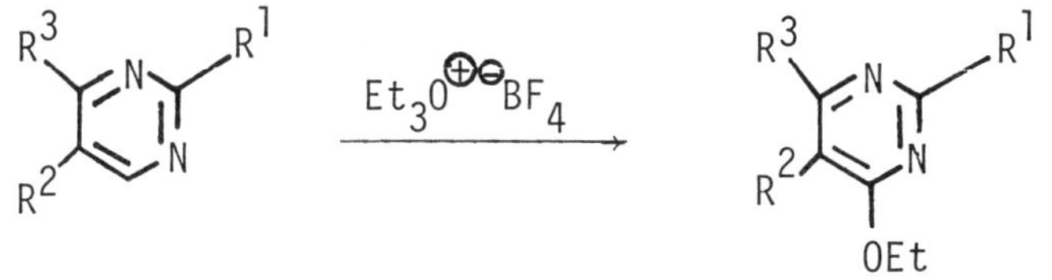

[Reaction: 2,4,5-trisubstituted pyrimidine + $Et_3O^{\oplus} BF_4^{\ominus}$ → 6-OEt pyrimidine]

R^1 = Me, OEt, OH, NH_2
R^2 = H, COOR
R^3 = H, Me, Ph

~60-75%

IV.I.1.b-7   R. H. Fischer and H. M. Weitz, Synthesis, 53 (1976).

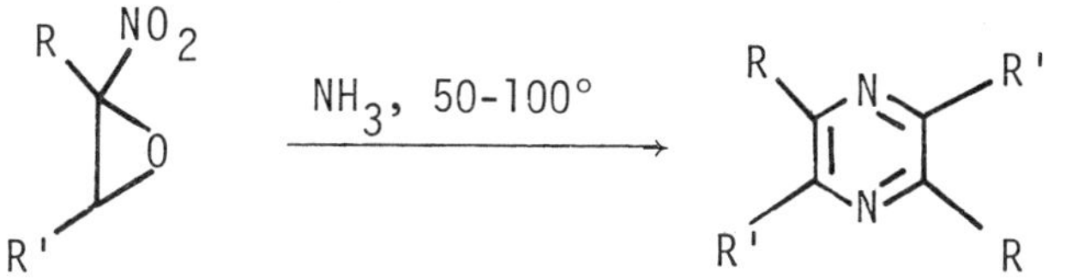

[Reaction: α-nitro epoxide + NH_3, 50-100° → tetrasubstituted pyrazine]

R,R' = alkyl, aryl, cyclic   44-94%

IV.I.1.b-8   T. Kato, A. Takada, and T. Ueda, Chem. Pharm. Bull., 24, 431 (1976).

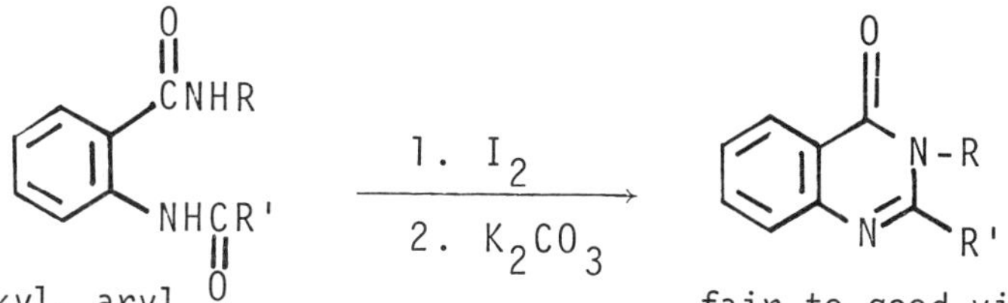

[Reaction: 2-(acylamino)benzamide 1. I_2 2. K_2CO_3 → 4(3H)-quinazolinone]

R = alkyl, aryl
R' = alkyl

fair to good yields (generally 30-60%)

IV.I.1.b-9   Y. F. Shealy and C. A. O'Dell,
J. Het. Chem., 13, 1041 (1976).

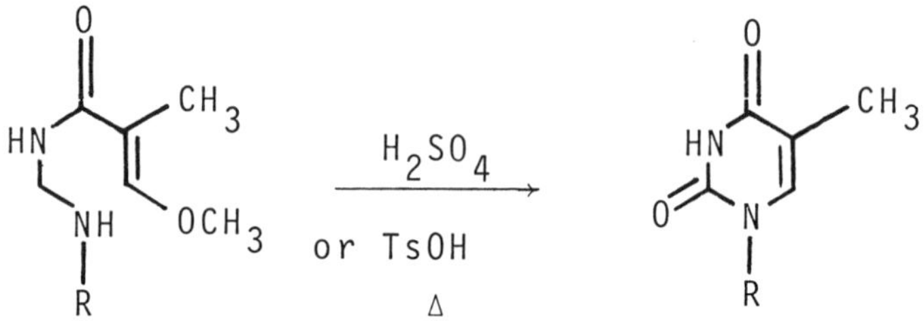

R = Ph, HOCH$_2$-cyclopentyl-OH

~85%

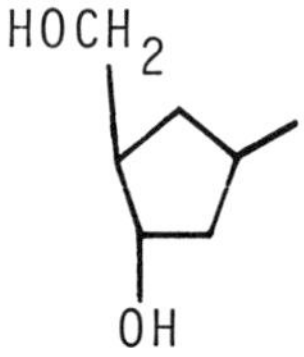

IV.I.1.b-10   J. R. Beck and R. P. Gajewski,
J. Het. Chem., 13, 605 (1976).

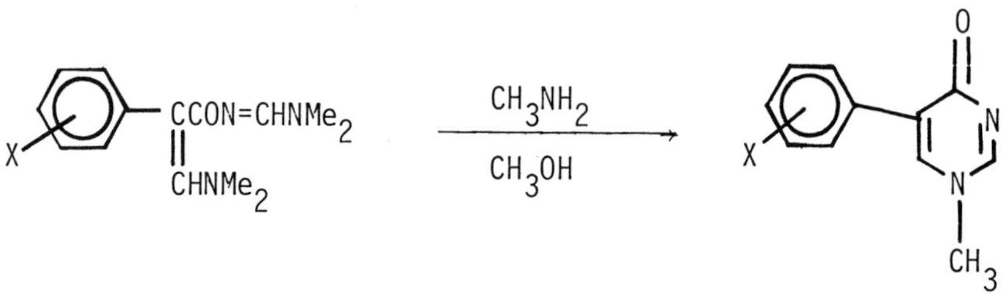

X = CF$_3$, Br, Cl, H

20-71%

IV.I.1.b-11 J.F.W. Keana et al., J. Org. Chem., 41, 2124 (1976).

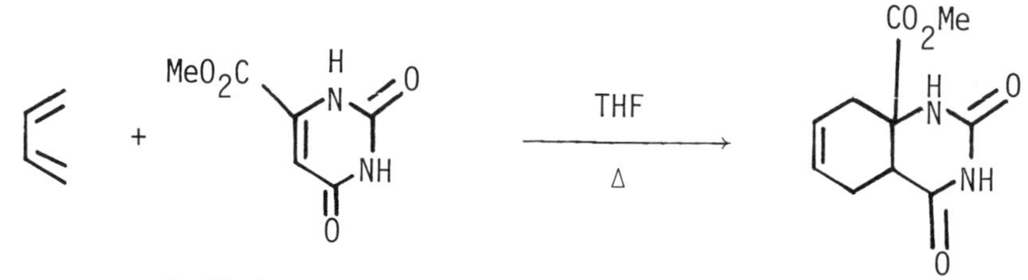

several dienes used.

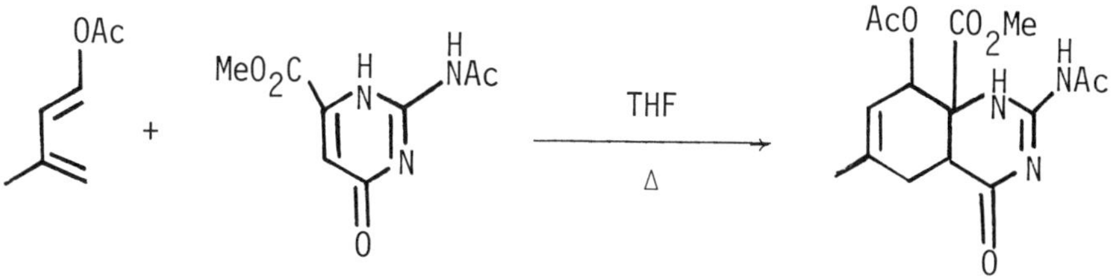

IV.I.1.b-12 K. Nagahara, K. Takagi, and T. Ueda, Chem. Pharm. Bull., 24, 1197 (1976).

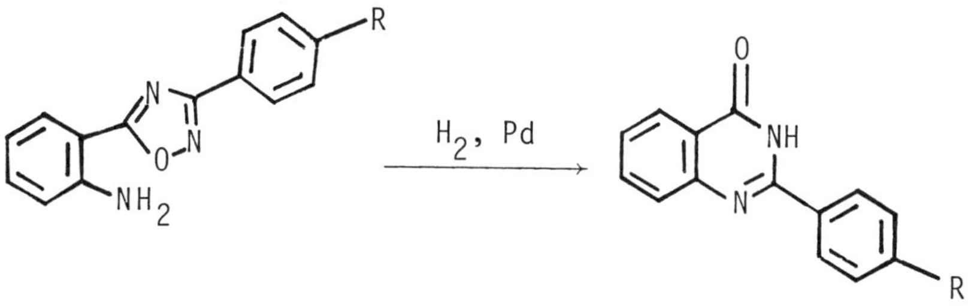

R = H, OCH$_3$ ~70%

IV.I.1.b-13   H. Wamhoff and B. Wehling, Chem. Ber., 109, 2983 (1976).

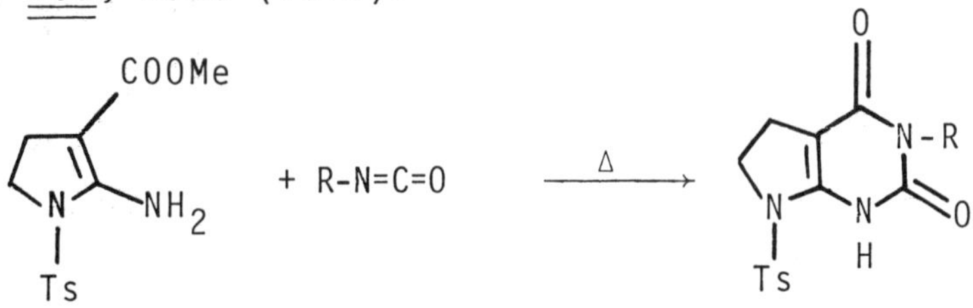

[pyrroline with COOMe and NH2, N-Ts] + R-N=C=O →Δ pyrrolo-pyrimidinedione (N-R, N-Ts)

R = Ph, Bz, CH$_2$OCH$_3$ ~50%

IV.I.1.b-14   R. F. Abdulla, Tetrahedron Lett., 521 (1976).

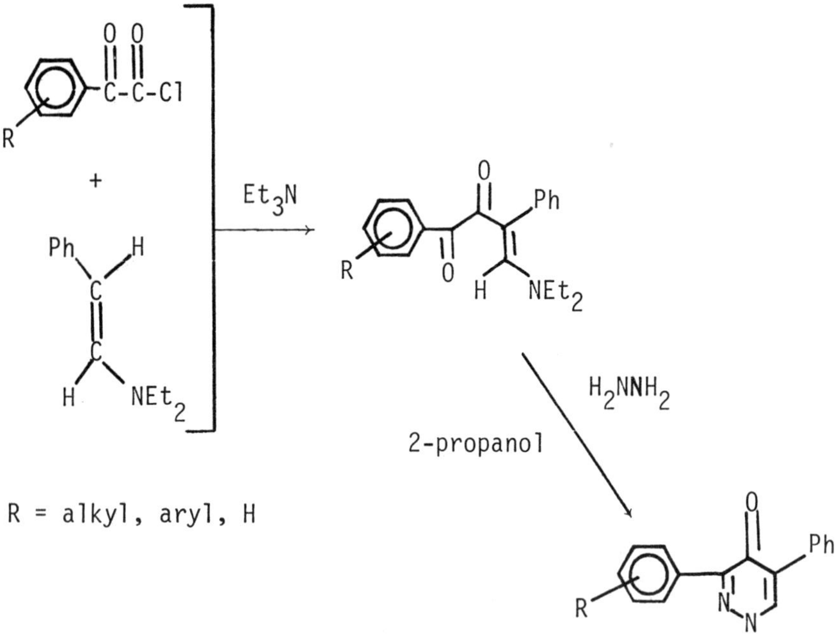

[R-C$_6$H$_4$-CO-CO-Cl] + [Ph-CH=CH-NEt$_2$] →Et$_3$N R-C$_6$H$_4$-CO-CO-C(Ph)=CH-NEt$_2$

R = alkyl, aryl, H

→ H$_2$NNH$_2$ / 2-propanol → pyridazinone (R-C$_6$H$_4$, Ph)

~60%

IV.I.1.c-1   M. Ogata and H. Matsumoto, *Chem. and Ind.*, 1067 (1976).

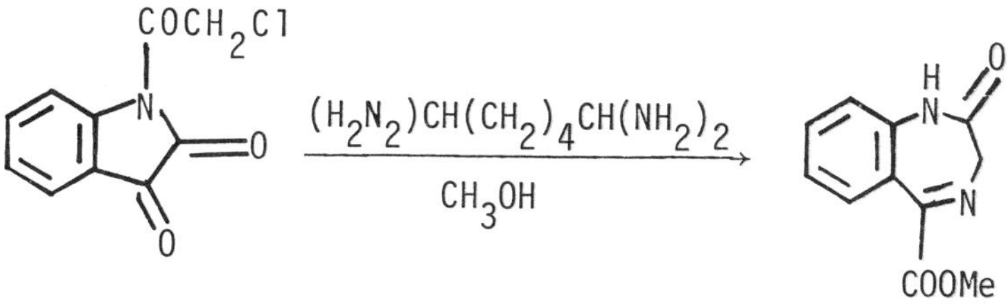

[Reaction: N-(chloroacetyl)isatin + $(H_2N_2)CH(CH_2)_4CH(NH_2)_2$ / CH_3OH → benzodiazepinone with COOMe, 51%]

IV.I.1.c-2   Y. Ohtsuka, *J. Org. Chem.*, **41**, 629 (1976).

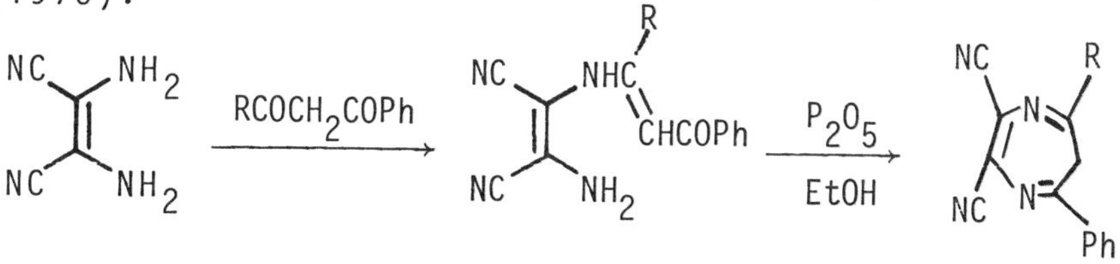

[Reaction: diaminodicyanoethylene + $RCOCH_2COPh$ → intermediate → P_2O_5/EtOH → pyrazine product]

18% (R=Me)
73% (R=H)

IV.I.2-1   M. Miocque *et al.*, *Comptes Rendus (C)*, **282**, 469 (1976).

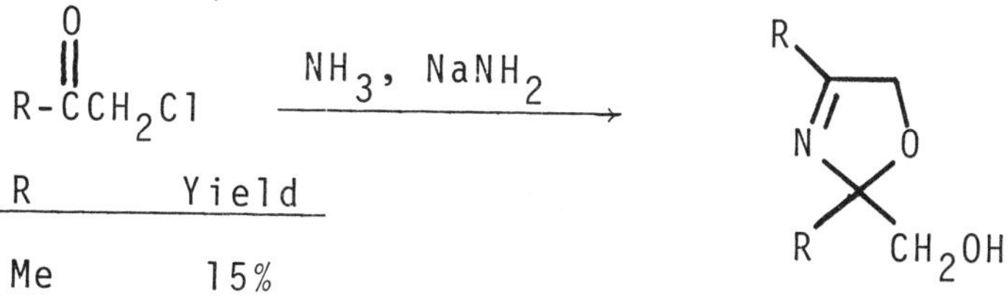

$R-\overset{O}{\underset{\|}{C}}CH_2Cl$ $\xrightarrow{NH_3,\ NaNH_2}$ [oxazoline product with CH_2OH]

R	Yield
Me	15%
Ph	62%
Bz	66%

IV.I.2-2   U. Schollkopf *et al.*, *Liebigs Ann. Chem.*, 183 (1976).

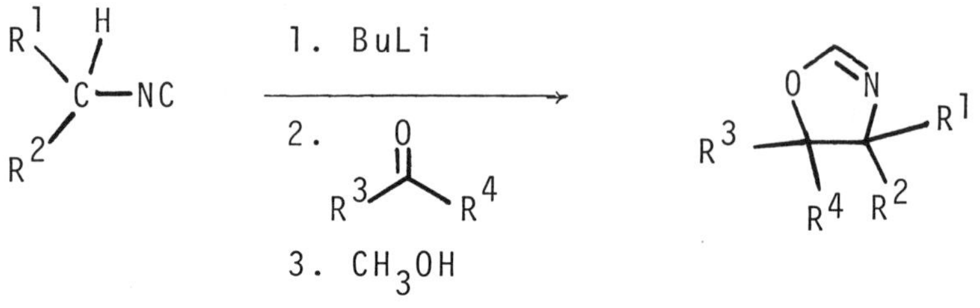

$$R^1R^2CH-NC \xrightarrow[\substack{2.\ R^3COR^4 \\ 3.\ CH_3OH}]{1.\ BuLi} \text{oxazoline}$$

R's = H, 1° alkyl, aryl, cyclic

~60-80%

IV.I.2-3   A. Wolloch and E. Zbiral, *Tetrahedron*, **32**, 1289 (1976).

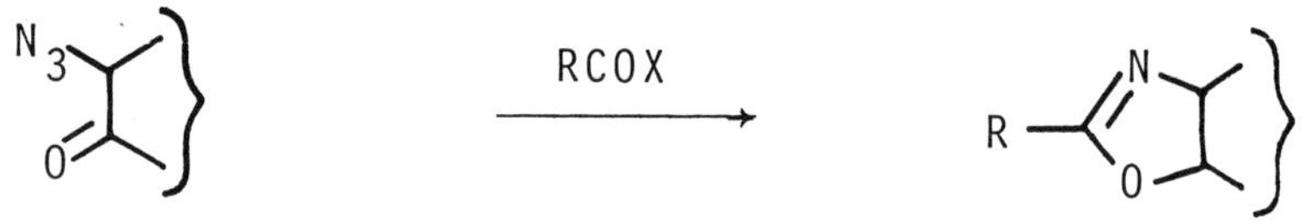

$$\text{N}_3\text{-cyclopentanone} \xrightarrow{RCOX} \text{oxazoline-fused}$$

X = Cl, Br   29%, R=Me
                 45%, R=Ph

IV.I.2-4   C. A. Park *et al.*, *J. Het. Chem.*, **13**, 449 (1976).

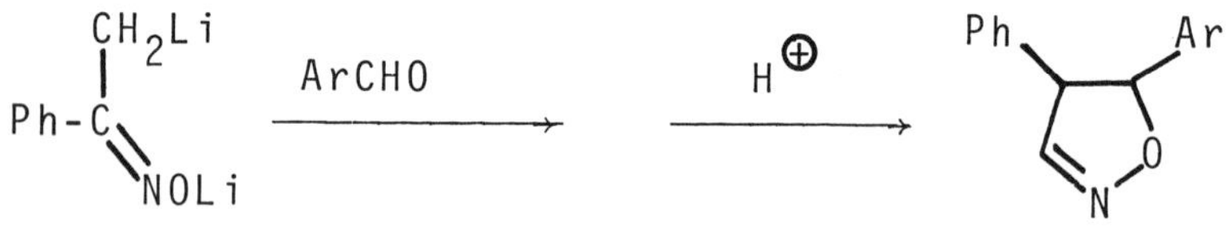

$$Ph-C(=NOLi)CH_2Li \xrightarrow{ArCHO} \xrightarrow{H^\oplus} \text{isoxazoline}$$

yields ~30-60%

IV.I.2-5   A. F. Cockerill et al., Synthesis, 591
(1976).

$$\underset{R^2\text{CHOH}}{\overset{R^1}{\underset{|}{>}}}C=O \;+\; R^3\text{NHCN} \xrightarrow{\text{NaOH, H}_2\text{O}} \underset{R^2}{\overset{R^1}{\diagdown}}\!\!\!\diagup\!\!\overset{N}{\underset{O}{\diagdown}}\!\!\!\diagup\text{NHR}^3$$

R^1, R^3 = 1°, 2° alkyl, H, aryl   ~40-90%
R^2 = H, Me

IV.I.2-6   S. P. McManus et al., J. Org. Chem., 41, 1642 (1976).

$$\underset{(\text{CH}_2)_n\!\!-\!\!\text{CHR'}}{\text{RCH}\overset{O}{\underset{}{\diagdown}}\!\!\!\overset{\|}{C}\!\!\!-\!O} \;+\; \underset{\text{HO}}{\overset{\text{Me}}{\diagdown}}\!\!\!\overset{\text{Me}}{\underset{\text{NH}_2}{\diagup}} \longrightarrow \underset{\text{RCH(CH}_2)_n\text{CHOH}}{\overset{O\diagup N}{\diagdown\!\!\diagup}\!\!-\!\!R'}$$

n = 1,2,3
R,R' = H, CH$_3$ 45-100% (VPC)

IV.I.2-7   A. I. Meyers and E. D. Mihelich, Angew. Chem. Int. Ed., 15, 270 (1976).

Review: "The Synthetic Utility of 2-Oxazolines"

IV.I.2-8   R. A. Kretchmer and P. J. Daly, J. Org. Chem., **41**, 192 (1976).

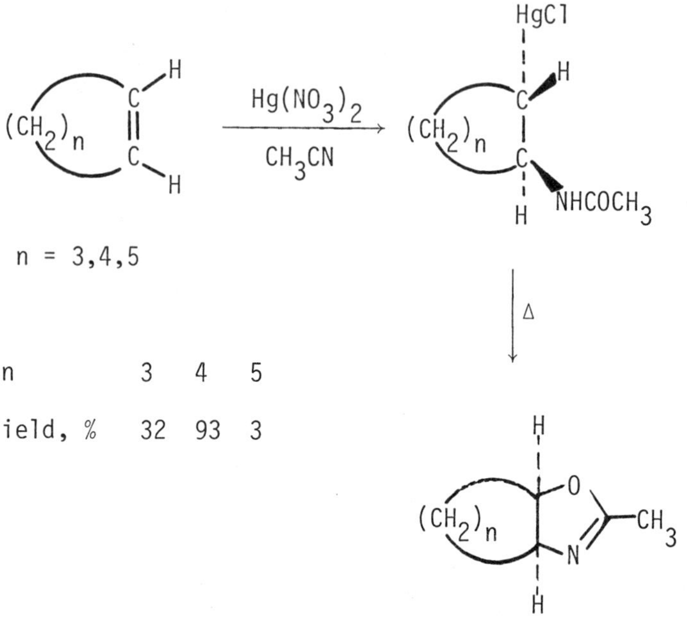

n = 3,4,5

n	3	4	5
Yield, %	32	93	3

IV.I.2-9   M. I. Shevchuk, S. T. Shpak, and A. V. Dombrovskii, J. Gen. Chem. (USSR), **45**, 2571 (1975).

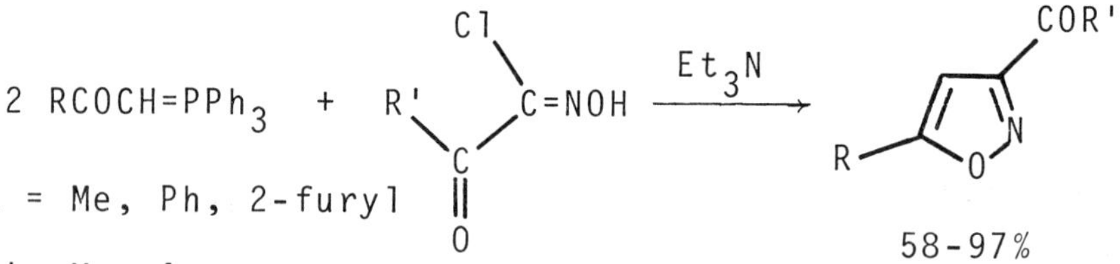

$$2\ RCOCH=PPh_3 + R'\underset{\underset{O}{\overset{\|}{C}}}{\overset{Cl}{C}}=NOH \xrightarrow{Et_3N} \text{[isoxazole]} \quad 58\text{-}97\%$$

R = Me, Ph, 2-furyl
R' = Me, Ar

IV.I.2-10   R. M. Sandifer et al., J. Het. Chem., 13, 607 (1976).

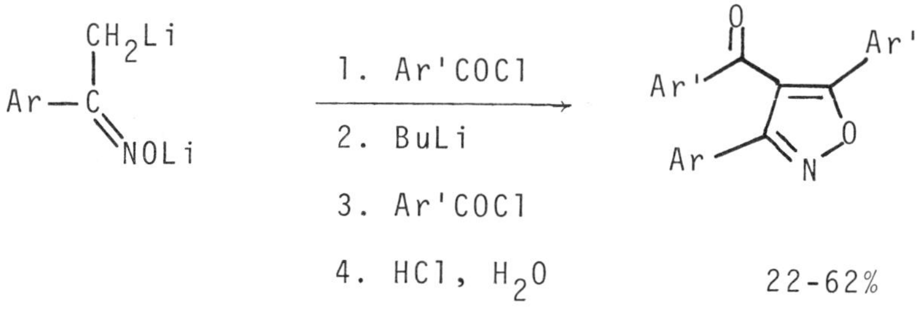

Ar—C(CH$_2$Li)=NOLi
$\xrightarrow{\begin{array}{l}1.\ Ar'COCl\\2.\ BuLi\\3.\ Ar'COCl\\4.\ HCl,\ H_2O\end{array}}$
[isoxazole: 3-Ar, 5-Ar', 4-COAr']

22-62%

IV.I.2-11/IV.I.3-1   N. Suzuki, T. Yamabayashi, and Y. Izawa, Bull. Chem. Soc. Japan, 49, 353 (1976).

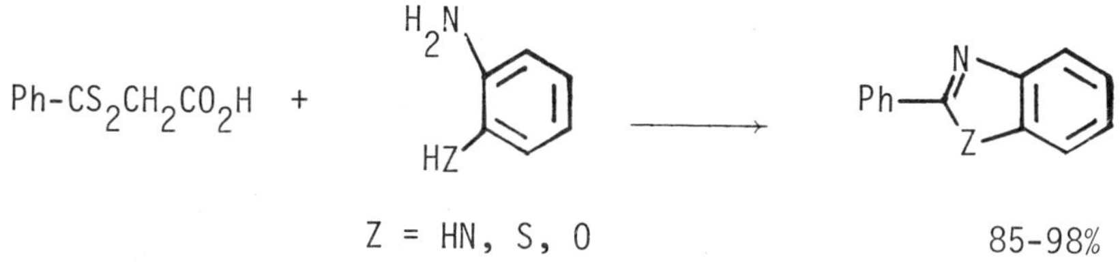

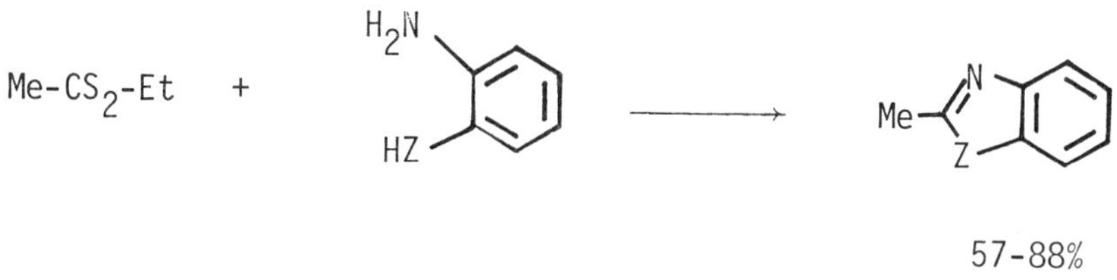

Ph-CS$_2$CH$_2$CO$_2$H + [2-HZ-aniline] ⟶ Ph—[benzazole, Z]

Z = HN, S, O   85-98%

Me-CS$_2$-Et + [2-HZ-aniline] ⟶ Me—[benzazole, Z]

57-88%

IV.I.2-12/IV.I.3-2   A. Botta, Liebigs Ann. Chem., 336 (1976).

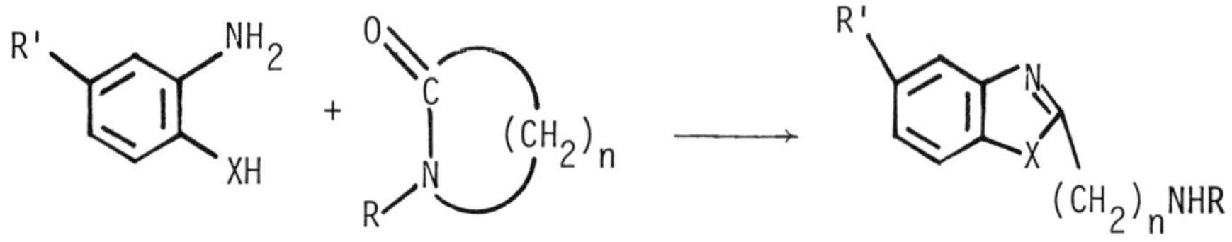

X = NH, O, S   R = H, Me, Ph   ~50-80%
R' = H, Me   n = 3,4,5,11

IV.I.3-3   P. Dubs and R. Stuessi, Synthesis, 696 (1976).

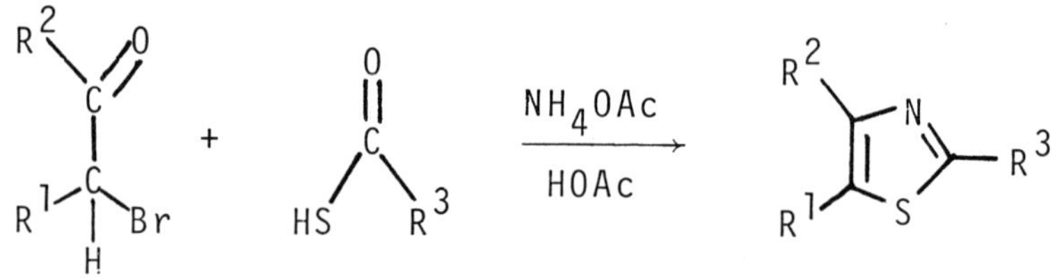

R^1, R^2 = H, CH_3
R^3 = CH_3, i-Bu   37-72%

IV.I.3-4   N. Suzuki and Y. Izawa, Bull. Chem. Soc. Japan, 49, 3155 (1976).

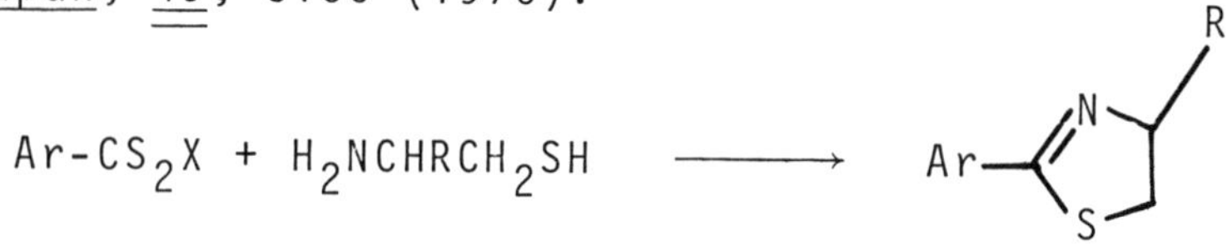

$Ar-CS_2X + H_2NCHRCH_2SH \longrightarrow$

Ar = Ph, p-Cl-Ph
R = H, Et, COOR, CH_2COOR generally ~70-80%

IV.I.3-5 W. Ried and L. Kaiser, Liebigs Ann. Chem., 395 (1976).

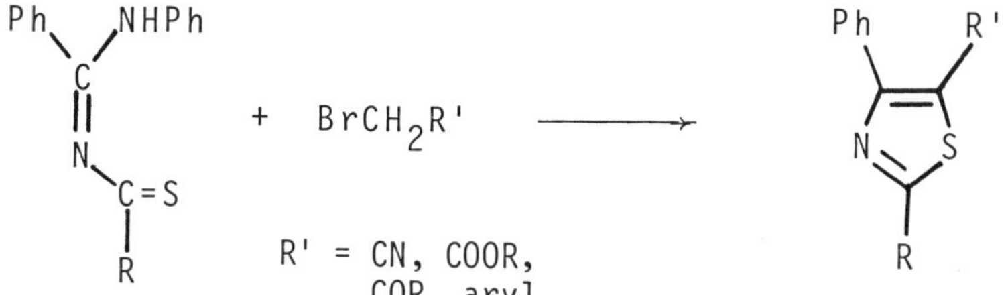

Ph\C(=NPh)–N=C(R)–S + BrCH$_2$R' ⟶ [thiazole: 2-R, 4-Ph, 5-R']

R' = CN, COOR, COR, aryl

R = OEt, morpholino

generally ~60-90%

IV.I.3-6 M. Ferrey, A. Robert, and A. Foucaud, Synthesis, 261 (1976).

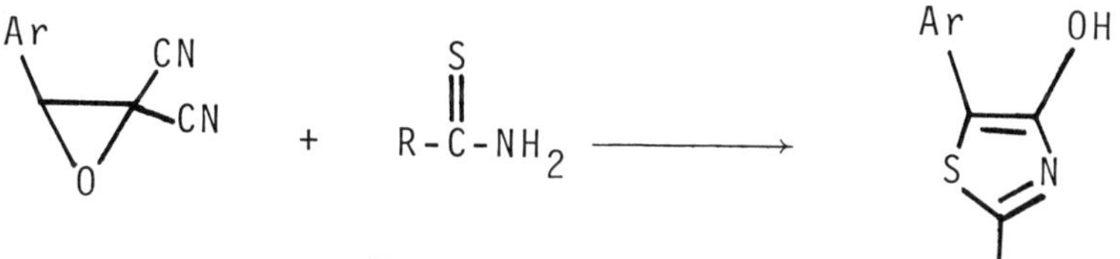

[Ar-epoxide-C(CN)$_2$] + R–C(=S)–NH$_2$ ⟶ [thiazole: 2-R, 4-OH, 5-Ar]

R = H, Me, Ph

36-55%

IV.I.3-7 R. D. Haugwitz et al., Synthesis, 336 (1976).

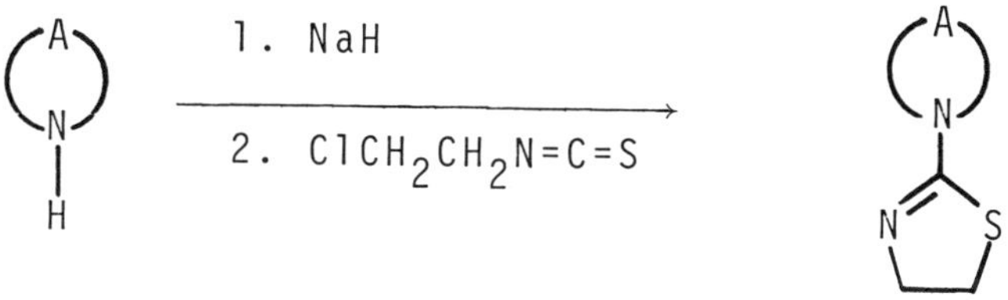

[A–NH heterocycle] $\xrightarrow{\text{1. NaH; 2. ClCH}_2\text{CH}_2\text{N=C=S}}$ [A–N-(thiazoline)]

A = various heterocycles

46-93%

IV.I.3-8   G. D. Hartman and L. M. Weinstock, Synthesis, 681 (1976).

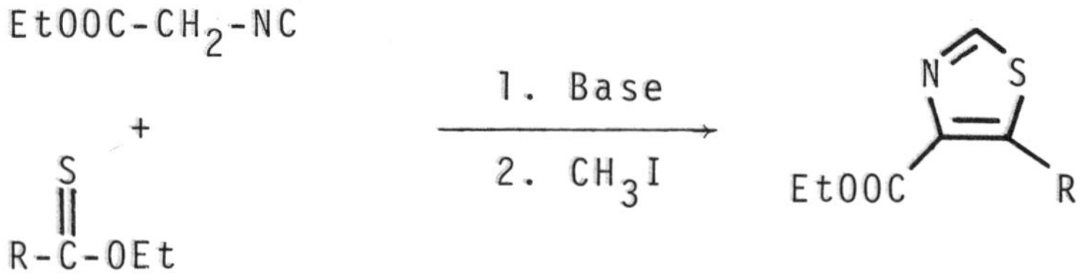

EtOOC-CH$_2$-NC

\+

R-C(=S)-OEt

1. Base
2. CH$_3$I

→ 4-EtOOC, 5-R thiazole

R = H, alkyl, Bz   68-92%

IV.I.3-9   M. J. Spitulnik, Synthesis, 730 (1976).

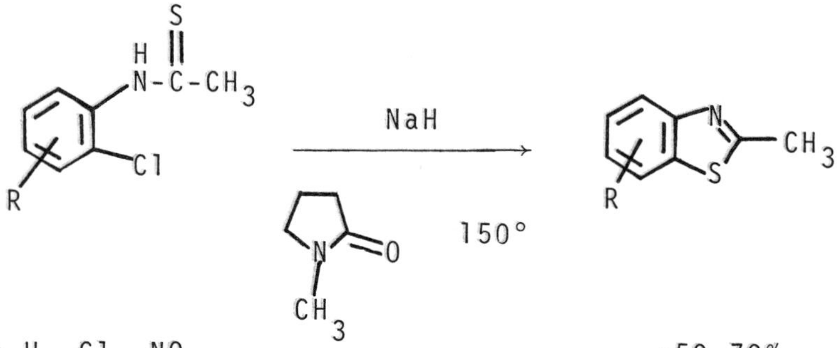

2-(N-H,N-C(=S)CH$_3$)-chloro-R-benzene + N-methylpyrrolidinone, NaH, 150° → 2-methyl-R-benzothiazole

R = H, Cl, NO$_2$ ~50-70%

IV.I.3-10   G. Westphal et al., J. Prakt. Chem., 318, 875 (1976).

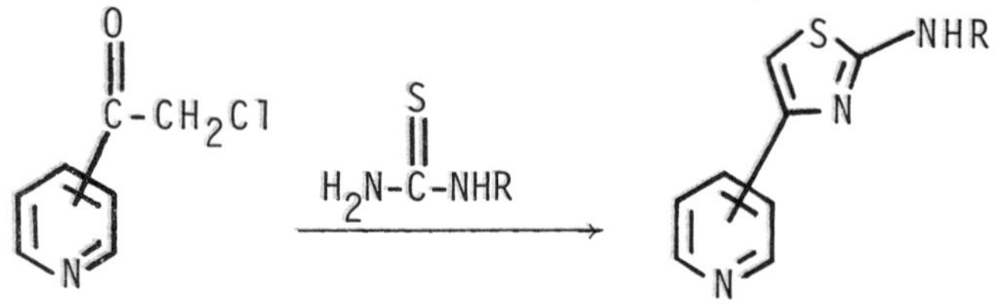

pyridyl-C(=O)-CH$_2$Cl + H$_2$N-C(=S)-NHR → 4-pyridyl-2-(NHR)-thiazole

R = 1° alkyl, aryl   ~30%

IV.I.3-11   M. Baudy and A. Robert, J. Chem. Soc., Chem. Commun., 23 (1976).

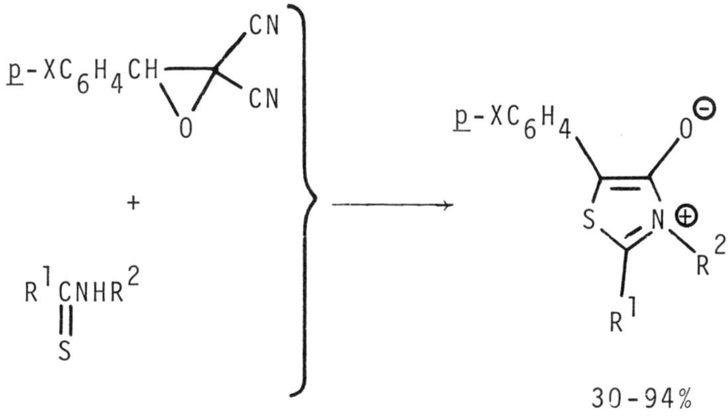

$\underline{p}\text{-}XC_6H_4CH\underset{O}{\overset{CN}{-}}CN$

+

$R^1\underset{S}{\overset{\|}{C}}NHR^2$

$\longrightarrow$

[product: thiazole with $\underline{p}\text{-}XC_6H_4$, $O^{\ominus}$, $N^{\oplus}\text{-}R^2$, R^1]

30-94%

X = H, Cl, NO_2
R^1 = Me, Ph
R^2 = Ph, Bz

IV.I.3-12   K. Peseke, Synthesis, 386 (1976).

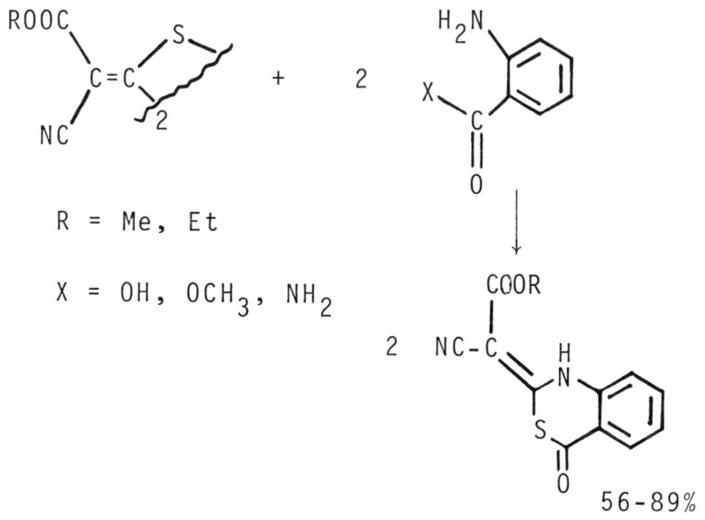

$\underset{NC}{\overset{ROOC}{>}}C=C\underset{2}{\overset{S}{<}}$ + 2 $X-\underset{O}{\overset{}{C}}-C_6H_4\text{-}NH_2$ (H$_2$N ortho)

R = Me, Et

X = OH, OCH_3, NH_2

↓

2 $NC\text{-}C(COOR)=$ [2H-benzo[e][1,3]thiazin-4(3H)-one]

56-89%

IV.I.4-1 S. Cabiddu, A. Maccioni, and M. Secci, *Synthesis*, 797 (1976).

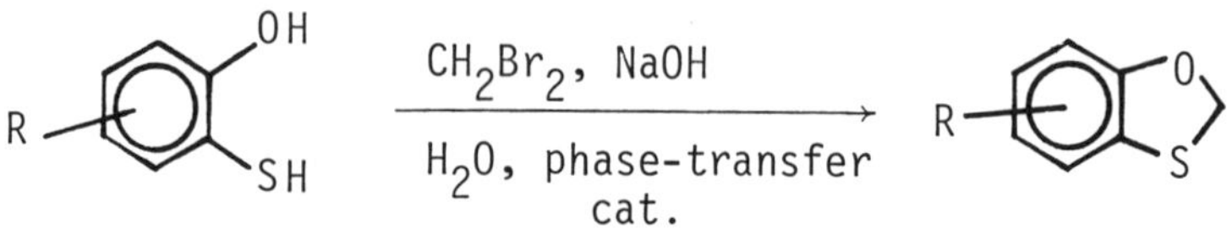

R—(C$_6$H$_3$)(OH)(SH) $\xrightarrow{\text{CH}_2\text{Br}_2,\ \text{NaOH}}_{\text{H}_2\text{O, phase-transfer cat.}}$ R—(benzo[1,3]oxathiole)

R = Me, Cl

70-81%

(C$_6$H$_4$)(OH)(SH) + (CH$_2$)$_n$C=O $\xrightarrow{\text{NaOH, DMSO}}$ spiro benzo[1,3]oxathiole-(CH$_2$)$_n$

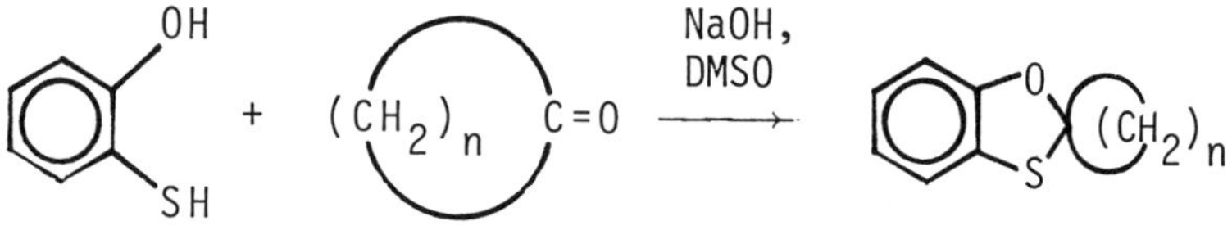

n = 4, 5, 6

65-71%

IV.I.5-1 N. E. Alexandrou and S. Adamopoulos, *Synthesis*, 482 (1976).

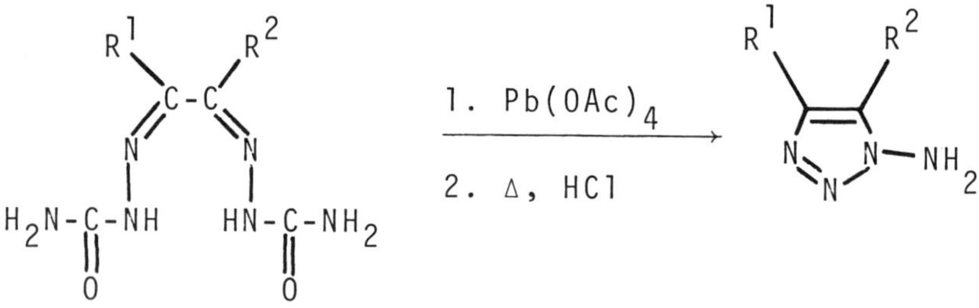

R^1(C=N-NH-C(=O)-NH$_2$)-C(R^2)(=N-NH-C(=O)-NH$_2$) $\xrightarrow[\text{2. }\Delta,\ \text{HCl}]{\text{1. Pb(OAc)}_4}$ R^1,R^2-triazole-NH$_2$

~20-30%

R^1, R^2 = H, Me, aryl

IV.I.5-2   J. Stein, J. Prakt. Chem., 318, 693

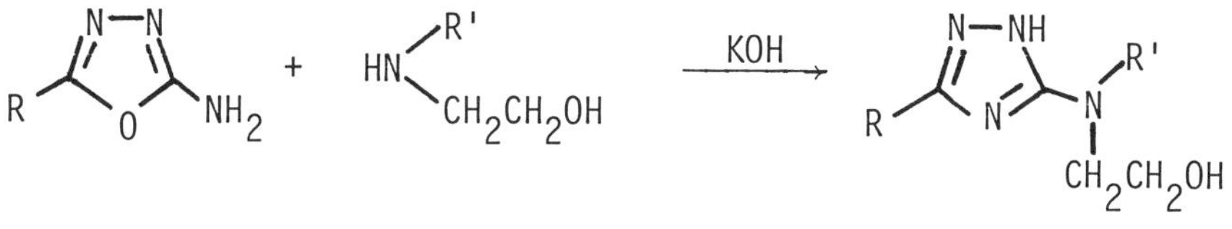

R = n-Bu, aryl

R' = CH_3, CH_2CH_2OH

40-95%

IV.I.5-3   S. Gorjan et al., Monat. fur Chem., 107, 1199 (1976).

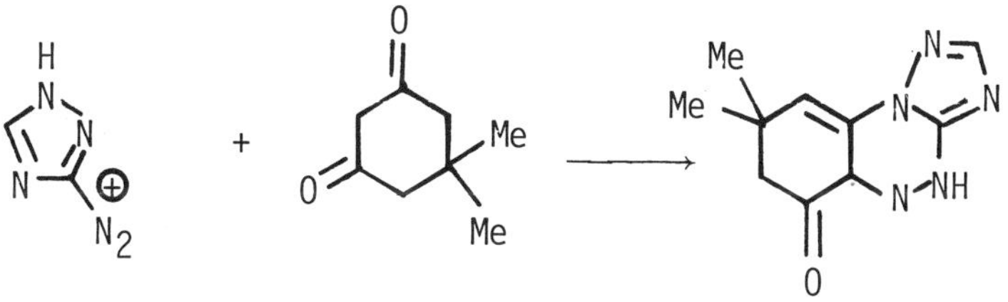

53%

IV.I.5-4   L. Garanti, A. Scandroglio and G. Zecchi, J. Het. Chem., 13, 1339 (1976).

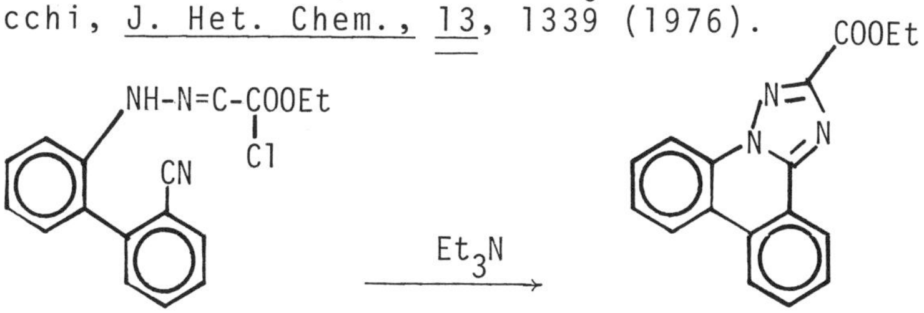

76%

IV.I.5-5   H. Neunhoeffer and V. Boehnisch,
Liebigs Ann. Chem., 153 (1976).

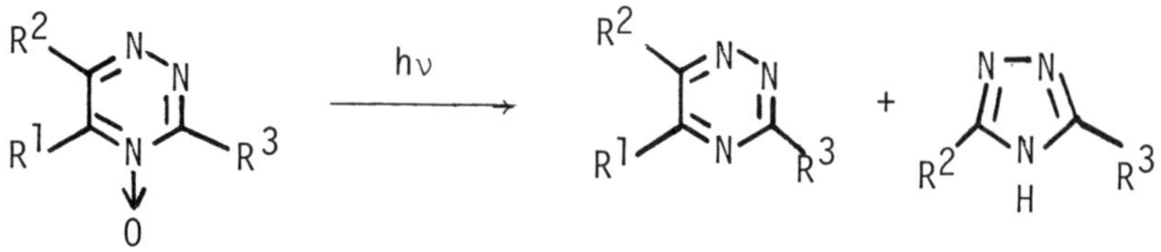

[Reaction: 1,2,4-triazine N-oxide with R², R¹, R³ substituents → hν → 1,2,4-triazine + 1,2,4-triazole (NH)]

48-82% total yield

IV.I.5-6   D. Alsofrom, H. Grossberg, and
H. Sheffer, J. Het. Chem., 13, 917 (1976).

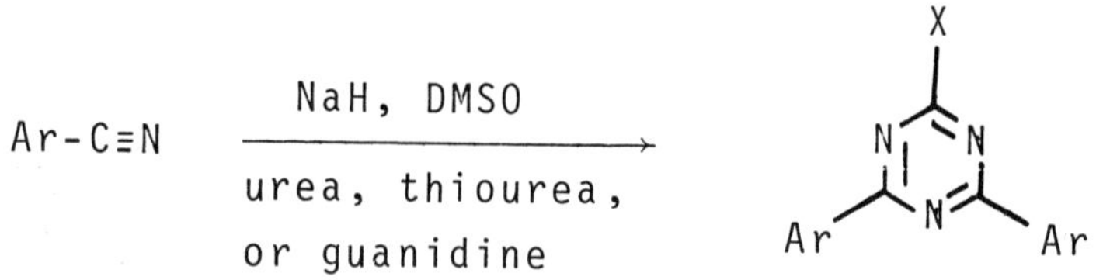

$$Ar-C\equiv N \xrightarrow[\text{urea, thiourea,}]{\text{NaH, DMSO}}$$
or guanidine

[Product: 2,4-diaryl-1,3,5-triazine with X substituent]

X = OH, SH, NH_2

35-93%

IV.I.6-1   T. Fuchigami and K. Odo, Bull. Chem.
Soc. Japan, 49, 3607 (1976).

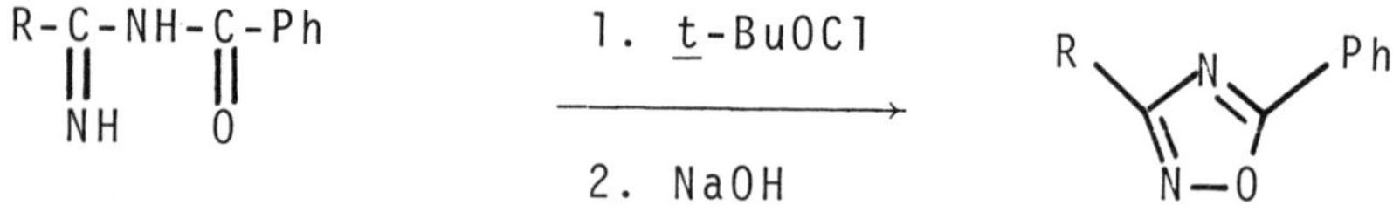

R-C-NH-C-Ph
‖   ‖
NH   O

1. t-BuOCl
───────────→
2. NaOH

[Product: 1,3,4-oxadiazole ring with R and Ph]

R = Ph, OR, N-piperidino

58-91%

IV.I.6-2   G. Rembarz, E. Fischer, and R. Behm, J. Prakt. Chem., 318, 479 (1976).

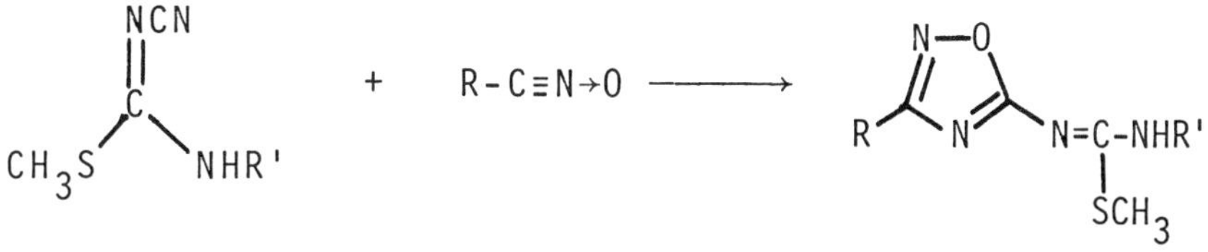

[Structure: CH₃S-C(=NCN)-NHR' + R-C≡N→O → 1,2,4-oxadiazole with R, N=C(NHR')(SCH₃) substituents]

R = Me, Ar
R' = H, 1°,2° alkyl, aryl

~40-70%

IV.I.6-3   I. T. Kay and I. T. Streeting, Synthesis, 38 (1976).

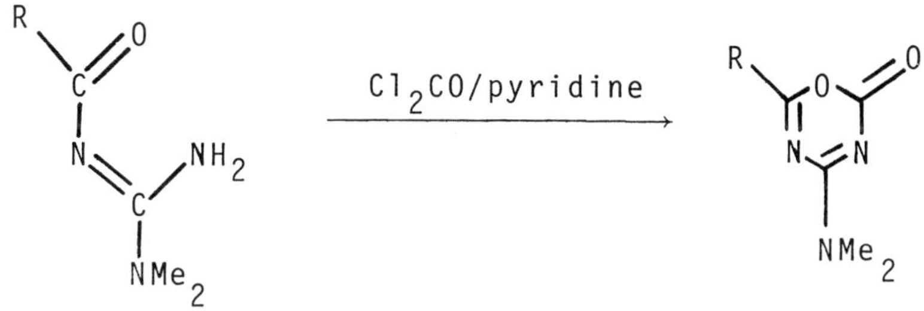

[Structure: R-C(=O)-N=C(NH₂)(NMe₂) + Cl₂CO/pyridine → cyclic product with NMe₂]

R = aryl, t-Bu

43-69%

IV.I.6-4   E. Lippmann and A. Könneche, Z. Chem., 16, 90 (1976).

Several methods of forming 2-aryl tetrazoles, e.g.:

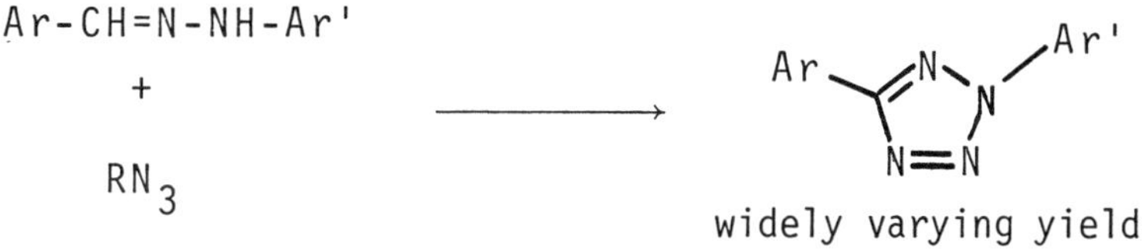

Ar-CH=N-NH-Ar'
+
RN₃

→ [tetrazole with Ar, Ar' substituents]

widely varying yields

IV.J-1   V. Y. Pochinok et al., Russ. Chem. Rev., 45, 183 (1976).

   Review: "Condensed Tetrazoles"

IV.J-2   M. Narita and C. U. Pittman, Jr., Synthesis, 489 (1976).

   Review: "Preparation of Tetrathiafulvalenes (TTF) and their Selenium Analogs - Tetraselenafulvalenes (TSeF)"

IV.J-3   W. D. Munslow and T. J. Delia, J. Het. Chem., 13, 675 (1976).

   Review: "Syntheses of the Isomeric Benzoquinazolines"

V.A-1   C. G. Kruse, N.L.J.M. Broekhof, and A. van der Gen, Tetrahedron Lett., 1725 (1976).

Protection of hydroxyl groups as 2'-O-tetrahydrofuranyl ether. Similar to THP ethers, but removed faster and more easily.

$$\underset{O}{\bigcirc} + ROH \xrightarrow[Et_3N]{SO_2Cl_2} \underset{O}{\bigcirc}\!\!\!-OR$$

Hydrolyzed under mildly acidic conditions.

V.A-2   E. J. Corey, J.-L. Gras, and P. Ulrich, Tetrahedron Lett., 809 (1976).

Use of the β-methoxyethoxymethyl (MEM) group for the protection of alcohols.

$$MEM = -CH_2OCH_2CH_2OCH_3$$

Stable to strong bases, reducing agents, many oxidizing agents, and mild acids.

$$RO^{\ominus} + MEM\text{-}Cl \longrightarrow RO\text{-}MEM$$

$$ROH + MEM\text{-}NEt_3^{\oplus}\ ^{\ominus}Cl \xrightarrow[reflux]{CH_3CN} RO\text{-}MEM$$

$$ROH + MEM\text{-}Cl + (\underline{i}\text{-}Pr)_2NEt \longrightarrow RO\text{-}MEM$$

Removed by $ZnBr_2$ or $TiCl_4$ in CH_2Cl_2.

V.A-3   E. Seymour and J.M.J. Fréchet, Tetrahedron Lett., 1149 (1976).

Use of a polystyrylboronic acid resin as a selective protecting group for hydroxyl functions in glycosides.

V.A-4   J.M.F. Fréchet and L. J. Nuyens, Can. J. Chem., 54, 926 (1976). Also T. M. Fyles and C. C. Leznoff, Can. J. Chem., 54, 935 (1976).

Use of polymers containing trityl chloride residues to block one primary alcohol functional group of a polyhydroxy alcohol. Cleaved by acid.

$$\text{(P)}-\text{C}_6\text{H}_4-\underset{\underset{O-(CH_2)_n-OH}{|}}{C}-Ph_2$$

V.A-5   P. M. Pojer and S. J. Angyal, Tetrahedron Lett., 3067 (1976).

$$R-OH \xrightarrow[Ac_2O]{DMSO, HOAc} R-OCH_2SCH_3 \xrightarrow[acetone/H_2O]{CH_3I, NaHCO_3} ROH$$

R = 1°, 2°, 3°

V.A-6   K. Yamada, et al., Tetrahedron Lett., 65 (1976).

Use of the methylthiomethyl ether function for protection of tertiary alcohols. Protected under basic conditions, stable to base. Removed under mildly acidic or neutral (Ag^+ or Hg^{++}) conditions.

V.A-7   J. P. Yardley and H. Fletcher, Synthesis, 244 (1976).

$$Ar\text{-}OH + H_2C(OCH_3)_2 \xrightarrow[\text{molecular sieves}]{\text{TsOH, } CH_2Cl_2} Ar\text{-}OCH_2OCH_3$$

60-80%
if no ortho-
substituents present

V.A-8   E. Nakamura et al., J. Am. Chem. Soc., 98, 2346 (1976).

Use of ethyl trimethylsilylacetate-tetra-n-butylammonium fluoride (ETSA-TBAF) for silylation of alcohols, e.g.:

$$Ph\text{-}CH_2CH_2CH_2\text{-}OH \xrightarrow{ETSA\text{-}TBAF} Ph\text{-}CH_2CH_2CH_2\text{-}OSiMe_3$$

92%

V.A-9   J. H. Clark, H. L. Holland, and J. M. Miller, Tetrahedron Lett., 3361 (1976).

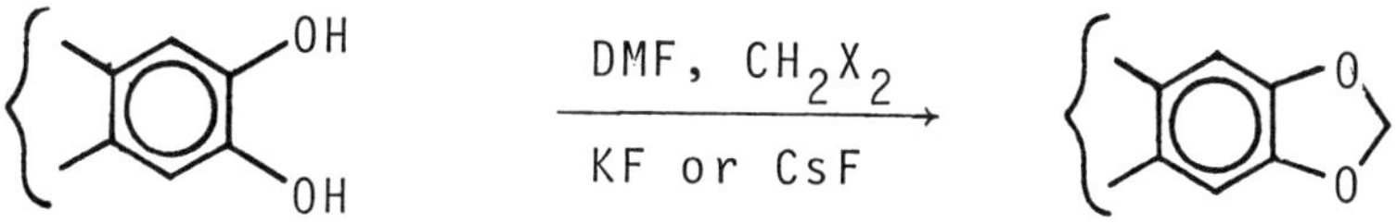

$$\text{Ar}(OH)_2 \xrightarrow[\text{KF or CsF}]{\text{DMF, } CH_2X_2} \text{Ar}(OCH_2O)$$

X = Cl, Br   70-90%

V.A-10   C. Hansson and B. Wickberg, Synthesis, 191 (1976).

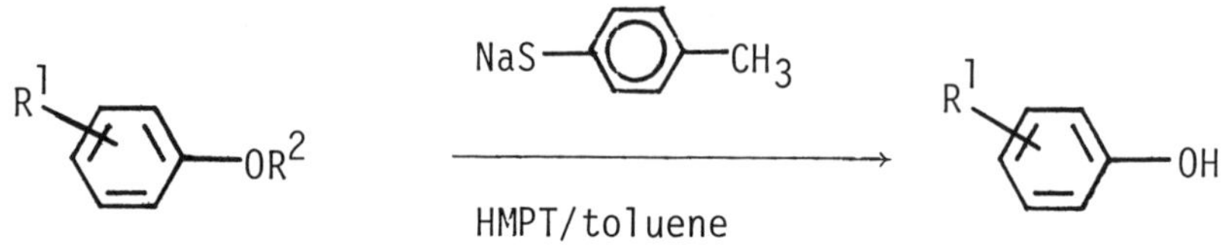

Ar(R^1)–OR^2 → (NaS–C$_6$H$_4$–CH$_3$, HMPT/toluene) → Ar(R^1)–OH

R^1 = Br, CHO
R^2 = Me, Bz

generally 80-90%

V.A-11   G. A. Olah, J. Welch, and T.-L. Ho, J. Am. Chem. Soc., 98, 6717 (1976).

$$R-O-Bz \xrightarrow{UF_6} R-OH$$

R=alkyl   44-69%

V.A-12   C. A. Smith and J. B. Grutzner, J. Org. Chem., 41, 367 (1976).

$$ROCH_3 \xrightarrow[\substack{CH_3CN \\ CH_3OCHI_2}]{HI} ROH$$

Widely varying yields

V.A-13   T. Kametani et al., J. Org. Chem., 41, 2545 (1976).

$$ArOR \xrightarrow{NaAlH_2(OC_2H_4OCH_3)_2} Ar-OH$$

R = Ph, CH=CMe$_2$

V.A-14 M. Node, H. Hori, and E. Fujita, J. Chem. Soc. Perkin I, 2237 (1976).

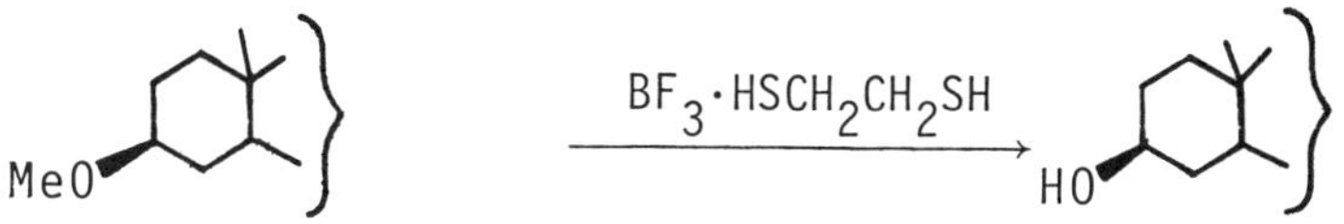

MeO-[cyclohexane] →[$BF_3 \cdot HSCH_2CH_2SH$]→ HO-[cyclohexane]

~60-90%
(retention of config.)

V.A-15 R. Boss and R. Scheffold, Angew. Chem. Int. Ed., 15, 558 (1976).

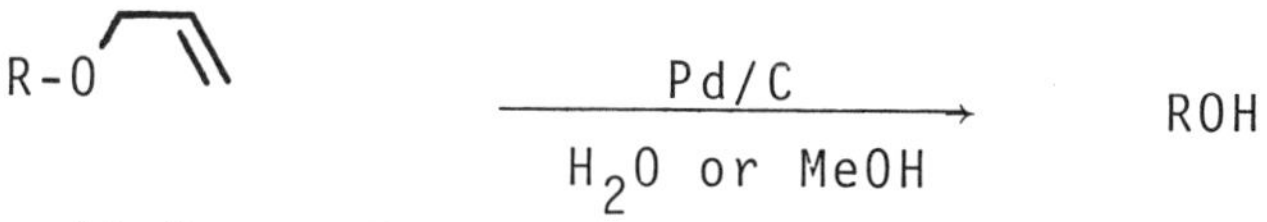

R-O-CH₂-CH=CH₂ →[Pd/C, H_2O or MeOH]→ ROH

R = alkyl, aryl   78-95%

R may contain a wide variety of functional groups.

V.A-16/V.B-1 F. D. King and D.R.M. Walton, Synthesis, 40 (1976).

Use of trimethylsilyl derivatives of phenols, anilines, and carboxylic acids in organocopper coupling reactions.

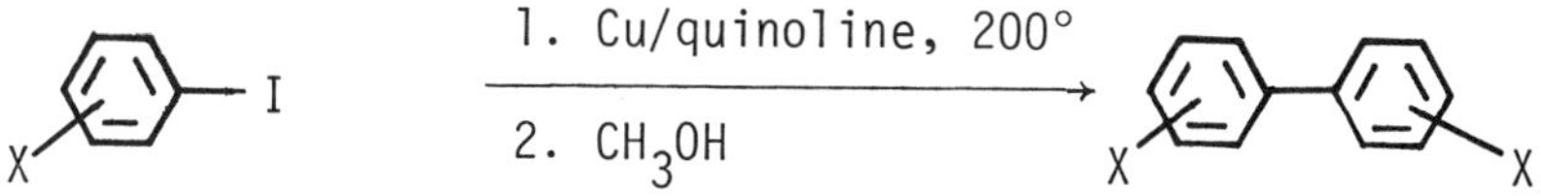

X-[phenyl]-I →[1. Cu/quinoline, 200°; 2. CH_3OH]→ X-[phenyl]-[phenyl]-X

X = $-OSiMe_3$ 55-70%
    $-N(SiMe_3)_2$
    $-COOSiMe_3$

V.B-2 H. Eckert and I. Ugi, Angew. Chem. Int. Ed., 15, 681 (1976).

Use of Cobalt(I)-phthalocyanine to cleave β-halogenated urethane protecting groups, e.g.:

$$R-NH-\overset{O}{\underset{\|}{C}}-OCH_2CH_2X \longrightarrow R-NH_2$$

X = Cl, Br   70-91%

V.B-3 S. Brandänge and L. Lindblom, Acta Chem. Scand. B, 30, 93 (1976).

Use of N-vinyl as an N-H protecting group in the synthesis of myosmine:

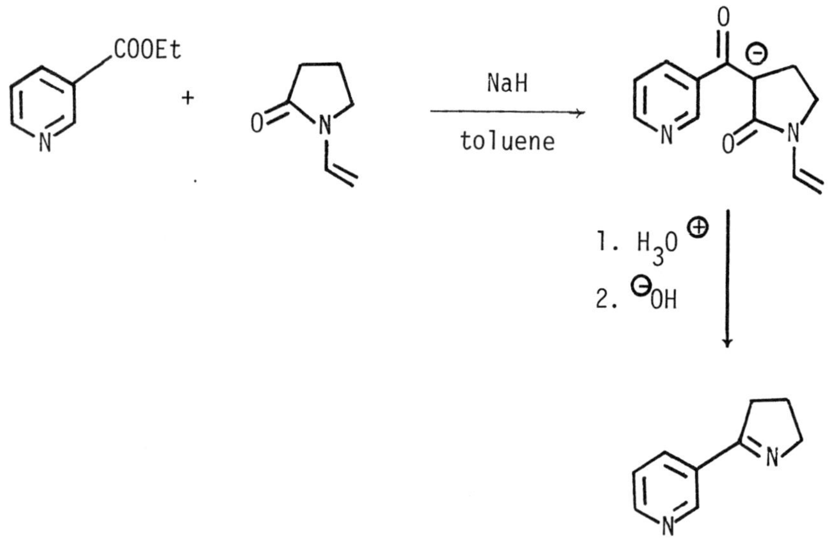

V.B-4   G. W. Kenner, G. A. Moore, and R. Ramage, Tetrahedron Lett., 3623 (1976).

Use of phosphinamides as amino protecting groups in peptide chemistry.

$Ph_2POCl + H_2NCHRCOOR' \xrightarrow{\text{morpholine-Me}} Ph_2PONHCHRCOOR'$

R' = Me, Bz

Cleaved with $HOAc/HCOOH/H_2O$.

V.B-5   A. Tun-Kyi and R. Schwyzer, Helv. Chim. Acta, 59, 1642 (1976).

Use of the (p-phenylazophenyl)-isopropyloxycarbonyl (AZOC) protecting group in peptide synthesis. Put on using AZOC-OPh or AZOC-N_3, and removed with mild acid.

V.B-6   A. P. Hope and B. Halpern, Aust. J. Chem., 29, 1591 (1976).

Use of the 2-hydroxyarylmethylene protecting group for N-terminii of amino acids in peptide synthesis. Removed under conditions (80% HOAc) which leave the t-butyloxycarbonyl group intact.

X = Cl, Me

V.B-7   B. Weinstein et al., Synth. Commun., 6, 17 (1976).

Use of the Tac (p-tosylaminocarbonyl) N-protecting group in peptide synthesis:

$$H_2NCHRCO_2H + TsNCO \longrightarrow TsNHCONHCHRCO_2H$$

Stable to most acids and bases, and to H_2/Pd. Removed by refluxing in ethanol or isopropanol.

V.B-8   H. Kunz, Liebigs Ann. Chem., 1674 (1976).

Use of the 2-phosphonioethoxycarbonyl function as an amine-protecting group in peptide synthesis, giving a higher water solubility than the corresponding triphenylphosphonium derivatives. Removed with .01 N NaOH.

V.B-9   H. Kunz, Chem. Ber., 109, 2670 (1976).

Use of the 2-(triphenylphosphonio)ethoxycarbonyl (Peoc) group as an amino protective function in peptide chemistry. Extremely stable to acids, but removed under mildly basic conditions.

V.B-10   H. Kunz, Chem. Ber., 109, 3693 (1976).

Use of the 2-(methylthio)ethoxycarbonyl (Mtc) group as an amino protecting group in peptide chemistry. Put on using Mtc-phenolate, and removed by methylation followed by basic hydrolysis.

V.B-11   M. Juillerat and J. P. Bargetzi, <u>Helv. Chem. Acta</u>, <u>59</u>, 855 (1976).

Use of NTBA or its methyl ester in cleaving the o-nitrophenylsulfenyl α-amino protecting group. Gives nearly quantitative yields under mild conditions.

$$\text{HS}-\underset{O_2N}{\underset{|}{\bigcirc}}-\underset{}{\overset{O}{\underset{\|}{C}}}-OH$$

NTBA

V.B-12   O. Nishimura and M. Fujino, <u>Chem. Pharm. Bull.</u>, <u>24</u>, 1568 (1976).

Use of the p-methoxybenzenesulfonyl (MBS) group to protect the ω-guanidino function in arginine in peptide synthesis. More easily removed than tosyl group, using methanesulfonic acid and BTFA.

V.B-13   M. Chorev and Y. S. Klausner, <u>J. Chem. Soc., Chem. Commun.</u>, 596 (1976).

Use of the benzyloxycarbonyl group to protect the indole ring of tryptophan in peptide synthesis. Achieved by acylation with p-nitrophenyl benzyl carbonate in the presence of unsolvated fluoride anion, generated by crown ethers. Removed by cat. H_2, H_2NNH_2, or liquid HF.

V.C-1   C. G. Kruse, N.L.J.M. Broekhof, and
A. Van der Gen, <u>Tetrahedron Lett.</u>, 1725 (1976).

Protection of thiol groups as 2'-S-tetrahydrofuranyl ethers. Similar to THP ethers, but removed faster and more easily (under mildly acidic conditions).

$$\text{(tetrahydrofuran)} + \text{RSH} \xrightarrow[\text{Et}_3\text{N}]{\text{SO}_2\text{Cl}_2} \text{(2-SR-tetrahydrofuran)}$$

V.C-2   C. Seidel <u>et al.</u>, <u>J. Prakt. Chem.</u>, <u>318</u>, 229 (1976).

Use of the S-ethylthio protecting group for cysteine in solid-phase peptide synthesis. Relatively stable to base, but not stable to acid.

V.C-3   S. Coyle and G. T. Young, <u>J. Chem. Soc., Chem. Commun.</u>, 980 (1976).

Use of the diphenyl-4-pyridlmethyl group as a protecting group for the thiol group of cysteine and the imidazole nitrogen of histidine in peptide synthesis. Removed by electrolytic reduction, Zn/HOAc, Hg(OAc)$_2$, and I$_2$/HOAc.

V.D-1   J. San Filippo, Jr. <u>et al.</u>, <u>J. Org. Chem.</u>, <u>41</u>, 586 (1976).

$$\text{RCO}_2\text{R'} \xrightarrow[\text{2. H}_3\text{O}^{\oplus}]{\text{1. KO}_2\text{-crown ether}} \text{RCO}_2\text{H} + \text{R'OH}$$

R,R' = alkyl, aryl, lactone

yields generally 35-95% (GLC)

V.D-2 T.-L. Ho and G. A. Olah, Angew. Chem. Int. Ed., 15, 774 (1976).

$$\underset{}{R-\overset{O}{\underset{\|}{C}}-OR'} \quad \xrightarrow[\text{2. } H_2O]{\text{1. } Me_3SiI} \quad RCO_2H$$

R = Ar, Bz, t-Bu, cyclohexyl

R' = Me, Et, Bz

∼ 60-90%

V.D-3 H. Eckert and I. Ugi, Angew. Chem. Int. Ed., 15, 681 (1976).

Use of Cobalt(I)-phthalocyanine to cleave β-halogenated alkyl ester protecting groups, e.g.:

$$R-\overset{O}{\underset{\|}{C}}-OCH_2CH_2X \quad \longrightarrow \quad R-COOH$$

X = Cl, Br

57-98%

V.D-4   J. W. Huffman and P. G. Harris, Synth. Commun., 6, 481 (1976).

$$R\text{-}COOCH_3 \xrightarrow{KH,\ DMSO} RCOOH$$

~50%

Useful for most hindered esters, but fails with methyl benzoate and methyl mesitoate.

V.D-5   G. Just and K. Grozinger, Synthesis, 457 (1976).

$$R\text{-}\overset{O}{\underset{\|}{C}}\text{-}O\text{-}CH_2CCl_3 \xrightarrow[pH\ 4.2\text{-}7.2]{Zn,\ THF/H_2O} RCOOH$$

R = alkyl, aryl, unsat., etc.

~80%

V.D-6   R. L. Prestidge, D.R.K. Harding, and W. S. Hancock, J. Org. Chem., 41, 2579 (1976).

Use of Cl—C$_6$H$_4$—CH$_2$-O- and O$_2$N—C$_6$H$_4$—CH$_2$-O-

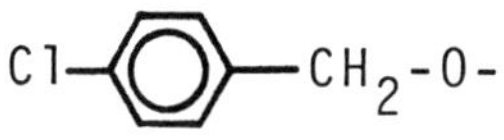

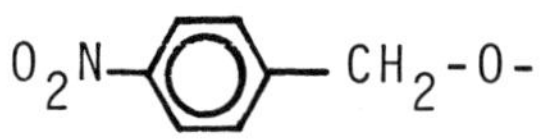

esters as carboxyl-protecting groups in solid-phase peptide synthesis.

PROTECTING GROUPS

V.E-1 E. J. Corey, E. J. Trybulski, and J. W. Suggs, Tetrahedron Lett., 4577 (1976).

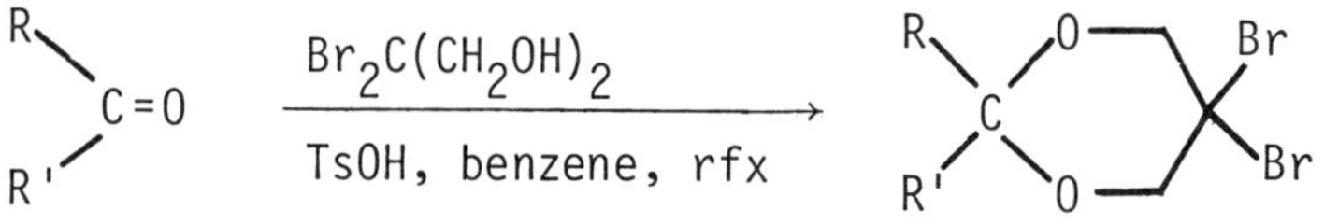

$$\underset{R'}{\overset{R}{>}}C=O \quad \xrightarrow{\underset{\text{TsOH, benzene, rfx}}{Br_2C(CH_2OH)_2}} \quad \underset{R'}{\overset{R}{>}}C\underset{O}{\overset{O}{<}}\underset{Br}{\overset{Br}{>}}$$

R,R' = H, alkyl   84-94%

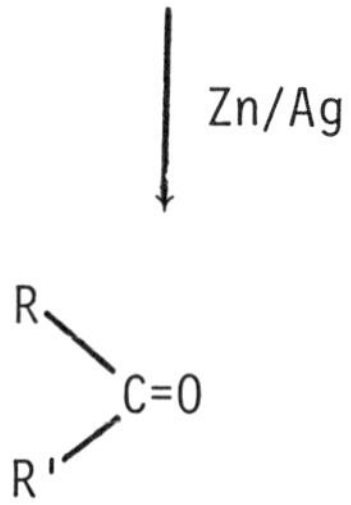

Stable to $NaBH_4$, RCO_3H, B_2H_6

$\downarrow$ Zn/Ag

$$\underset{R'}{\overset{R}{>}}C=O$$

>90%

V.E-2 J. B. Paine III, R. B. Woodward, and D. Dolphin, J. Org. Chem., 41, 2826 (1976).

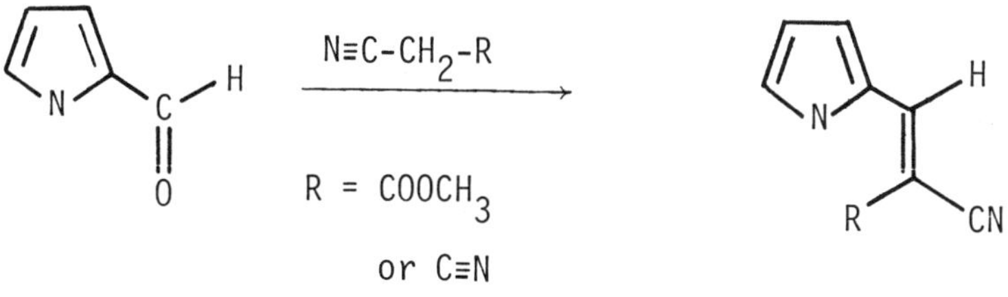

[pyrrole-2-CHO] $\xrightarrow{N\equiv C-CH_2-R}$ [pyrrole-2-C(R)=C(H)(CN)]

R = $COOCH_3$ or $C\equiv N$

Stable to acids, most oxidants, reductants

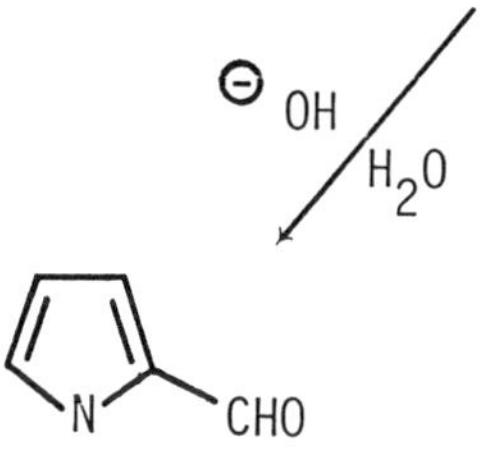

$\xrightarrow{\overset{\ominus}{OH}, H_2O}$ [pyrrole-2-CHO]

V.E-3   E. Nakamura et al., J. Am. Chem. Soc., 98, 2346 (1976).

Use of ethyl trimethylsilylacetate-tetra-n-butylammonium fluoride (ETSA-TBAF) for silylation of ketones, e.g.:

cyclohexanone →(ETSA-TBAF)→ 1-(trimethylsilyloxy)cyclohexene   98%

V.E-4   E. J. Corey and S. Knapp, Tetrahedron Lett., 3667 (1976).

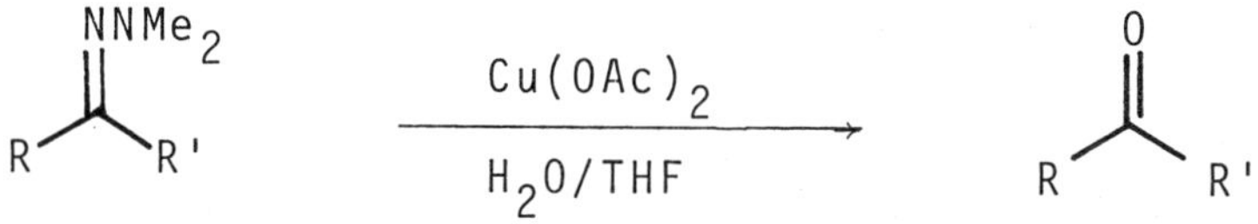

$R\text{-}C(\text{=NNMe}_2)\text{-}R'$ →($Cu(OAc)_2$, H_2O/THF)→ $R\text{-}C(\text{=O})\text{-}R'$

R,R' = H, alkyl, vinyl, etc.   generally >95%

V.E-5   E. J. Corey and D. Enders, Tetrahedron Lett., 3 (1976).

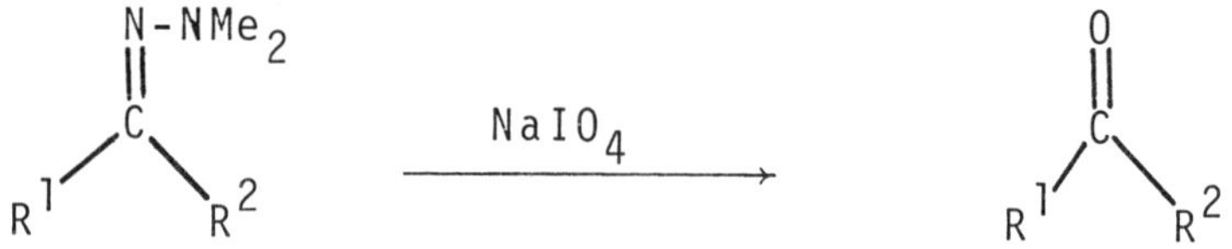

$R^1\text{-}C(\text{=N-NMe}_2)\text{-}R^2$ →($NaIO_4$)→ $R^1\text{-}C(\text{=O})\text{-}R^2$

R^1, R^2 = H, alkyl   >90% isolated

V.E-6   G. A. Olah et al., Synthesis, 808, 809 (1976).

$$\underset{R'}{\overset{R}{>}}C=N-NTs \quad \xrightarrow{\text{MoOCl}_3 \text{ or WF}_6} \quad \underset{R'}{\overset{R}{>}}C=O$$

R,R' = alkyl, aryl   ~80-90%

V.E-7   T.-L. Ho and G. A. Olah, Synthesis, 611 (1976).

$$\underset{R^2}{\overset{R^1}{>}}C=N-NHTos \quad \xrightarrow{\text{Na}_2\text{O}_2} \quad \underset{R^2}{\overset{R^1}{>}}C=O$$

cyclohexanone, acetophenone, etc.   69-73%

V.E-8   C. E. Sacks and P. L. Fuchs, Synthesis, 456 (1976).

$$\underset{R^1 \quad R^2}{\overset{\text{NNHTs}}{C}} \quad \xrightarrow{\text{acetone/H}_2\text{O} \atop \text{BF}_3 \cdot \text{Et}_2\text{O}} \quad \underset{R^1 \quad R^2}{\overset{O}{C}}$$

R^1, R^2 = alkyl, aryl, cyclic   80-97%

V.E-9   M. Hojo and R. Masuda, Synthesis, 678 (1976).

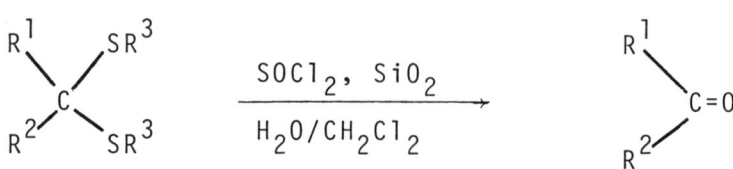

R^1, R^2 = H, alkyl, aryl   quantitative by NMR

V.E-10   E. Fujita, Y. Nagao, and K. Kaneko, Chem. Pharm. Bull., 24, 1115 (1976).

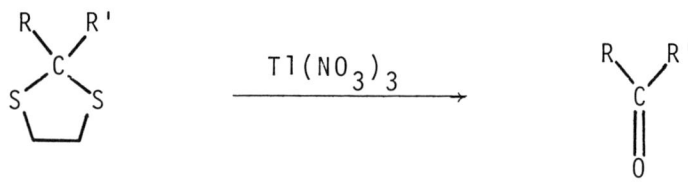

R,R' = H, aryl, 1°,2°,3°   73-99%
       alkyl, cyclic

V.F-1   K. K. Ogilvie, S. L. Beaucage, and D. W. Entwistle, Tetrahedron Lett., 1255 (1976).

Use of tetrabutylammonium fluoride in THF for the removal of phosphate-protecting groups in nucleotide synthesis.

V.G-1 G. Evans, B.F.G. Johnson and J. Lewis, J. Organometal. Chem., 102, 507 (1975).

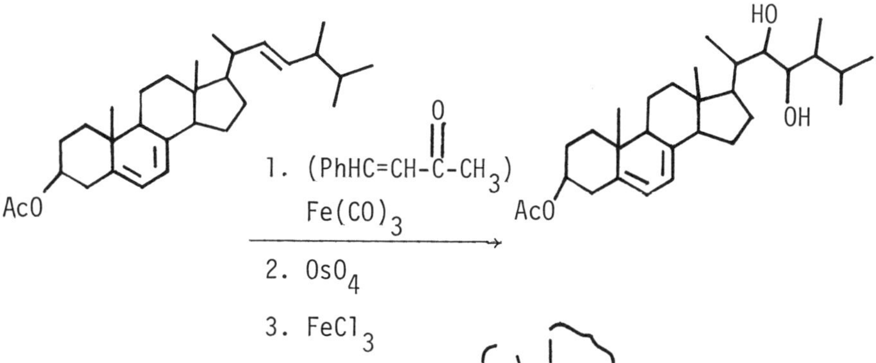

1. (PhHC=CH-C(=O)-CH$_3$) Fe(CO)$_3$
2. OsO$_4$
3. FeCl$_3$

Fe protects the diene system as [Fe(CO)$_3$ complex]

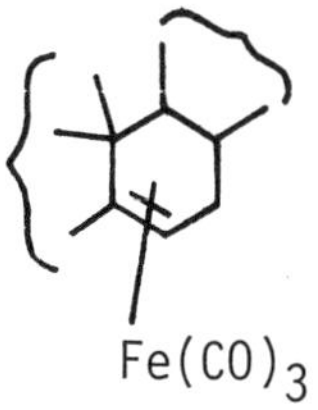

V.G-2 T. C. Jain, C. M. Banks, and J. E. McCloskey, Tetrahedron, 32, 765 (1976).

Reversible dimethylamine addition as a protecting reaction for α,β-unsaturated methylene groups of γ-lactones.

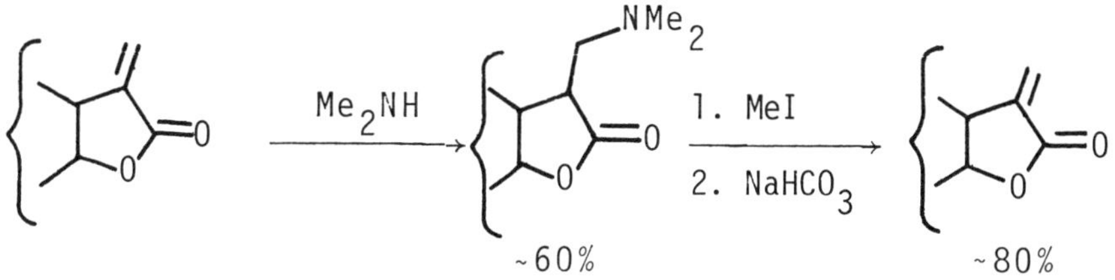

Me$_2$NH → ~60%
1. MeI
2. NaHCO$_3$ → ~80%

V.H-1   V. G. Mairanovsky, Angew. Chem. Int. Ed.,
15, 281 (1976).

Review: "Electro-Deprotection --
         Electrochemical Removal of
         Protecting Groups"

Protecting groups discussed:

Tosyl, benzoyl, trityl, benzhydryl, benzyl, phenyl, cinnamyl, benzyloxycarbonyl, and benzylidene.

VI.A.1-1   A. I. Meyers et al., J. Am. Chem. Soc., 98, 567 (1976).

$$R^1CH_2-\underset{N}{\overset{O}{\diagup}}\overset{Ph}{\underset{OCH_3}{\diagdown}}\overset{H}{\underset{H}{>}} \quad \xrightarrow[\text{2. } R^2X]{\text{1. n-BuLi or LDA}} \quad R^1R^2CH\text{-}COOH$$

3. $H_3O^{\oplus}$

30-85% yields
50-90% ee

Full paper, many examples.

VI.A.1-2   A. I. Meyers and K. Kamata, J. Am. Chem. Soc., 98, 2290 (1976).

$$\underset{H}{\overset{H}{>}}C=C\underset{\underset{Li\leftarrow O}{|}}{\overset{Ph}{\diagup}}\underset{CH_3}{\overset{O}{\diagdown}}\overset{H}{\underset{H}{>}} \quad \xrightarrow[\text{2. }H_3O^\oplus]{\text{1. }R^2\overset{R^1}{\diagdown}CHI \ (\pm)} \quad R^2\overset{H}{\underset{R^1}{-}}\overset{|}{C}\text{-}COOH$$

R^1R^2 = alkyl

25-89% yields
30-49% ee

VI.A.1-3   L. Cassar, M. Foa, and A. Gardano, J. Organometal. Chem., 121, C55 (1976).

$$R\text{-}X \quad \xrightarrow[\underline{n}\text{-}Bu_4NI, \ [Pd(PPh_3)_2Cl_2]]{PPh_3, \ NaOH, \ CO} \quad R\text{-}COOH$$

RX = "Aryl, benzyl, vinyl, and heterocyclic halides"

up to 83%
Few details.

VI.A.1-4   T. Mukaiyama and T. Tanaka, <u>Chem. Lett.</u>, 303 (1976).

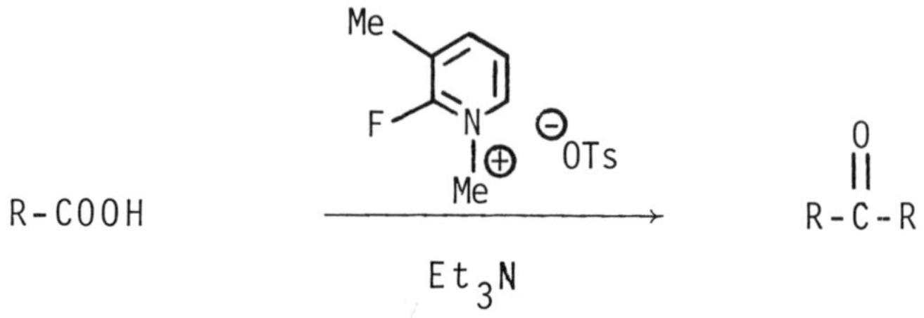

R-COOH → R-C(=O)-R

reagent: 2-fluoro-3-methyl-1-methylpyridinium tosylate; Et₃N

R = Ph, Bz, alkyl, cinnamyl, etc.

VI.A.2-1   Y. Yamamoto <u>et al.</u>, <u>J. Chem. Soc., Chem. Commun.</u>, 672 (1976).

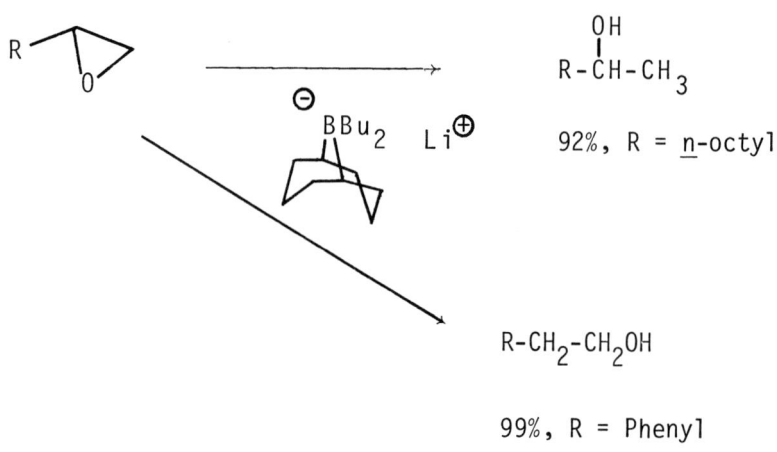

R-CH(OH)-CH₃   92%, R = n-octyl

R-CH₂-CH₂OH   99%, R = Phenyl

VI.A.2-2   B. Kokel, A. Lefebvre, and C. Lespagnol, <u>Comptes Rendus (C)</u>, <u>282</u>, 1125 (1976).

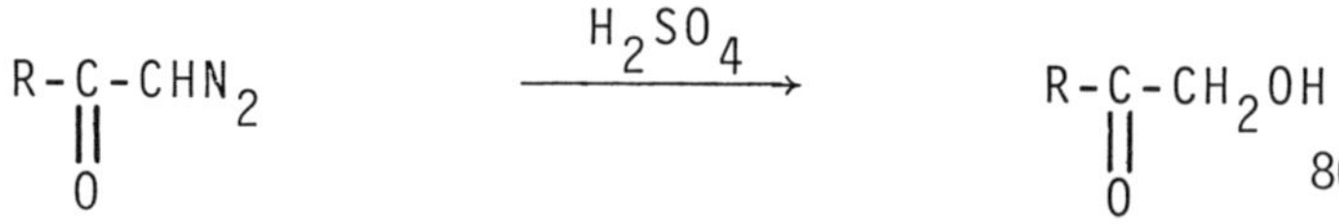

R-C(=O)-CHN₂ $\xrightarrow{H_2SO_4}$ R-C(=O)-CH₂OH   80-85%

Few details; TFA esters may also be obtained.

VI.A.2-3   T. Kametani et al., J. Org. Chem., 41, 2545 (1976).

$$\begin{array}{c} CH_3 \\ C=C \\ CH_3 O-Ar \end{array} \quad \overset{H}{}$$

or

$PhCH_2-O-Ar$

$\xrightarrow{NaAlH_2(OC_2H_4OCH_3)_2}$ Ar-OH

40-50%

VI.A.2-4   E. Negishi, S. Baba, and A. O. King, J. Chem. Soc., Chem. Commun., 17 (1976).

$R^1C \equiv CH \quad \xrightarrow[2.\ \underline{n}\text{-BuLi}]{1.\ \underline{i}\text{-Bu}_2AlH} \quad \left[\begin{array}{c} R H \\ C=C \\ H Al(\underline{i}\text{-Bu})_2(\underline{n}\text{-Bu}) \end{array} \right]^{\ominus} \quad Li^{\oplus}$

R^1 = $\underline{n}$-Bu, $\underline{t}$-Bu, cyclohexyl

$\begin{array}{c} R^1 H \\ C=C R^2 \\ H CH_2\overset{|}{C}HOH \end{array}$

57-82% (glc)

$\longleftarrow$

$\overset{R^2}{\underset{O}{\triangle}}$

R^2 = H, Me, Ph

VI.A.2-5   G. Deleris, J. Dunoguès, and R. Calas, Tetrahedron Lett., 2449 (1976).

$$R\text{-CHO} + R'\text{-CH=CH-CH}_2\text{-SiMe}_3 \xrightarrow[\text{2. H}_2\text{O}]{\text{1. AlCl}_3} R\text{-CHOH-CH(R')-CH=CH}_2$$

$$R\text{-CHO} + \text{Me}_3\text{Si-C}\equiv\text{C-SiMe}_3 \xrightarrow[\text{2. H}_2\text{O}]{\text{1. AlCl}_3} R\text{-CHOH-C}\equiv\text{C-SiMe}_3$$

VI.A.2-6   G. N. Barber and R. A. Olofson, Tetrahedron Lett., 3783 (1976).

[Reaction scheme: $R^1R^3C=CR^2R^4$ alkene reacts with $ClCH_2CH_2OCHCl_2$ / LiTMP to give a cyclopropane bearing R^1,R^2,R^3,R^4 and an OCH_2CH_2Cl substituent, 55-81%. Then 1. n-BuLi, 2. $H^{\oplus}$ gives the cyclopropane with an OH group.]

R's = Me, vinyl, OEt, cyclic

VI.A.2-7   J. F. Ruppert and J. D. White, J. Org. Chem., 41, 550 (1976).

$$R^1R^2C=O + R^3\text{-CH=CH-CH}_2\text{-Br} \xrightarrow[\text{2. H}_3\text{O}^{\oplus}]{\text{1. Zn}} R^1R^2C(OH)\text{-CH}_2\text{-CH=CH-}R^3$$

R^1, R^2 = H, alkyl, aryl, cyclic; R^3 = H, alkyl   74-97% (isolated)

USEFUL SYNTHETIC PREPARATIONS

VI.A.2-8   A. Debal, T. Cuvigny, and M. Larchevêque, Synthesis, 391 (1976).

$$\underset{R^2}{\overset{R^1}{>}}CHCN \xrightarrow[2.\ Cl(CH_2)_nOR^3]{1.\ LiNEt_2} R^1-\underset{R^2}{\overset{CN}{\underset{|}{C}}}-(CH_2)_nOR^3 \xrightarrow[2.\ H_3O^{\oplus}]{1.\ Na,\ HMPT,\ \underline{t}-BuOH} $$

$R^1, R^2 = 1°$ alkyl R^3 = TMS or THP

$$\underset{R^2}{\overset{R^1}{>}}CH(CH_2)_nOH$$

~50-80%

VI.A.2-9   A. B. Levy and S. J. Schwartz, Tetrahedron Lett., 2201 (1976).

$$R_3B + \underset{OMe}{\overset{}{Li-C=CH_2}} \xrightarrow[3.\ H_2O_2,\ NaOH]{\substack{1.\ -80°\ to\ r.t. \\ 2.\ 2N\ HCl}} R_2\overset{OH}{\underset{|}{C}}-CH_3$$

R = alkyl, cyclic   87-100% (VPC)

VI.A.3-1   A. G. Anderson, Jr. et al., Synthesis, 398 (1976).

$$ROH \xrightarrow[ZnCl_2]{PCl_3,\ DMF} R-Cl$$

R = 1°, 2° alkyl

generally 60-87%;
poor yields with hindered alcohols

VI.A.3-2   R. M. Carman and I. M. Shaw, Aust. J. Chem., 29, 133 (1976).

$$R-OH + PCl_5 \xrightarrow[CHCl_3]{CaCO_3} R-Cl$$

R = 3° alkyl   Retention of configuration. "Good" yields.

VI.A.3-3   D. R. Hepburn and H. R. Hudson, J. Chem. Soc. Perkin I, 754 (1976).

$$R-\underset{OH}{\overset{H}{\underset{*}{C}}}-R' \xrightarrow[CH_3CN]{[Me_2NCHX]^{+-}X} R-\underset{X}{\overset{H}{\underset{*}{C}}}-R'$$

R,R' = alkyl   X = Cl or Br   Chemical yields ~70-90%
~75% ee (retention)

VI.A.3-4   R. O. Hutchins, D. Masilamani, and C. A. Maryanoff, J. Org. Chem., 41, 1071 (1976).

Use of PBr$_3$ at low temperature (4°C) and long reaction times (1-2 days) to convert optically active alcohols to the corresponding bromides. Yields 60-75%, with 84-94% optical purity (on 1-phenylethanol, 2-octanol, and 2-pentanol).

USEFUL SYNTHETIC PREPARATIONS

VI.A.3-5   P. E. Sonnet, Synth. Commun., 6, 21 (1976).

$$R\text{-}O\text{-}THP \xrightarrow[X = Cl, Br]{PPh_3X_2} RX$$

R = 1° alkyl   80%

VI.A.3-6   P. Place, M.-L. Roumestant, and J. Gore, Bull. Soc. Chim. France, 169 (1976).

$$R\text{-}OTs \xrightarrow[MgX_2]{Et_2O} R\text{-}X$$

R = 1°, 2° alkyl   27-98%
     X = Br, I

VI.A.3-7   H. C. Brown et al., J. Am. Chem. Soc., 98, 1290 (1976).

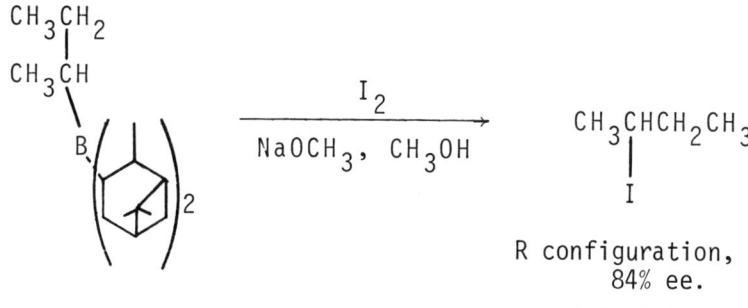

$$\xrightarrow[NaOCH_3,\ CH_3OH]{I_2}$$

CH$_3$CHCH$_2$CH$_3$
         |
         I

R configuration, 84% ee.
49% yield

329

VI.A.3-8 M. Zupan and A. Pollak, *J. Org. Chem.*, **41**, 4002 (1976).

$$\underset{Ph}{\overset{Ph}{}}C=C\underset{R}{\overset{H}{}} \xrightarrow[CH_2Cl_2/CF_3COOH]{XeF_2} Ph_2C-\underset{F}{\overset{H}{\underset{|}{\overset{|}{C}}}}-R$$
$$ F$$

R = H, CH$_3$ \qquad\qquad\qquad\qquad 95%

VI.A.3-9 N. R. DeLue and H. C. Brown, *Synthesis*, 114 (1976).

$$H_2C=CH(CH_2)_8COOCH_3 \xrightarrow[2.\ I_2,\ NaOCH_3]{1.\ B_6H_6,\ THF} I-CH_2(CH_2)_9-COOCH_3$$

68% isolated

cyclohexene $\xrightarrow{\text{same conditions}}$ iodocyclohexane

72% (VPC)

VI.A.3-10 J. Kollonitsch, S. Marburg, and L. M. Perkins, *J. Org. Chem.*, **41**, 3107 (1976).

$$-\underset{SH}{\overset{|}{C}}-\underset{NH_2}{\overset{|}{C}}-CO_2H \xrightarrow[\text{liquid HF}]{CF_3OF} -\underset{F}{\overset{|}{C}}-\underset{NH_2}{\overset{|}{C}}-CO_2H$$

Widely varying yields.

USEFUL SYNTHETIC PREPARATIONS

VI.A.3-11 M. P. Doyle and B. Sigfried, J. Chem. Soc., Chem. Commun., 433 (1976).

$$RCH_2NH_2 \xrightarrow[CuX_2]{R'ONO} RCHX_2$$

R = unbranched alkyl
R' = isopentyl or $\underline{t}$-butyl
X = Cl, Br

38-79%

VI.A.3-12 M. P. Doyle, B. Siegfried, and J. J. Hammond, J. Am. Chem. Soc., 98, 1627 (1976).

$$RCH_2NH_2 \xrightarrow[2.\ (CuX_2 \cdot NO)_2,\ CH_3CN]{1.\ CuX_2} RCHX_2$$

R = Ph, Bz, alkyl, etc.
X = Cl, Br

14-58%

VI.A.3-13 G. Cainelli and F. Manescalchi, Synthesis, 472 (1976).

$$R-X \xrightarrow{Resin-\overset{\oplus}{N}Me_3\overset{\ominus}{F}} R-F$$

X = Br, OTs, OMes

R = 1°,2° alkyl; may contain carbonyl

60-90%

Reaction fails when R = cyclohexyl

VI.A.3-14   J. A. Miller and M. J. Nunn, J. Chem. Soc. Perkin I, 416 (1976).

$$\text{R-Cl} + \text{NaI} \xrightarrow[\text{ZnCl}_2]{\text{CS}_2} \text{R-I}$$

R = 3° alkyl, generally very hindered   >90%

VI.A.3-15   W. E. Willy, D. R. McKean, and B. A. Garcia, Bull. Chem. Soc. Japan, 49, 1989 (1976).

$$\text{R-Cl} \xrightarrow[\text{N-methylpyrrolidinone}]{\text{NaBr, EtBr}} \text{R-Br}$$

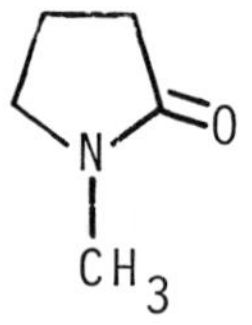

R = primary alkyl   70-93%

VI.A.3-16   A. L. Fridman and N. A. Kolobov, J. Org. Chem. (USSR), 12, 228 (1976).

$$\underset{\text{R = alkyl, aryl}}{\text{R-CO-CHN}_2} \xrightarrow{\text{X-I} \atop (\text{X = Cl, Br})} \text{R-CO-CHIX}$$

~70%

USEFUL SYNTHETIC PREPARATIONS

VI.A.3-17 P. E. Sonnet and J. E. Oliver, J. Org. Chem., 41, 3279 (1976).

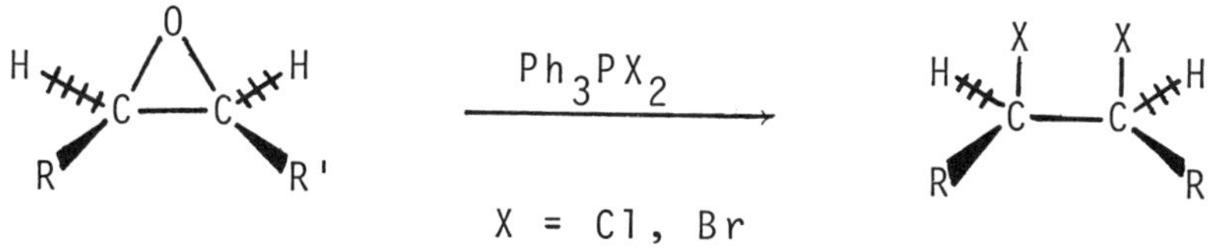

$$\text{epoxide (R,R'=alkyl)} \xrightarrow{Ph_3PX_2,\ X=Cl,Br} \text{vicinal dihalide}$$

yields ~60-90%

VI.A.3-18 D.H.R. Barton et al., J. Am. Chem. Soc., 98, 3036 (1976).

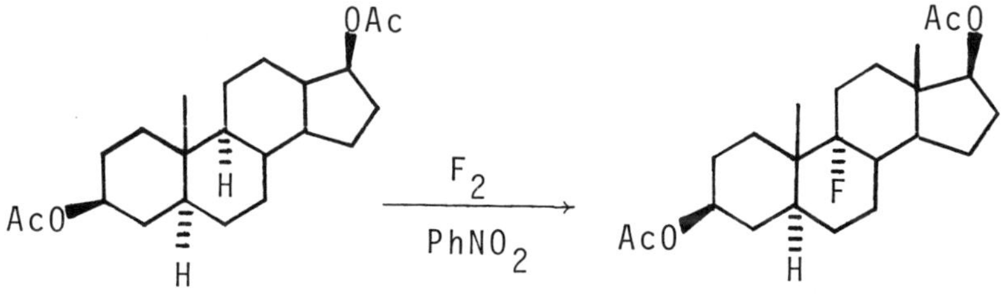

$$\text{steroid enol acetate} \xrightarrow[PhNO_2]{F_2} \text{fluorinated steroid}$$

50%

VI.A.3-19 J. Pfab, Tetrahedron Lett., 943 (1976).

$$R^1R^2CClNO \xrightarrow{Br_2,\ h\nu} R^1R^2CClBr$$

R^1, R^2 = Me, Et, t-Bu, cyclic

75-92%

VI.A.3-20 F. de Reinach-Hirtzbach and T. Durst, Tetrahedron Lett., 3677 (1976).

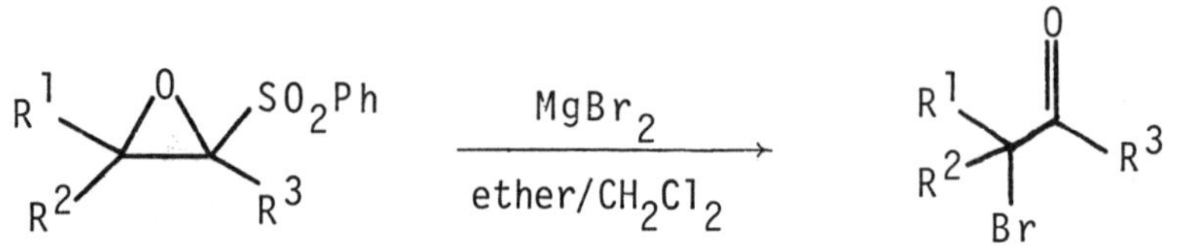

R^1, R^2 = H, Me, Ph, cyclic

R^3 = H, Me, cyclic

80-95%

VI.A.3-21 P. Wieland, Helv. Chim. Acta, 59, 1027 (1976).

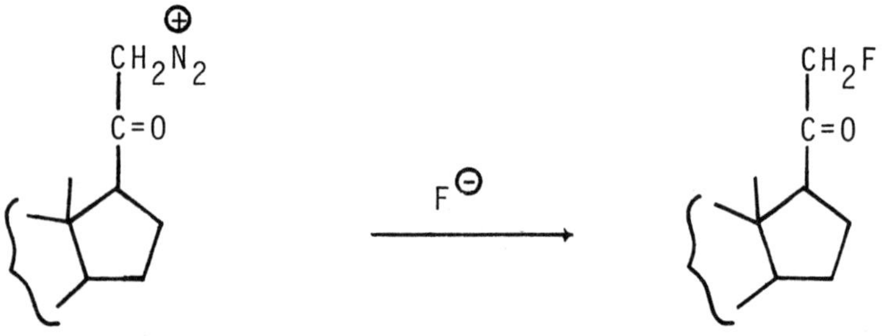

60%

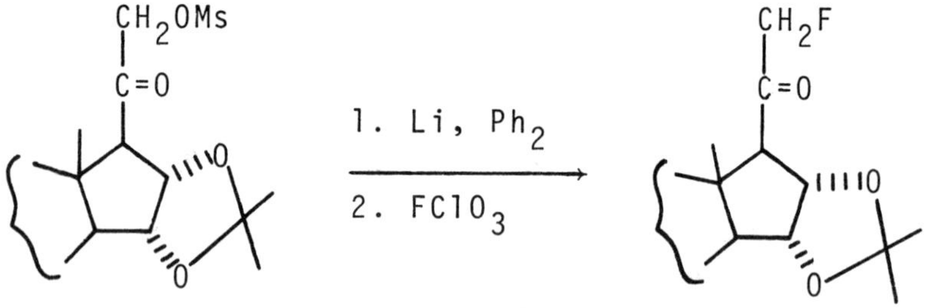

~70%

VI.A.3-22 V. E. Platonov and G. G. Yakobson, Synthesis, 374 (1976).

Review: "The Application of Thermolytic Reactions for the Syntheses of Fluoro-Organic Compounds"

VI.A.3-23 G. G. Yakobson and V. M. Vlasov, Synthesis, 652 (1976).

Review: "Recent Synthetic Methods for Polyfluoroaromatic Compounds"

VI.A.4-1 J. H. Hall and M. Gisler, J. Org. Chem., 41, 3769 (1976).

$$R-C\equiv N \xrightarrow{KOH,\ t\text{-BuOH}} R-\overset{O}{\underset{}{C}}-NH_2$$

R = Bz, Et, Ph, etc.   ~80-90%

VI.A.4-2 T. Mukaiyama, Y. Aikawa, and S. Kobayashi, Chem. Lett., 57 (1976).

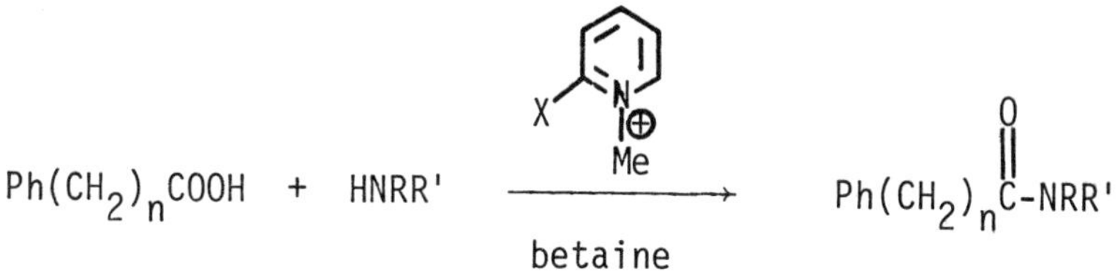

$$Ph(CH_2)_n COOH + HNRR' \xrightarrow[\text{betaine}]{} Ph(CH_2)_n \overset{O}{\underset{}{C}}-NRR'$$

n = 0,1,2   X = F, Cl
R = H, Bu
R' = Bu, t-Bu, Ph, Bz

VI.A.4-3 K. Takeuchi, M. Nojima, and N. Tokura,
J. Chem. Soc. Perkin I, 2205 (1976).

$$ROH \xrightarrow[CH_3CN]{SO_2Cl_2} RNHCCH_3 \;(\text{with } C=O)$$

R = n-hexyl, bornyl, norbornyl, etc.

Widely varying yields

VI.A.4-4 H. Schuttenberg and R. C. Schulz,
Angew. Chem. Int. Ed., 15, 777 (1976).

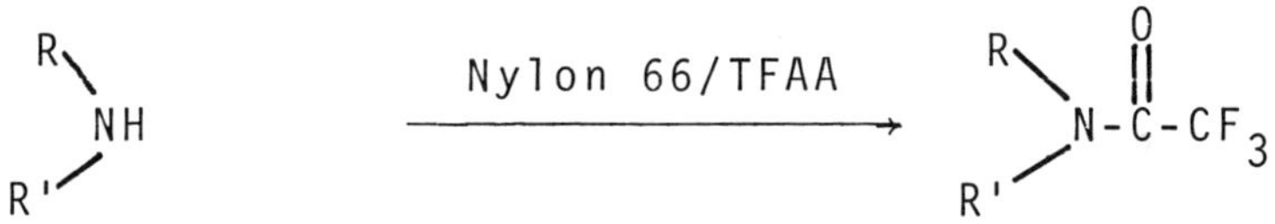

$$R_2NH \xrightarrow{\text{Nylon 66/TFAA}} R_2N-C(=O)-CF_3$$

aryl amines, piperidine, hexylamine, hexamethylene diamine

~90% except for p-nitroaniline, 40%

VI.A.4-5 N. Ishikawa and S. Shin-Ya, Chem. Lett., 673 (1976).

$$R-NH_2 \xrightarrow[TMED]{Ph-C(=O)-CF(CF_3)_2} R-NH-C(=O)-Ph$$

R = Bu, Ph

99%

VI.A.4-6   U. Schöllkopf and H. Beckhaus, <u>Angew. Chem. Int. Ed.</u>, <u>15</u>, 293 (1976).

$$\text{Li-}\overset{\text{O}}{\underset{\|}{\text{C}}}\text{-N(CH}_2\text{OMe)}_2 \;+\; \underset{R^2}{\overset{R^1}{\diagdown}}\!\!\!\!C=O \;\longrightarrow\; \xrightarrow{H_3O^{\oplus}} \; R^1\text{-}\underset{R^2}{\overset{HO}{\underset{|}{\text{C}}}}\text{-}\overset{O}{\underset{\|}{\text{C}}}\text{-NH}_2$$

R^1, R^2 = H, alkyl, aryl   44-88%

VI.A.4-7   J. Auerbach, M. Zamore, and S. M. Weinreb, <u>J. Org. Chem.</u>, <u>41</u>, 725 (1976).

$$R\text{-}\overset{O}{\underset{\|}{\text{C}}}\text{-NH}_2 \;+\; HCHO$$

$$\updownarrow$$

$$R\text{-}\overset{O}{\underset{\|}{\text{C}}}\text{-NHCH}_2\text{OH} \quad \xrightarrow[\text{H}_2\text{-Pd/C-TFA}]{\text{Et}_2\text{SiH-TFA} \atop \text{or}} \quad R\text{-}\overset{O}{\underset{\|}{\text{C}}}\text{-NHCH}_3$$

Yields generally 80-90%

VI.A.4-8   S. R. Jones and J. M. Mellor, <u>Synthesis</u>, 32 (1976).

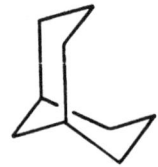
$\xrightarrow[\text{2. } CH_3CN/H_3O^{\oplus}]{\text{1. } Pb(OAc)_4, CF_3COOH}$
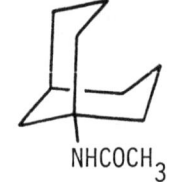

NHCOCH$_3$

78%

adamantane $\xrightarrow[\text{2. Reagent}]{\text{1. } Pb(OAc)_4, CF_3COOH}$ [adamantyl]-X

Reagent	X	Yield (%)
KCN	-NHCOH	82
CH$_3$CN	-NHCOCH$_3$	85
phenol	<u>p</u>-OHC$_6$H$_4$	81
ethyl acetoacetate	-CHCOOCH$_3$ | COOEt	87
<u>n</u>-C$_4$H$_9$SH	-S-C$_4$H$_9$	97

VI.A.5-1   T.-L. Ho, M. Henninger and G. A. Olah, <u>Synthesis</u>, 815 (1976).

$$Ar-N_3 \xrightarrow[H_2O/THF]{VCl_2} Ar-NH_2$$

~70-90%

VI.A.5-2   Y. Watanabe *et al.*, *Bull. Chem. Soc. Japan*, **49**, 1378 (1976).

$$\begin{array}{c} R \\ \diagdown \\ R' \diagup \end{array} N-H \quad \xrightarrow[CO]{KHFe(CO)_4, \; HCHO} \quad \begin{array}{c} R \\ \diagdown \\ R' \diagup \end{array} N-CH_3$$

R, R' = alkyl, aryl, H   Widely varying yields.

VI.A.5-3   K. B. Sharpless *et al.*, *J. Am. Chem. Soc.*, **98**, 269 (1976). See also *J. Org. Chem.*, **41**, 176 (1976).

$$\left\{ \begin{array}{c} =\!\!\!\diagdown \\ \diagup CH \\ | \end{array} \right\} \quad \xrightarrow{Se(NR)_2} \quad \left\{ \begin{array}{c} =\!\!\!\diagdown \\ \diagup C-NHR \\ | \end{array} \right\}$$

R = $\underline{t}$-Bu, Ts

Isolated yields generally 40-60%

29 examples cited

VI.A.5-4   D. Enders *et al.*, *Synthesis*, 548 (1976).

$$\begin{array}{c} R^1 \diagdown \diagup R^2 \\ N \\ | \\ NO \end{array} \quad \xrightarrow[MeOH, \; 200°]{H_2/Ra-Ni} \quad \begin{array}{c} R^1 \diagdown \diagup R^2 \\ N \\ | \\ H \end{array}$$

R^1, R^2 = Ph, 1°, 2°, 3° alkyl   70-95%

VI.A.5-5   F.C.M. Chen and N. L. Benoiton, Can. J. Chem., 54, 3310 (1976).

$$\text{R-NH}_2 \text{ or } R_2\text{NH} \xrightarrow[\text{MeOH}]{\text{CH}_3\text{I, NaHCO}_3} \overset{\oplus}{\text{RNMe}_3} \text{I}^\ominus \text{ or } R_2\overset{\oplus}{\text{NMe}}_2 \text{I}^\ominus$$

75-96%

VI.A.5-6   S. Coulton, G. A. Moore, and R. Ramage, Tetrahedron Lett., 4005 (1976).

$$\text{RNH}_2 \xrightarrow[\text{Et}_3\text{N/CH}_2\text{Cl}_2]{\text{Ph}_2\text{POCl}} \text{Ph}_2\overset{\text{O}}{\overset{\|}{\text{P}}}\text{-NHR} \xrightarrow[\text{2. MeI}]{\text{1. NaH, THF}} \text{Ph}_2\overset{\text{O}}{\overset{\|}{\text{P}}}\text{-NMeR}$$

R = Bz, PhCH$_2$CH$_2$, n-Hex, cyclohexyl

$$\downarrow \begin{array}{c} \text{TsOH} \\ \text{MeOH} \end{array}$$

$$\text{RNH}_2\overset{\oplus}{\text{Me}} \quad \overset{\ominus}{\text{OTs}}$$

Overall yields ~50%

VI.A.5-7   H. Berbalk, K. Eichinger, and E. Decker, Monat. für Chem., 107, 401 (1976).

2-chlorocyclopentanone + ArNHR $\xrightarrow[\text{K}_2\text{CO}_3]{80°, \text{dioxane}}$ 2-(N-Ar-N-R-amino)cyclopentanone

R = H, Me
Ar = Cl, CF$_3$, C$_6$H$_4$OCF$_3$

~50%

USEFUL SYNTHETIC PREPARATIONS

VI.A.5-8   S. Yamada, N. Ikota, and K. Achiwa, *Tetrahedron Lett.*, 997, 1001 (1976).

$$R^1\text{-}\underset{\underset{}{}}{\overset{\overset{O}{\|}}{C}}\text{-}CH_3$$
+
$$R^2\text{-}\underset{NH_2}{CH}\text{-}COOR^3$$

pure (L)

R^1 = Ph, Bz

1. $-H_2O$
2. H_2, Pd/C
3. $\underline{t}$-BuOCl
4. Base, H_2O

$$R^1\text{-}\overset{*}{\underset{NH_2}{C}}HCH_3$$

Predominantly (S)
Chemical yield up to 63%
ee up to 81%

VI.A.5-9   L. E. Overman, *J. Am. Chem. Soc.*, **98**, 2901 (1976).

$$R^1R^2C=CR\text{-}\underset{OH}{CR^3R^4}$$

1. CCl_3CN
2. Thermolysis
3. $NaOH/H_2O$

$$R^1R^2\underset{NH_2}{C}\text{-}CR=CR^3R^4$$

R's = H, alkyl, aryl

23-83% overall

Full paper, many examples cited.

VI.A.5-10   J. C. Arnould, J. Cossy and J. P. Pète, *Tetrahedron Lett.*, 3919 (1976).

R = Ts, α,β-naphthyl
R'= Et, Bz, $\underline{i}$-Pr, allyl

$\xrightarrow{h\nu}$
benzene

30-70%

VI.A.5-11 R. Chaabouni, A. Laurent, and B. Marquet, Tetrahedron Lett., 757 (1976).

[Reaction: 2-phenyl-2-methyl-vinyl aziridine; R = H, CH$_3$]

1. B$_2$H$_6$
2. H$_2$O$_2$, $^\ominus$OH

→ product (75-79%)

[Reaction: 1-vinyl-bicyclic aziridine]

1. B$_2$H$_6$
2. H$_2$O$_2$, $^\ominus$OH

→ product (51%)

VI.A.5-12 E. Schmitz, Russ. Chem. Rev., **45**, 16 (1976).

Review: "Electrophilic Amination"

VI.A.6-1 M. Ali, N. H. Khan, and A. A. Siddiqui, Synth. Commun., **6**, 227 (1976).

[Oxazolone starting material, Ph at C-2]

$\xrightarrow{\text{H}_2,\ \text{Pd/C}}_{\text{EtOH, NH}_3}$ RR'CH–CH–CONH$_2$
                                                                    |
                                                                 NHCOPh

65-98%

R, R' = alkyl, aryl

USEFUL SYNTHETIC PREPARATIONS

VI.A.6-2 I. Tabushi, Y. Yabushita, and T Nakajima, <u>Tetrahedron Lett.</u>, 4343 (1976).

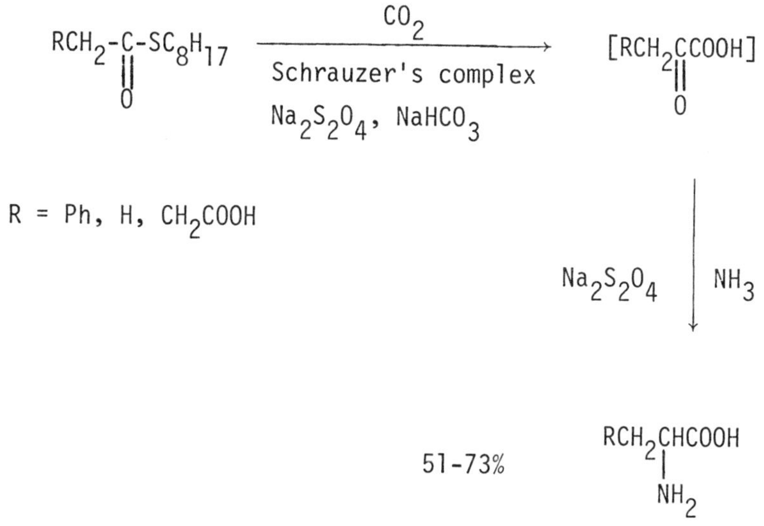

$$RCH_2-\underset{O}{\underset{\|}{C}}-SC_8H_{17} \xrightarrow[Na_2S_2O_4,\ NaHCO_3]{CO_2,\ Schrauzer's\ complex} [RCH_2\underset{O}{\underset{\|}{C}}COOH]$$

R = Ph, H, CH$_2$COOH

$$\downarrow Na_2S_2O_4 \mid NH_3$$

51-73% $RCH_2\underset{NH_2}{\underset{|}{C}HCOOH}$

VI.A.6-3 D. Ben-Ishai, I. Sataty, and Z. Bernstein, <u>Tetrahedron</u>, <u>32</u>, 1571 (1976).

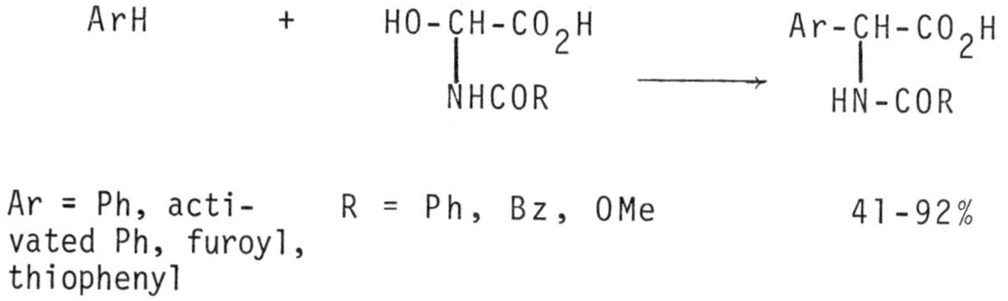

$$ArH\ +\ \underset{NHCOR}{\underset{|}{HO-CH-CO_2H}} \longrightarrow \underset{HN-COR}{\underset{|}{Ar-CH-CO_2H}}$$

Ar = Ph, activated Ph, furoyl, thiophenyl   R = Ph, Bz, OMe   41-92%

VI.A.6-4 S. Yamada, T Oguri, and T. Shioiri, J. Chem. Soc., Chem. Commun., 136 (1976).

$NH_2CH_2CO_2Bu^t$ $\xrightarrow{\underline{I}}$ [pinanone-derived imine with $CH_2CO_2Bu^t$ and O-H···N hydrogen bond]

$\underline{I}$ = [pinanone with OH and =O]

1. $LiNPr^i_2$
2. RX
3. $H_3O^{\oplus}$

↓

$\underline{I}$ + $H_2NCHCO_2Bu^t$
          |
          R

(predominantly D isomer)

RX	Overall Yield	Optical Yield
MeI	52	83
Bu^iI	50	83
$PhCH_2Br$	79	72

VI.A.6-5 N. Takaishi et al., J. Am. Chem. Soc., **98**, 5400 (1976).

$RCH=C\begin{smallmatrix}COOH\\R'\end{smallmatrix}$ $\xrightarrow[\text{catalyst}]{H_2}$ $RCH_2-CH\begin{smallmatrix}CO_2H\\R'\end{smallmatrix}$

R = H, Ph
R' = Ph, $NHCOCH_3$

Yields ~100% (NMR)
52-86% ee

catalyst = polymer-supported Rh-DIOP

VI.A.6-6 H. Poisel and U. Schmidt, Angew. Chem. Int. Ed., 15, 294 (1976).

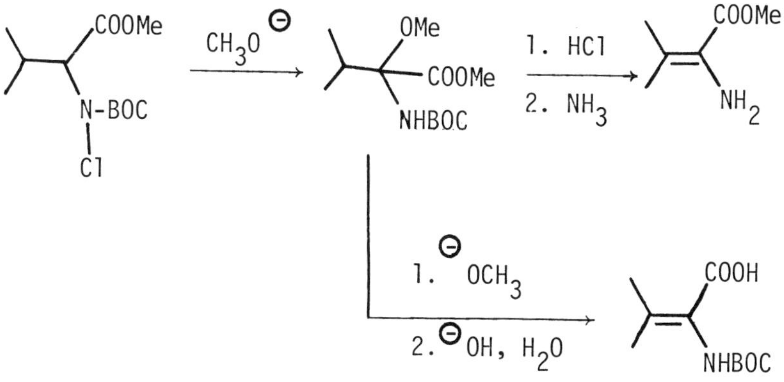

$$\underset{\substack{\text{N-BOC}\\|\\\text{Cl}}}{\overset{\text{COOMe}}{\bigvee}} \xrightarrow{CH_3O^{\ominus}} \underset{\text{NHBOC}}{\overset{\text{OMe}}{\bigvee}\text{COOMe}} \xrightarrow[2.\ NH_3]{1.\ HCl} \underset{NH_2}{\overset{\text{COOMe}}{\bigvee}}$$

$$\xrightarrow[2.\ ^{\ominus}OH,\ H_2O]{1.\ ^{\ominus}OCH_3} \underset{\text{NHBOC}}{\overset{\text{COOH}}{\bigvee}}$$

yields ~60% overall

VI.A.7-1 G. W. Griffin, R. L. Smith, and A. Manmade, J. Org. Chem., 41, 338 (1976).

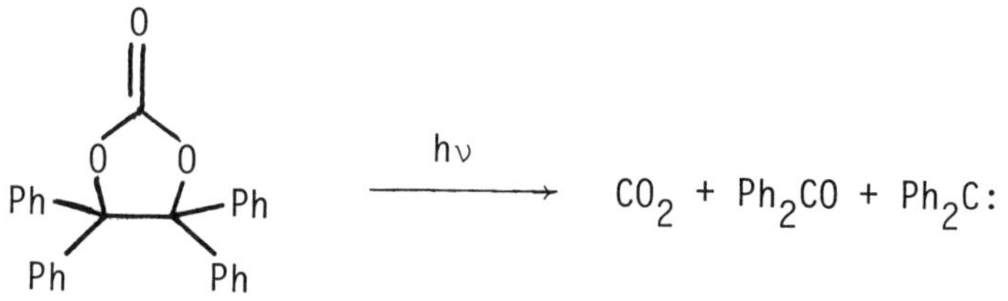

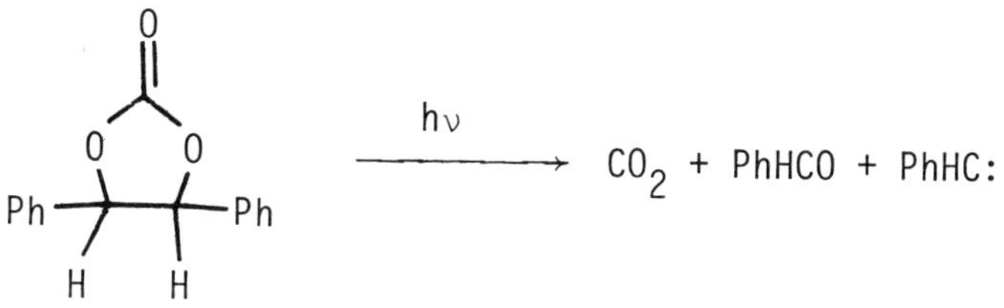

Ph,Ph / Ph,Ph dioxolanone $\xrightarrow{h\nu}$ CO_2 + Ph_2CO + Ph_2C:

Ph,H / Ph,H dioxolanone $\xrightarrow{h\nu}$ CO_2 + PhHCO + PhHC:

VI.A.7-2   M. Sekiya *et al.*, Chem. Pharm. Bull., **24**, 369 (1976).

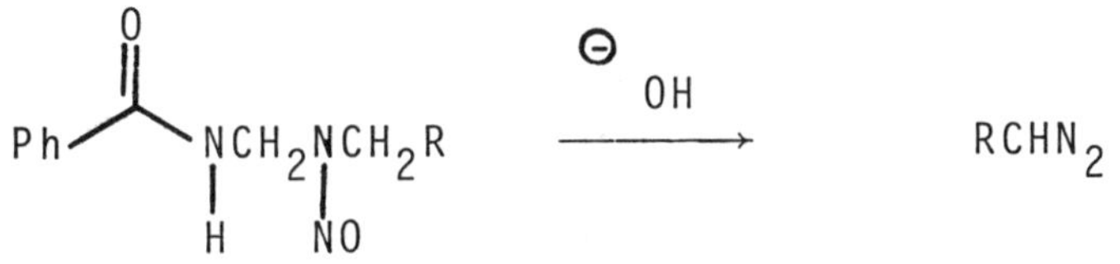

$$Ph-C(=O)-N(H)-CH_2-N(NO)-CH_2-R \xrightarrow{{}^{\ominus}OH} RCHN_2$$

R = Me, Et, *n*-Bu   75%, 65%, 44%, respectively

VI.A.7-3   R. A. Olofson, K. D. Lotts, and G. N. Barber, Tetrahedron Lett., 3779 (1976).

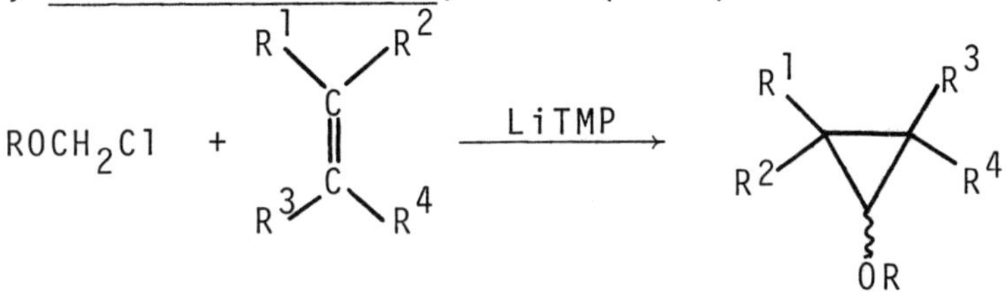

$$ROCH_2Cl + \underset{R^3R^4}{\overset{R^1R^2}{C=C}} \xrightarrow{LiTMP} \text{cyclopropane}(R^1, R^2, R^3, R^4, OR)$$

R = Me, Et, *t*-Bu, *i*-Pr   ~40-60%
$R^1, \ldots, R^4$ = Me, vinyl, cyclic, etc.

VI.A.7-4   T. L. Gilchrist and D. J. Pearson, J. Chem. Soc. Perkin I, 1257 (1976).

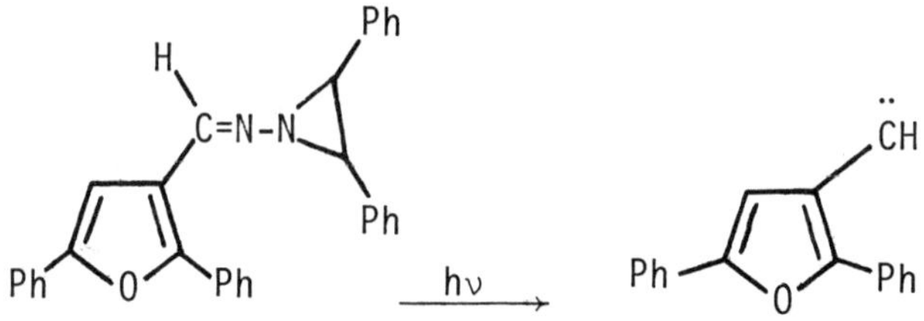

[aziridine with C=N-N linkage to furan(Ph,Ph), Ph substituents] $\xrightarrow{h\nu}$ [furan(Ph,Ph)-CH:]

USEFUL SYNTHETIC PREPARATIONS

VI.A.7-5 M. Makosza, <u>Pure and Appl. Chem.</u>, <u>43</u>, 439 (1975).

Review: "Two-phase reactions in the chemistry of carbanions and halocarbenes. A useful tool in organic synthesis."

VI.A.7-6 C. Wentrup, <u>Fortschritte der Chem. Forschung</u>, <u>62</u>, 173 (1976).

Review: "Rearrangements and Interconversions of Carbenes and Nitrenes"

VI.A.8-1 K. Nyberg, <u>Synthesis</u>, 545 (1976).

$$\text{(cyclic N-CH(OCH}_3\text{)-C(=O)-R)} \xrightarrow{NH_4Br} \text{(cyclic N-CH=...-C(=O)-R)}$$

~70-90%

VI.A.8-2 H. Bohme and K.-H. Weisel, <u>Chem. Ber.</u>, <u>109</u>, 2908 (1976).

$$\underset{\substack{\| \; | \\ O \; Cl}}{Me-C-CH-CN} \xrightarrow{ArNH_2} \underset{\substack{/ \; | \\ ArNH \; Cl}}{Me-C=C-CN}$$

~50-70%

VI.A.8-3   E. Elkik and M. Imbeaux-Oudotte,
Comptes Rendus (C), 282, 935 (1976).

$$R-\overset{O}{\underset{\|}{C}}-CHCl-CHO \ + \ NR'_2 \ \longrightarrow \ R-\overset{O}{\underset{\|}{C}}-CCl=CHNR'_2$$

R = Ph, OEt R'_2 = Me_2,   35-48%
              Et_2, $(CH_2)_5$

VI.A.8-4   J. Ficini, Tetrahedron, 32, 1449 (1976).

Review: "Ynamine: a versatile tool in organic synthesis."

VI.A.9-1   Y. Tamura et al., Synthesis, 35 (1976).

$$Ph-\overset{NTs}{\underset{\|}{S}}-CH_2R^1 \quad \xrightarrow[\substack{2. \ R^2\\ \diagdown \\ \quad C=O \\ R^3 \diagup}]{1. \ NaH/DMSO} \quad \underset{R^3}{\overset{R^2}{\diagup}}\!\!\!\triangle\!\!\!\underset{R^1}{\diagdown}O$$

54-65%

R^1 = H, CH_3
R^2, R^3 = H, alkyl, aryl, cyclic

USEFUL SYNTHETIC PREPARATIONS

VI.A.9-2   D. Van Ende and A. Krief, Tetrahedron Lett., 457 (1976).

$$R^1R^2C=O \xrightarrow[\begin{array}{c}3.\ -\overset{O}{\overset{\|}{C}}-\overset{|}{C}=C\diagdown\end{array}]{\begin{array}{c}1.\ CH_3SeH\\ 2.\ \underline{n}\text{-BuLi}\end{array}} CH_3Se-\underset{R^2}{\overset{R^1}{\underset{|}{\overset{|}{C}}}}-\underset{OH}{\overset{|}{C}}-\overset{|}{C}=C\diagdown$$

$$\downarrow \begin{array}{c}1.\ CH_3I\\ 2.\ \underline{t}\text{-BuOK, DMSO}\end{array}$$

$$\underset{R^2}{\overset{R^1}{\diagdown}}\!\!\underset{O}{\overset{C-C}{\triangle}}\!\!\diagdown C=C\diagdown$$

Yields up to 80% (isolated)

VI.A.10-1   J. Fairhurst and D. C. Horwell, Synth. Commun., 6, 89 (1976).

$$R\text{-COOH} \xrightarrow{Et_2SO_4,\ DBN} R\text{-COOEt}$$

R = β-keto, β-cyano alkyl, etc.   76-95%

Particularly useful for thermally unstable acids.

VI.A.10-2   H. Normant and C. Laurenco, Comptes Rendus (C), 283, 483 (1976).

$$R\text{-COOK} + R'X \xrightarrow{CH_3CN,\ TMEDA} R\text{-COOR}'$$

R = hindered alkyl   R'X = BzCl, n-BuBr   ~80-90%

VI.A.10-3   T. Mukaiyama, H. Toda, and S. Kobayashi, Chem. Lett., 13 (1976).

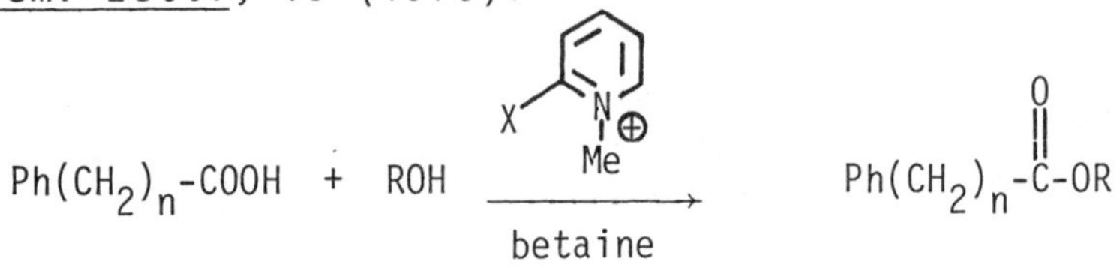

$$Ph(CH_2)_n\text{-COOH} + ROH \xrightarrow[\text{betaine}]{} Ph(CH_2)_n\text{-C(=O)-OR}$$

betaine = 2-methyl-N-substituted pyridinium (X = F, Cl)

n = 1, 2   X = F, Cl

R = Ph, alkyl, Bz, cinnamyl, crotyl

VI.A.10-4   U. Weber, Z. Naturforsch (B), 31, 1157 (1976).

$$R\text{-COOH} \xrightarrow[\text{BuOH, pyridine}]{R'OH,\ DCC} R\text{-COOR'}$$

(N-protected amino acid)   R' = Me, Bz   ~80-90%

VI.A.10-5   S. Takimoto et al., Bull. Chem. Soc. Japan, 49, 2335 (1976).

$$R\text{-C(=O)-Cl} + R'\text{-OH} \xrightarrow{AgCN} R\text{-C(=O)-OR'}$$

R = hindered alkyl or aryl   R' = 1°, 2°, or 3° alcohol   57-100% (VPC)

USEFUL SYNTHETIC PREPARATIONS

VI.A.10-6   B. M. Trost et al., Tetrahedron Lett., 3477 (1976).

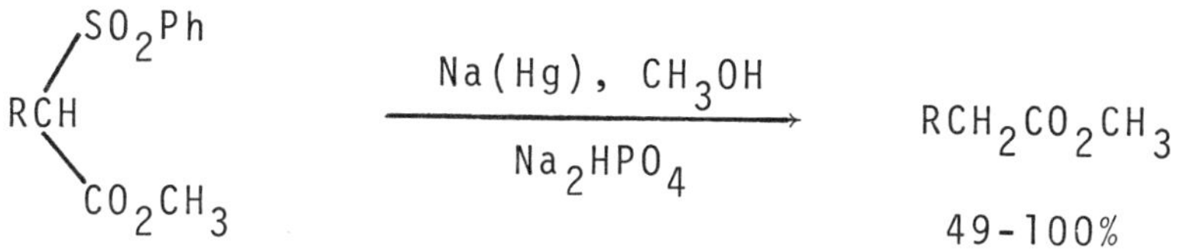

$$\underset{CO_2CH_3}{\overset{SO_2Ph}{RCH}} \xrightarrow[Na_2HPO_4]{Na(Hg),\ CH_3OH} RCH_2CO_2CH_3$$

49-100%

R = complicated alkyl

VI.A.10-7   K. R. Fountain and M. Pierschbacher, J. Org. Chem., 41, 2039 (1976).

A phosgeneless synthesis of diaryl carbonates, e.g.:

$$Ph-O-CO_2CHCl_2 + PhO^{\ominus} \xrightarrow{NaH} Ph-OCOOPh$$

35-70%

VI.A.10-8   T. Ogawa and M. Matsui, J. Am. Chem. Soc., 98, 1629 (1976).

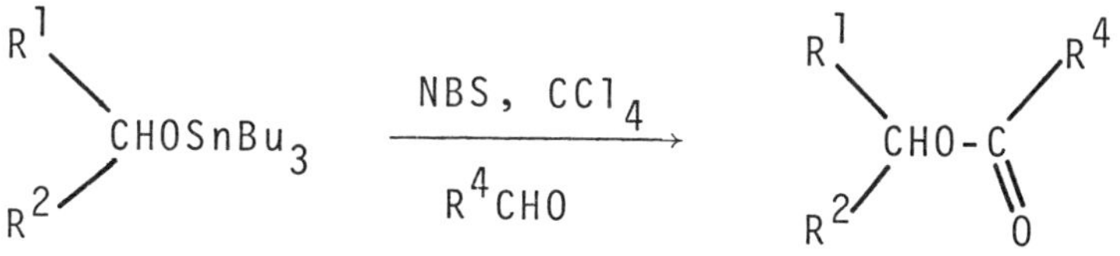

$$\underset{R^2}{\overset{R^1}{\diagdown}}CHOSnBu_3 \xrightarrow[R^4CHO]{NBS,\ CCl_4} \underset{R^2}{\overset{R^1}{\diagdown}}CHO-\underset{O}{\overset{R^4}{\overset{\|}{C}}}$$

R's = H, alkyl, aryl   yields up to 95%

VI.A.10-9   P. Catsoulacos, Bull. Soc. Chim. France, 642 (1976).

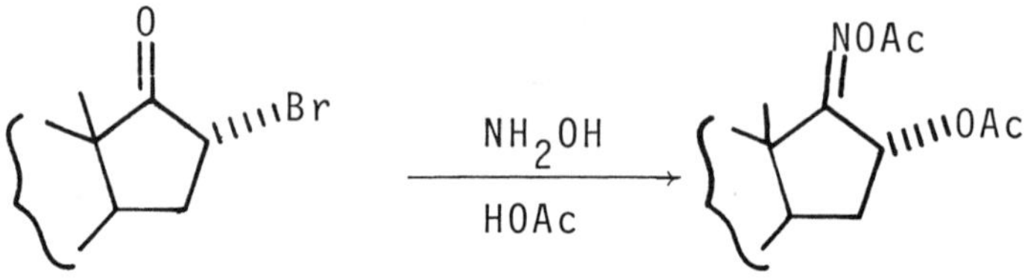

[reaction: bicyclic cyclopentanone with Br substituent + NH_2OH / $HOAc$ → bicyclic oxime with NOAc and OAc substituents]

~50-85%

VI.A.10-10   T. Francis and M. P. Thorne, Can. J. Chem., 54, 24 (1976).

$$Ar\text{-}NCO + ROH \xrightarrow{Sn(OCOC_7H_{15})_2} Ar\text{-}\underset{H}{\overset{}{N}}\text{-}\underset{}{\overset{O}{C}}\text{-}OR$$

R = 3° alkyl   ~70-95%

VI.A.10-11   M. Julia and B. Badet, Bull. Soc. Chim. France, 525 (1976).

$BrCH_2CO_2Et$
+
$PhSO_2\text{-}RR'C^{\ominus}$

→

$PhSO_2\text{-}RR'C\text{-}CH_2CO_2Et$

$\xrightarrow{Na_2CO_3,\ EtOH}$

$RR'C=CHCO_2Et$

R,R' = H, alkyl, vinyl, aryl

30-90%

VI.A.10-12 E. Yoshii, T. Koizumi, and T. Kawazoe, Chem. Pharm. Bull., 24, 1957 (1976).

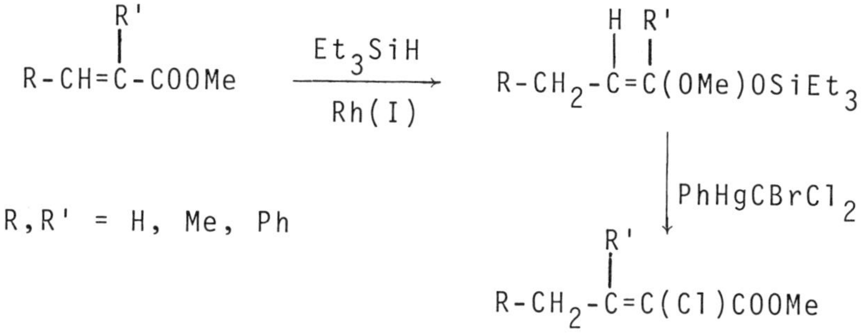

$$R-CH=C(R')-COOMe \xrightarrow{Et_3SiH, Rh(I)} R-CH_2-C(R')=C(OMe)OSiEt_3 \xrightarrow{PhHgCBrCl_2} R-CH_2-C(R')=C(Cl)COOMe$$

R,R' = H, Me, Ph

52-92%

VI.A.10-13 G. M. Rubottom, J. M. Gruber, and K. Kincaid, Synth. Commun., 6, 59 (1976).

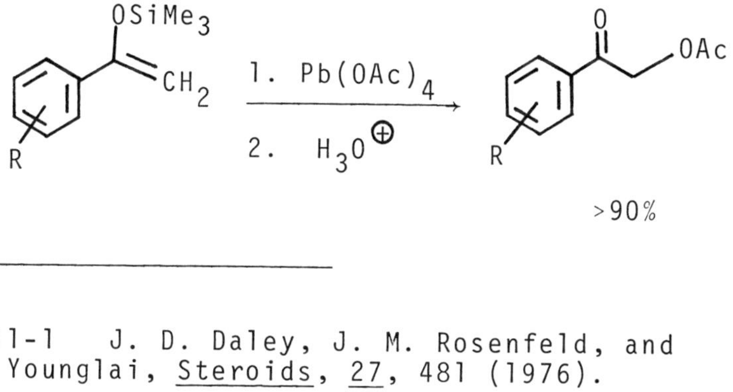

Ar-C(OSiMe$_3$)=CH$_2$ $\xrightarrow{\text{1. Pb(OAc)}_4 \quad \text{2. } H_3O^{\oplus}}$ Ar-C(O)-CH$_2$-OAc

(Ar = substituted phenyl with R)

>90%

VI.A.11-1 J. D. Daley, J. M. Rosenfeld, and E. V. Younglai, Steroids, 27, 481 (1976).

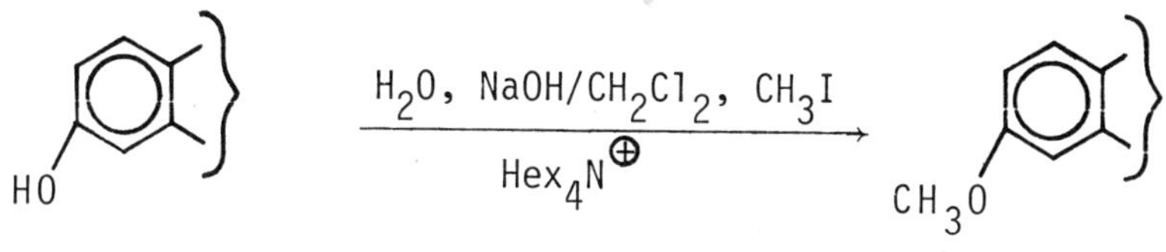

HO-Ar $\xrightarrow{H_2O, \text{NaOH/}CH_2Cl_2, CH_3I, \text{Hex}_4N^{\oplus}}$ CH$_3$O-Ar

quantitative

VI.A.11-2   J. E. Shaw, D. C. Kunerth, and S. B. Swanson, J. Org. Chem., **41**, 732 (1976).

$$\text{R-C}_6\text{H}_4\text{-Cl} \xrightarrow[\text{HMPA}]{\text{NaOMe, 90°}} \text{R-C}_6\text{H}_4\text{-OMe}$$

R = H, Cl   50-80%

                                       13% for R = p-CH$_3$

VI.A.12-1   A. W. Burgstahler, L. O. Weigel, and C. G. Shaefer, Synthesis, **6**, 767 (1976).

$$\text{R-CO-Cl} \xrightarrow[\substack{\text{2,6-dimethyl-}\\\text{pyridine}}]{\text{H}_2/\text{Pd-C, THF}} \text{R-CHO}$$

R = alkyl, aryl   77-96% isolated

VI.A.12-2   G. Cardillo, M. Orena, and S. Sandri, J. Chem. Soc., Chem. Commun., 190 (1976).

$$\text{R-CH}_2\text{X} + \text{KO-Cr(=O)}_2\text{-OK} \xrightarrow[\text{crown ethers}]{\text{HMPT}} \text{R-CHO}$$

halide	Yield (%, by glc)
γ,γ-dimethylallyl Br	78
Geranyl Br	82
Farnesyl Br	80
Benzyl Br	80
Octyl Br	20

VI.A.12-3   I. Degani and R. Fochi, J. Chem. Soc. Perkin I, 323 (1976).

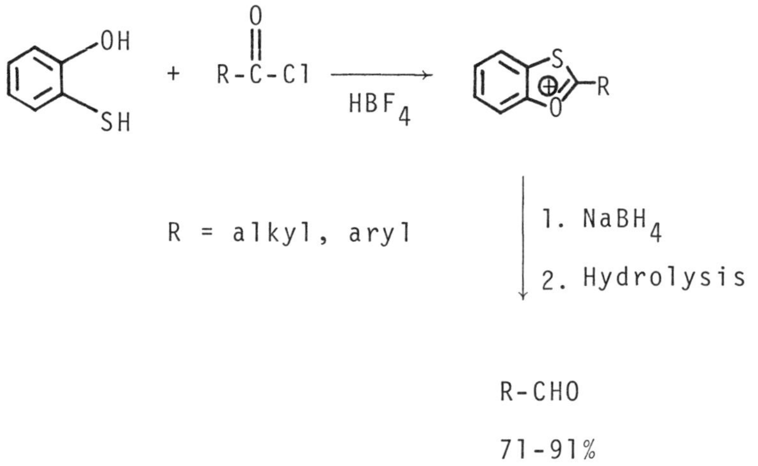

[structure: 2-hydroxythiophenol] + $R\text{-}\overset{O}{\overset{\|}{C}}\text{-}Cl$ $\xrightarrow{HBF_4}$ [benzoxathiolium salt with R]

R = alkyl, aryl

1. $NaBH_4$
2. Hydrolysis

R-CHO

71-91%

VI.A.12-4   G. Doleschall, Tetrahedron, 32, 2549 (1976).

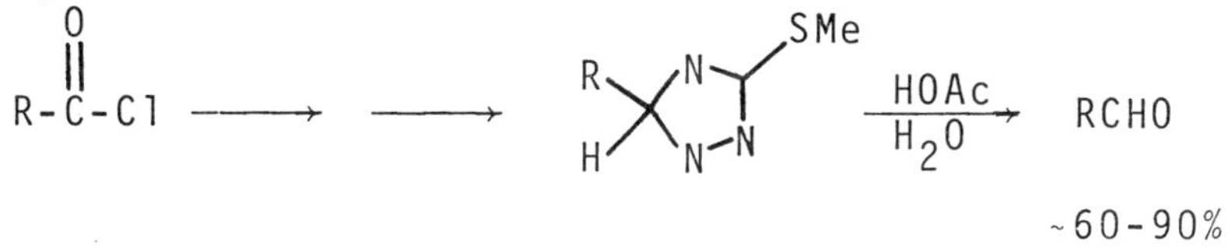

$R\text{-}\overset{O}{\overset{\|}{C}}\text{-}Cl$ ⟶ ⟶ [triazolidine with SMe, R, H, N-N] $\xrightarrow[H_2O]{HOAc}$ RCHO

~60-90%

R = 1°,2°,3° alkyl, aryl, cinnamoyl

VI.A.12-5   C. A Scott, D. G. Smith, and D.J.H. Smith, Synth. Commun., 6, 135 (1976).

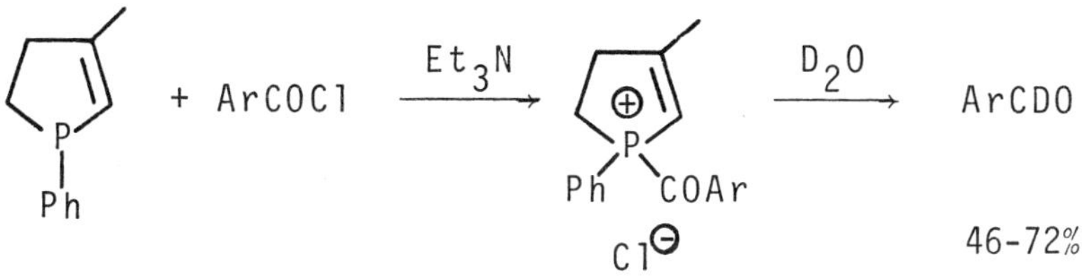

[methylphospholene-Ph] + ArCOCl $\xrightarrow{Et_3N}$ [phospholium with Ph, COAr, $Cl^\ominus$] $\xrightarrow{D_2O}$ ArCDO

46-72%

VI.A.12-6 K. Sachdev and H. S. Sachdev. *Tetrahedron Lett.*, 4223 (1976).

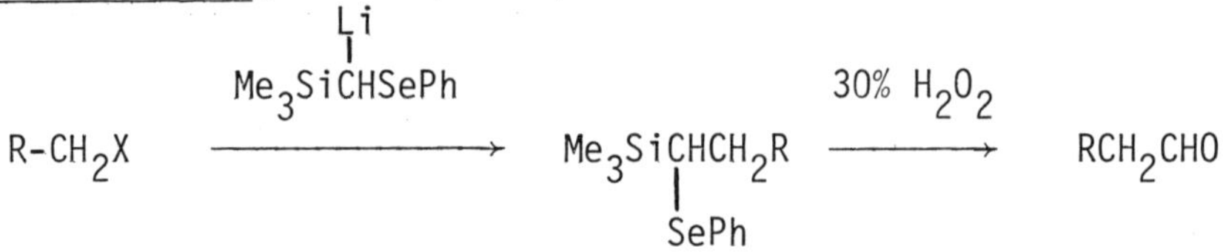

$$R\text{-}CH_2X \xrightarrow{Me_3SiCHSePh \atop Li} Me_3SiCHCH_2R \atop SePh \xrightarrow{30\% \ H_2O_2} RCH_2CHO$$

X = Br, I
R = 1° alkyl

>66%

VI.A.12-7 J. Nakayama, *J. Chem. Soc. Perkin I*, 540 (1976).

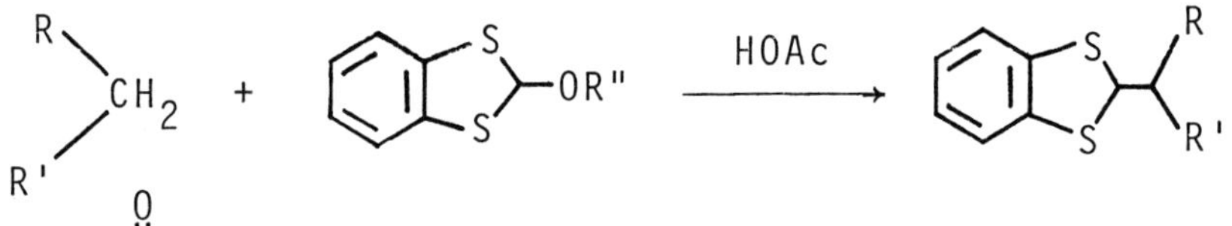

R,R' = $-\overset{O}{\underset{\|}{C}}-R$, CN, SPh   R" = alkyl

~50-80%

VI.A.12-8 P. Bakuzis *et al.*, *J. Org. Chem.*, **41**, 2769 (1976).

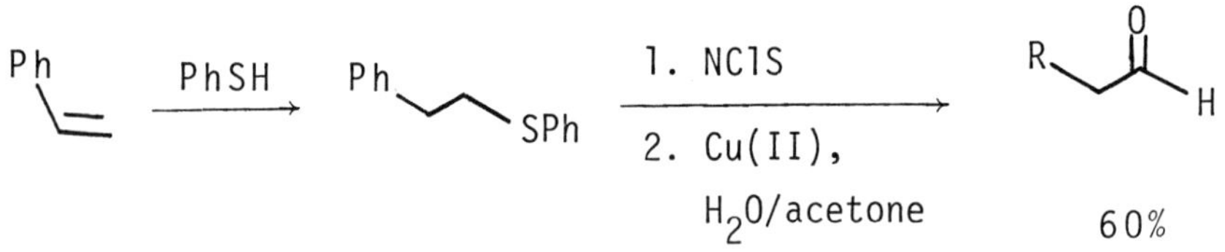

$$Ph\text{-}CH=CH_2 \xrightarrow{PhSH} Ph\text{-}CH_2CH_2SPh \xrightarrow[\text{2. Cu(II), } H_2O/\text{acetone}]{\text{1. NCIS}} R\text{-}CH_2CHO$$

60%

VI.A.12-9 H. Pauling, D. A. Andrews, and N. C. Hindley, <u>Helv. Chim. Acta</u>, 59, 1233 (1976).

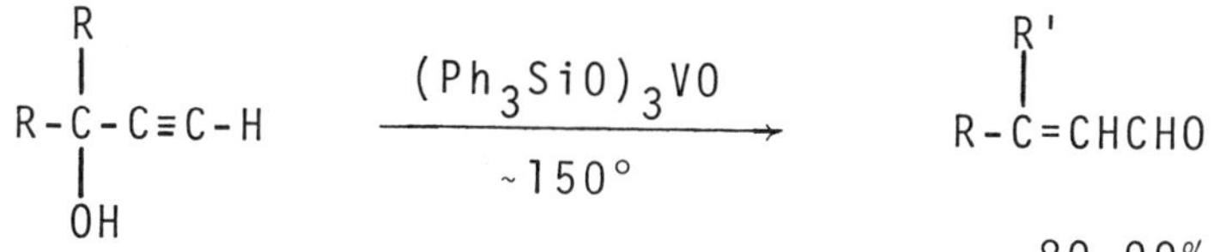

$$R-\underset{\underset{OH}{|}}{\overset{\overset{R}{|}}{C}}-C \equiv C-H \xrightarrow[\sim 150°]{(Ph_3SiO)_3VO} R-\underset{}{\overset{\overset{R'}{|}}{C}}=CHCHO$$

~80-90%

R' = H, CH_3
R = alkyl, aryl

VI.A.12-10 J. H. Babler and M. J. Coghlan, <u>Synth. Commun.</u>, 6, 469 (1976).

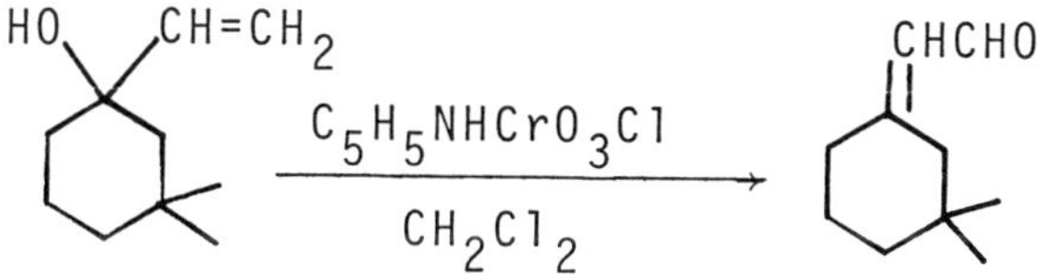

$$\xrightarrow[CH_2Cl_2]{C_5H_5NHCrO_3Cl}$$

85%

(mixture of E and Z)

VI.A.12-11 J. Hocker, H. Giesecke, and R. Merten, <u>Angew. Chem. Int. Ed.</u>, 15, 169 (1976).

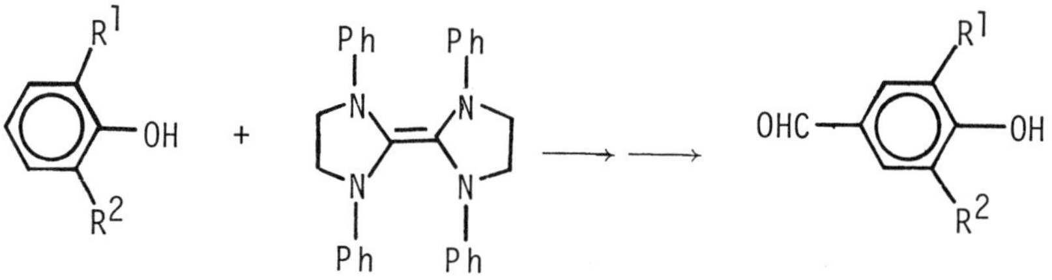

26-84%

R^1, R^2 = H, Cl, CH_3, <u>t</u>-Bu

VI.A.12-12   A. Rahman and A. Basha, *J. Chem. Soc., Chem. Commun.*, 594 (1976).

$$\text{R-C}_6\text{H}_4\text{-C(O)-N(piperidine)} \xrightarrow{POCl_3} \text{Ar-C(Cl)=N}^{\oplus}\text{(piperidine)} \xrightarrow[\text{2. }H_2O]{\text{1. Zn}} \text{ArCHO}$$

90-95%

R = H, Cl, Br

VI.A.12-13   C. A. Bertelo and J. Schwartz, *J. Am. Chem. Soc.*, **98**, 262 (1976).

$$\underset{R^2}{\overset{R^3\;\;\;R^4}{\text{R}^1\text{-C=C-C=CH}_2}} \xrightarrow{(C_5H_5)_2Zr(H)Cl} \underset{R^2}{\overset{R^3\;\;\;\;\;\;R^4}{\text{R}^1\text{-C=C-CH-CH}_2\text{-Zr(Cl)(C}_5H_5)_2}}$$

R's = H, CH$_3$

$$\downarrow \begin{array}{l}\text{1. CO} \\ \text{2. H}_3\text{O}^{\oplus}\end{array}$$

$$\underset{R^2}{\overset{R^3\;\;\;\;\;\;R^4}{\text{R}^1\text{-C=C-CH-CH}_2\text{-CHO}}}$$

59-98% (VPC, based on Zr complex)
Five examples cited.

VI.A.12-14   G. Cahiez, et al., Tetrahedron Lett., 3155 (1976).

$$RMgBr \xrightarrow{MnI_2} RMnI \xrightarrow{R'COCl} R-\underset{\underset{O}{\|}}{C}-R'$$

R,R' = Me, n-alkyl, i-Pr, t-Bu   60-90%

VI.A.12-15   G. Cardillo, M. Orena, and S. Sandri, Tetrahedron Lett., 3985 (1976).

$$\underset{R'}{\overset{R}{>}}CHX \xrightarrow{Polymer} \underset{R'}{\overset{R}{>}}C=O$$

BzCl, geranyl Br,   ~95%
Ph_2CHBr, etc.

$$Polymer = \underset{}{(P)}-\underset{}{\bigcirc}-\overset{\oplus}{N}Me_3 \; HCrO_4^{\ominus}$$

VI.A.12-16   R. Kirchoff, Tetrahedron Lett., 2533 (1976).

$$\underset{R^1}{\overset{H}{>}}\underset{R^2}{\overset{NO_2}{<}} \xrightarrow[DMF/H_2O]{VCl_2, \; HCl} \underset{R^1}{\overset{O}{\underset{\|}{C}}}\underset{}{R^2}$$

R^1 = H, alkyl   24% (heptaldehyde)
R^2 = alkyl, benzyl   53-71% (ketones)

VI.A.12-17 Y. Masuyama, Y. Ueno and M. Okawara, Tetrahedron Lett., 2967 (1976).

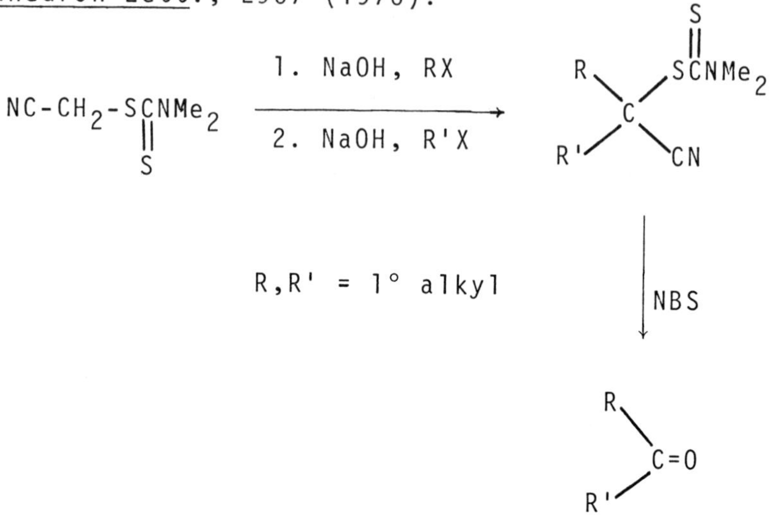

$$NC-CH_2-\underset{\underset{S}{\|}}{S}CNMe_2 \xrightarrow[\text{2. NaOH, R'X}]{\text{1. NaOH, RX}} \underset{R'}{\overset{R}{>}}C\underset{CN}{\overset{SCNMe_2 \; (S)}{<}}$$

R,R' = 1° alkyl

↓ NBS

$$\underset{R'}{\overset{R}{>}}C=O$$

overall ~60%

VI.A.12-18 J. Tsuji, I. Shimizu, and K. Yamamoto, Tetrahedron Lett., 2975 (1976).

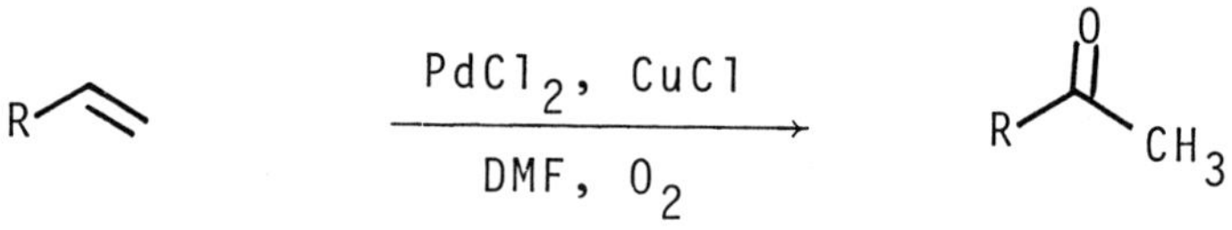

$$R-CH=CH_2 \xrightarrow[\text{DMF, } O_2]{PdCl_2, \; CuCl} R-CO-CH_3$$

R = alkyl, often γ-keto

58-77%

USEFUL SYNTHETIC PREPARATIONS

VI.A.12-19   P. C. Traas, H. Boelens, and H. J. Takken, <u>Tetrahedron Lett.</u>, 2287 (1976).

$$R^1R^2C(NNHTs)\text{-}CH_2 \xrightarrow{\text{1. BuLi} \atop \text{2. DMF} \atop \text{3. } H_2O} R^1R^2C\text{=}CHCHO$$

Yields generally 50-60%

VI.A.12-20   L. M. Stephenson and L. C. Falk, <u>J. Org. Chem.</u>, **41**, 2928 (1976).

$$R^1\text{-}\underset{\underset{O}{\|}}{C}\text{-}\underset{\underset{O}{\|}}{C}\text{-}R^2 \xrightarrow{(CH_3O)_3P} (CH_3O)_3P\underset{O}{\overset{O}{\diagdown}}\underset{R^2}{\overset{R^1}{\diagup}} \xrightarrow{H_2/\text{cat.}} R^1CH_2\underset{\underset{O}{\|}}{C}R^2$$

R^1, R^2 = alkyl, aryl

VI.A.12-21   R. C. Larock and J. C. Bernhardt, <u>Tetrahedron Lett.</u>, 3097 (1976).

$$\underset{H}{\overset{R}{\diagdown}}C\text{=}C\underset{HgCl}{\overset{H}{\diagup}} + Cl\text{-}\underset{\underset{}{\overset{O}{\|}}}{C}\text{-}R' \xrightarrow{AlCl_3} \underset{H}{\overset{R}{\diagdown}}C\text{=}C\underset{\underset{O}{\|\atop}}{\overset{H}{\diagup}}R'$$

R = alkyl, aryl
R' = alkyl, vinyl

~90-100%

VI.A.12-22 L. Barsky et al., J. Org. Chem., 41, 3651 (1976).

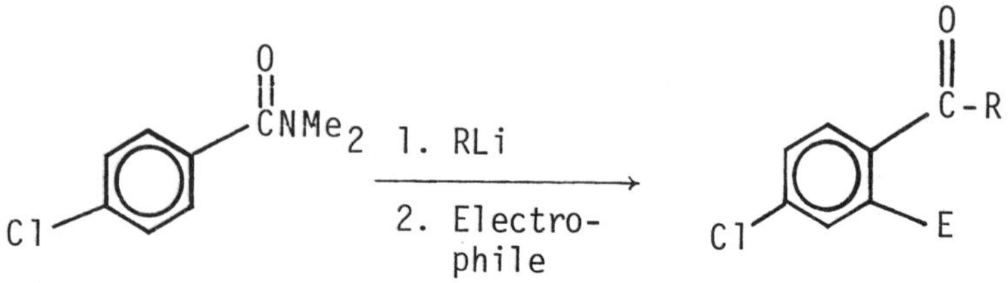

R	Electrophile	E	Yield (%)
Bu	$(CH_3S)_2$	CH_3S	56
Ph	$(CH_3S)_2$	CH_3S	53
Me	Ph_2CO	Ph_2COH	47
Me	DMF	CHO	56
Me	MeI	CH_3	81 (NMR)

Several additional examples.

VI.A.12-23 P. T. Lansbury and R. W. Britt, J. Am. Chem. Soc., 98, 4577 (1976).

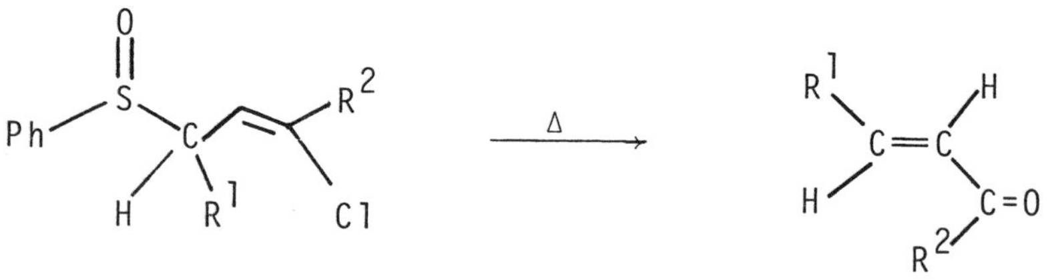

R^1, R^2 = H, alkyl

VI.A.12-24 B. Corbel et al., Tetrahedron Lett., 835 (1976).

$(EtO)_2\underset{\underset{O}{\|}}{P}-\underset{}{N}(CH_3)-CH=C=CH_2 \xrightarrow[\text{2. RX}]{\text{1. }\underline{n}\text{-BuLi}} (EtO)_2\underset{\underset{O}{\|}}{P}-\underset{}{N}(CH_3)-\underset{\underset{R}{}}{C}=C=CH_2$

R = CH_3, Bz

$\downarrow H_3O^{\oplus}$

$R-\underset{\underset{O}{\|}}{C}-CH=CH_2$

yields generally 50-70%

VI.A.12-25 J. S. Walia and A. S. Walia, J. Org. Chem., 41, 3765 (1976).

$Ph-C\equiv C-\underset{\underset{}{\|}}{\overset{O}{C}}-CH_3 \xrightarrow[\text{NaCN}]{CH_3OH} Ph-\underset{\underset{OMe}{}}{\overset{OMe}{C}}-CH_2-\underset{\underset{}{\|}}{\overset{O}{C}}-CH_3$

86%

$H-C\equiv C-\underset{\underset{}{\|}}{\overset{O}{C}}-OCH_3 \xrightarrow[\text{NaCN}]{CH_3OH} (MeO)_2CHCH_2COOCH_3$

75%

VI.A.12-26   J. I. Okogun, J. Chem. Soc. Perkin I, 2241 (1976).

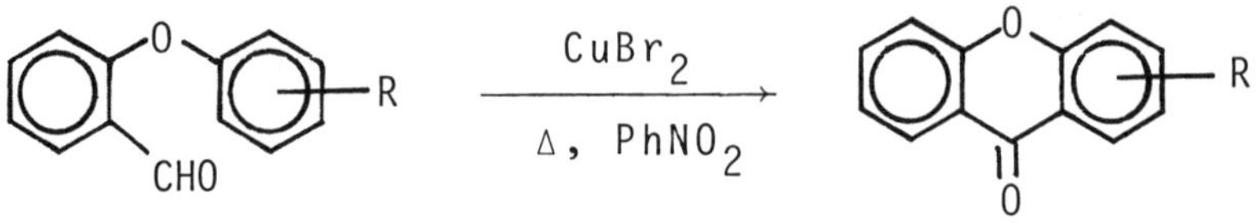

[2-(2-formylphenoxy)phenyl]-R →(CuBr$_2$, Δ, PhNO$_2$)→ xanthone-R

R = H, OMe, Me, NO$_2$ low yields (<50%)

VI.A.12-27   S. Nimgirawath, E. Ritchie, and W. C. Taylor, Aust. J. Chem., 29, 339 (1976).

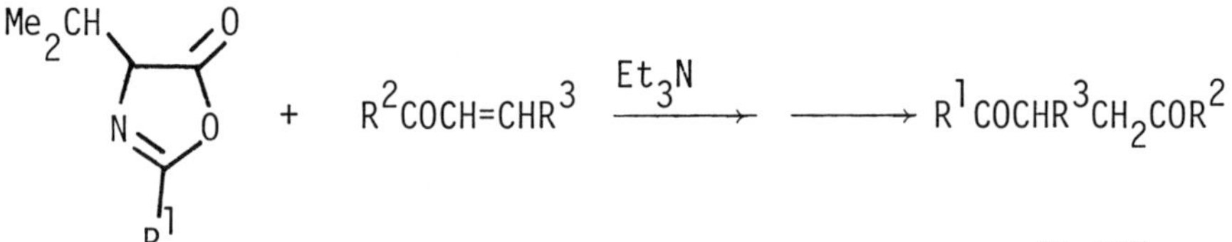

[2-isopropyl-4-R^1-oxazolin-5-one] + R^{2}COCH=CHR^3 →(Et$_3$N)→ R^{1}COCHR^{3}CH$_2$COR^2

25-35%

R^1,R^2 = alkyl, aryl
R^3 = H, Ph

VI.A.12-28   R. C. Cookson and R. M. Lane, J. Chem. Soc., Chem. Commun., 804 (1976).

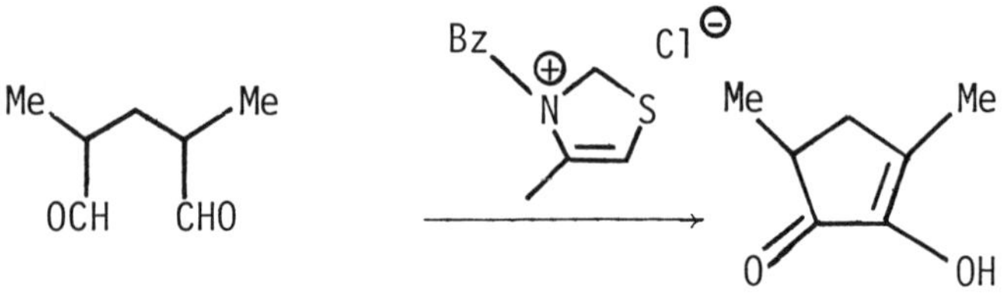

Me-CH(OCH)-CH(CHO)-Me + [N-Bz-4-methylthiazolium Cl$^\ominus$] → 2,5-dimethyl-3-hydroxycyclopent-2-enone

88%

USEFUL SYNTHETIC PREPARATIONS

VI.A.12-29   H. D. Scharf, P. Friedrich, and
A. Linckens, Synthesis, 256 (1976).

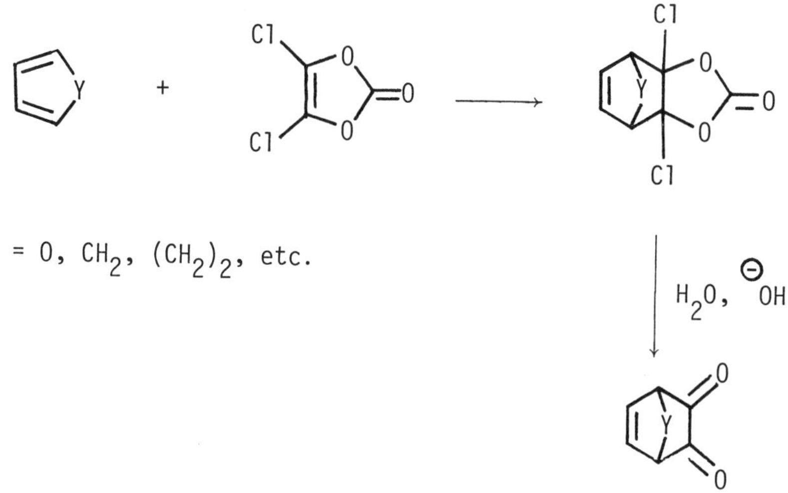

Y = O, CH_2, $(CH_2)_2$, etc.

70-80%
for second step

VI.A.12-30   K. Oguar, M. Yamashita, and
G. Tsuchihashi, <u>Tetrahedron Lett.</u>, 759 (1976).

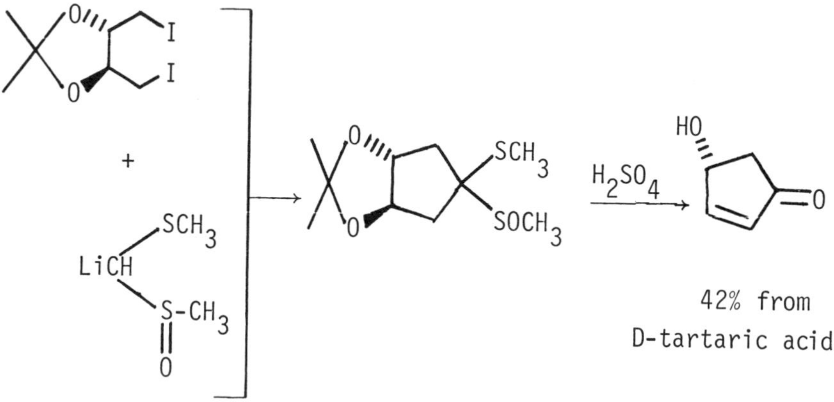

42% from
D-tartaric acid

VI.A.12-31   J. M. Conia, <u>Pure and Appl. Chem.</u>, <u>43</u>, 317 (1975).

> Review: "The Cyclopropanation of Silyl Enol Ethers. A Powerful Synthetic Tool."

Particularly useful in the formation of alkylated ketones, e.g.:

$$\text{(OSiMe}_3\text{-cyclohexene)} \xrightarrow[\text{Zn/Ag}]{\text{CH}_2\text{I}_2} \text{(OSiMe}_3\text{-bicyclic)} \xrightarrow{^{\ominus}\text{OH}} \text{(methylcyclohexenone)}$$

~80% overall

VI.A.12-32   G. Rio and A. Lecas-Nawracka, <u>Bull. Soc. Chim. France</u>, 317 (1976).

> Review: "Preparation of Saturated and Unsaturated 1,4-Diketones"

VI.A.13-1   J. Streith, C. Fizet, and H. Fritz, <u>Helv. Chem. Acta</u>, <u>59</u>, 2786 (1976).

$$\text{R-CHO} \xrightarrow[\text{H}_2\text{O}]{\text{NH}_2\text{OSO}_3\text{H}} \text{R-C} \equiv \text{N}$$

~70-90%

R = alkyl, cyclic alkyl, aryl, heterocyclic

VI.A.13-2   J. B. Hendrickson, K. W. Bair, and P. M. Keehn, Tetrahedron Lett., 603 (1976).

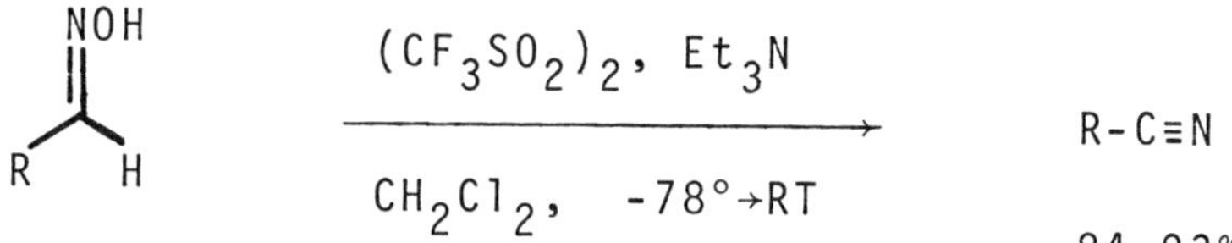

$$\underset{RH}{C}=NOH \xrightarrow[CH_2Cl_2,\ -78°\to RT]{(CF_3SO_2)_2,\ Et_3N} R-C\equiv N$$

84-93%

R = alkyl, aryl, cinnamyl, etc.

VI.A.13-3   M. R. Czarny, Synth. Commun., 6, 285 (1976).

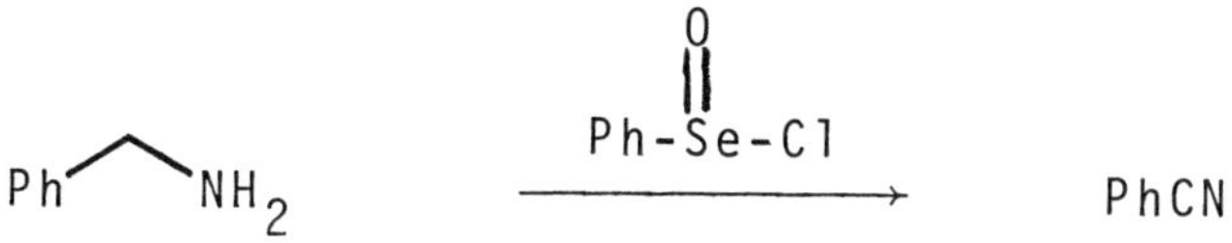

$$Ph\text{-}CH_2\text{-}NH_2 \xrightarrow{Ph\text{-}Se(O)\text{-}Cl} PhCN$$

85%

VI.A.13-4   W. P. Reeves and M. R. White, Synth. Commun., 6, 193 (1976).

$$R\text{-}Br + NaCN \xrightarrow{Bu_3N} R\text{-}CN$$

R = unhindered 1° alkyl   >90%

VI.A.13-5   J. B. Bapat et al., Tetrahedron Lett., 2691 (1976).

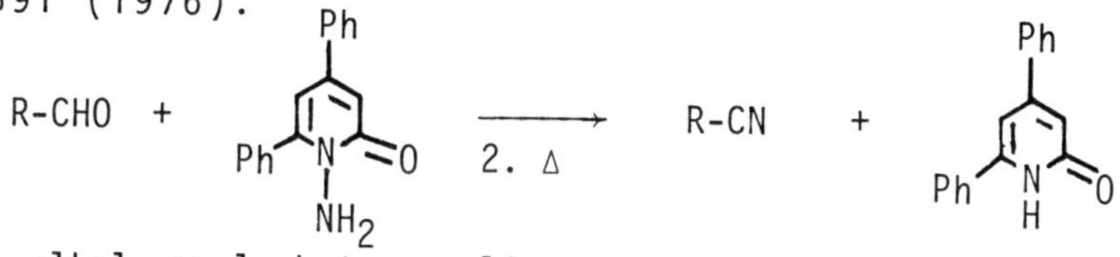

R-CHO + [4,6-diphenyl-1-amino-2-pyridone] ⟶ R-CN + [4,6-diphenyl-2-pyridone]

2. Δ

R = alkyl, aryl, heterocyclic   50-90%

VI.A.13-6   A. Ahmad, Synthesis, 418 (1976).

$$\underset{HO}{\overset{R}{\underset{\|}{C}}}\underset{N}{\overset{COOH}{}} + PhNCO \longrightarrow R-C\equiv N$$

generally >90%

R = 1°, 2° alkyl, aryl

VI.A.13-7   T. Cuvigny et al., Synthesis, 237, 238 (1976).

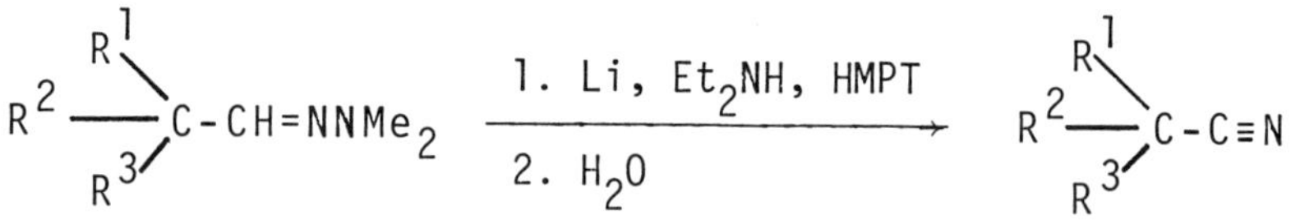

$$R^2\underset{R^3}{\overset{R^1}{-}}C-CH=NNMe_2 \xrightarrow[\text{2. } H_2O]{\text{1. Li, Et}_2\text{NH, HMPT}} R^2\underset{R^3}{\overset{R^1}{-}}C-C\equiv N$$

71-96%

R's = H, n-alkyl

VI.A.13-8   A. Antonowa and S. Hauptmann, Z. Chem., 16, 17 (1976).

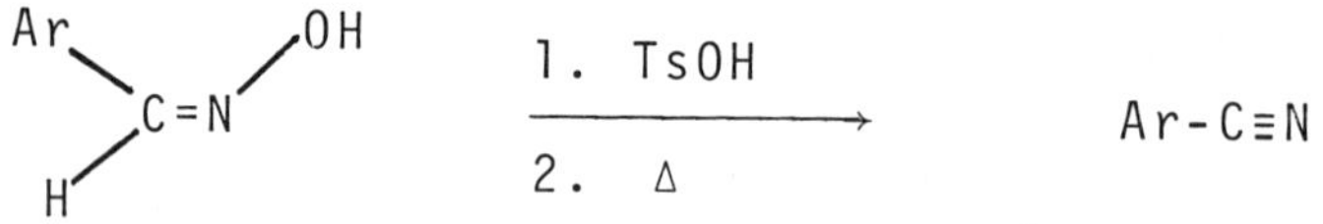

$$\underset{H}{\overset{Ar}{}}C=N\overset{OH}{} \xrightarrow[\text{2. } \Delta]{\text{1. TsOH}} Ar-C\equiv N$$

Widely varying yields, ~50-80% for subst. Ph

VI.A.13-9   R. S. Glass and R. C. Hoy, *Tetrahedron Lett.*, 1781 (1976).

$$Ar\text{-}CHO \longrightarrow ArCH=NTs \xrightarrow[HMPA]{NaCN} ArC\equiv N$$

generally 70-90%

VI.A.13-10   N. Suzuki et al., *J. Chem. Soc. Perkin I*, 1901 (1976).

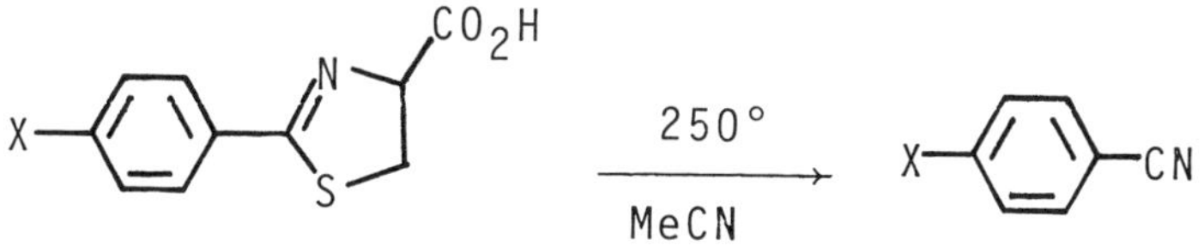

X = H, Cl

~100%

VI.A.13-11   M. A. Schwartz et al., *J. Org. Chem.*, **41**, 2502 (1976).

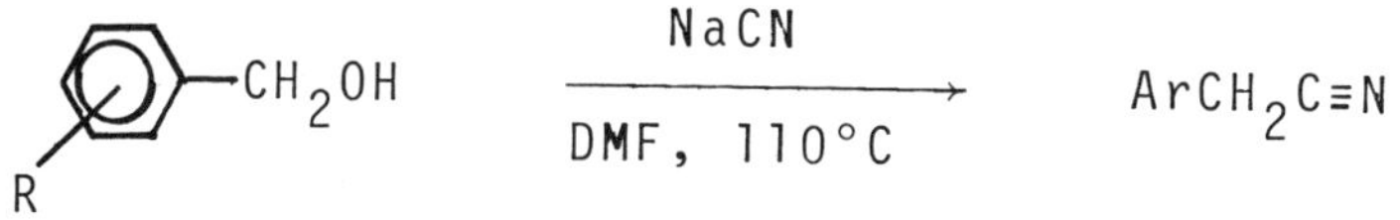

$$Ar\text{-}CH_2OH \xrightarrow[DMF, 110°C]{NaCN} ArCH_2C\equiv N$$

R = OH, OCH$_3$

0-79%

VI.A.13-12 G. W. Gokel, S. A. DiBiase, and B. A. Lipisko, Tetrahedron Lett., 3495 (1976).

$$CH_3CN + R-\underset{\underset{}{\|}}{\overset{\overset{O}{\|}}{C}}-R' \xrightarrow[\text{crown ether}]{KOH} RR'C=CH-CN$$

R,R' = alkyl, aryl, H   50-86% isolated

VI.A.13-13 D. S. Watt, J. Am. Chem. Soc., 98, 271 (1976).

$$R^1\underset{CN}{\overset{OOAc}{-C-}}CH_2CH_2\underset{R^3}{\overset{}{CH}}-R^2 \xrightarrow{h\nu} R^1\underset{}{\overset{O}{\overset{\|}{C}}}CH_2CH_2\underset{R^3}{\overset{CN}{C}}R^2$$

R^1, R^2, R^3 = alkyl, aryl, etc.   9-50%

VI.A.13-14 J. Tsuji, H. Takayanagi and Y. Toshida, Chem. Lett., 147 (1976).

$$\underset{}{\text{o-C}_6H_4(NH_2)_2} \xrightarrow[Cu_2Cl_2]{O_2, \text{ pyridine}} \text{o-C}_6H_4(CN)_2$$

90%

VI.A.13-15   G. Höfle and B. Lange, *Angew. Chem. Int. Ed.*, 15, 113 (1976).

$$\underset{\underset{R}{|}}{\underset{NH}{\overset{N=N}{\underset{N}{\diagdown}}\hspace{-0.5em}\overset{}{\underset{}{\diagup}}NH}} \xrightarrow{NaOBr} R-NC + 2N_2$$

75-92%

R = Ph, n-Bu, Bz

VI.A.14-1   Y. Takahashi et al., *Synthesis*, 616 (1976).

$$R-CH=CH_2 \longrightarrow (RCH_2CH_2)_3B \xrightarrow[\substack{\text{electrolysis} \\ CH_3NO_2}]{Et_4\overset{\oplus}{N} \overset{\ominus}{I}} RCH_2CH_2NO_2$$

25-50%

VI.A.14-2   M. W. Barnes and J. M. Patterson, *J. Org. Chem.*, 41, 733 (1976).

$$\underset{RCR'}{\overset{OH}{\overset{|}{N}}} \xrightarrow[\substack{2.\ O_3,\ CH_2Cl_2}]{1.\ Cl_2,\ CH_2Cl_2} \underset{\underset{NO_2}{|}}{\overset{\overset{Cl}{|}}{R-C-R'}} \xrightarrow[NaOH-H_2O]{H_2-Pd/C} \underset{}{\overset{\overset{NO_2}{|}}{R-CH-R'}}$$

R,R' = alkyl

Yields ~50-80%

VI.A.14-3   M. E. Kurz, and R.T.Y. Chen, *J. Chem. Soc., Chem. Commun.*, 968 (1976).

$$X-Ph-H + CH_3NO_2 \xrightarrow{Mn(OAc)_3} X-Ph-CH_2NO_2$$

~77% for X = H, Me, OMe
0-20% for X = NO_2, Cl

VI.A.14-4   N. Kornblum et al., *J. Org. Chem.*, **41**, 1560 (1976).

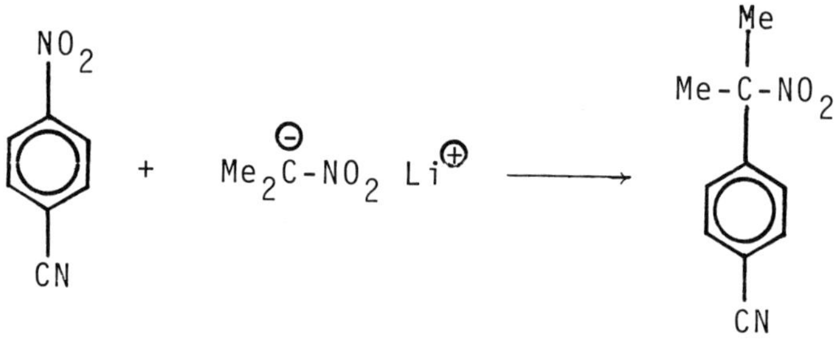

[4-nitro-benzonitrile] + $Me_2\overset{\ominus}{C}-NO_2$ $Li^{\oplus}$ ⟶ [4-(2-nitro-2-propyl)benzonitrile]

82%

VI.A.15-1   M. J. Robins et al., *J. Am. Chem. Soc.*, **98**, 7381 (1976).

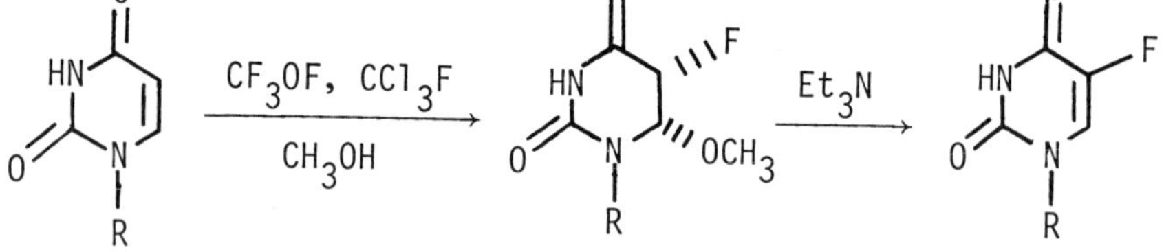

[uracil derivative] $\xrightarrow[CH_3OH]{CF_3OF, CCl_3F}$ [5-fluoro-6-methoxy-5,6-dihydrouracil] $\xrightarrow{Et_3N}$ [5-fluorouracil]

R = H, CH_3, sugar

~80%

VI.A.15-2   U. Reichman et al., Synthesis, 533 (1976).

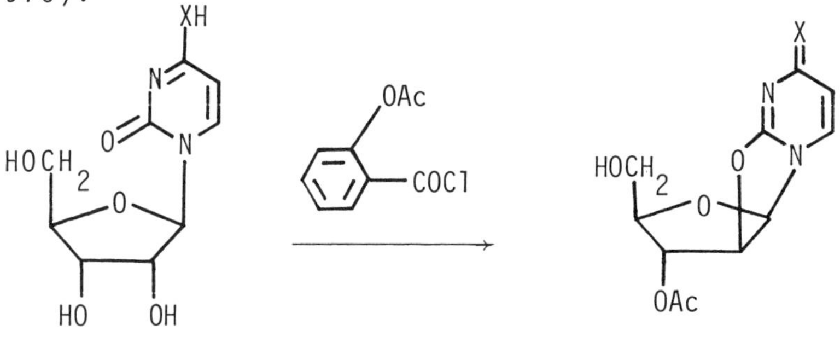

X = NH, O   ~65%

VI.A.15-3   D. E. Bergstrom and J. L. Ruth, J. Am. Chem. Soc., 98, 1587 (1976).

5-chloromercuriuridine $\xrightarrow{\begin{array}{l}1.\ Li_2PdCl_4/MeOH/ethylene \\ 2.\ NaBH_4 \\ 3.\ H_2/Pd\end{array}}$ 5-ethyluridine   68%

Reactions with allyl chloride and methyl acrylate are also described. (Yields 78% and 57%).

VI.A.15-4   M. Sekiya, J. Suzuki, and Y. Terao, Chem. Pharm. Bull., 24, 1331 (1976).

NCCHCN
 |
 N=NPh
$\xrightarrow[NH_3,\ HCONH_2]{H_2,\ Ra-Ni,\ 150°}$

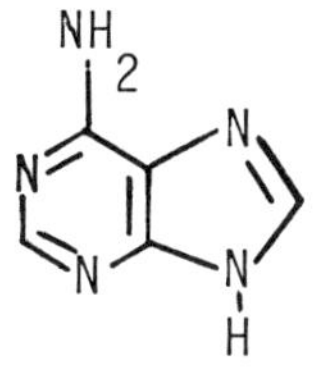

Similar procedures can be used to form xanthine and guanine.

VI.A.15-5 F. Yoneda and T. Nagamatsu, J. Chem. Soc. Perkin I, 1547 (1976).

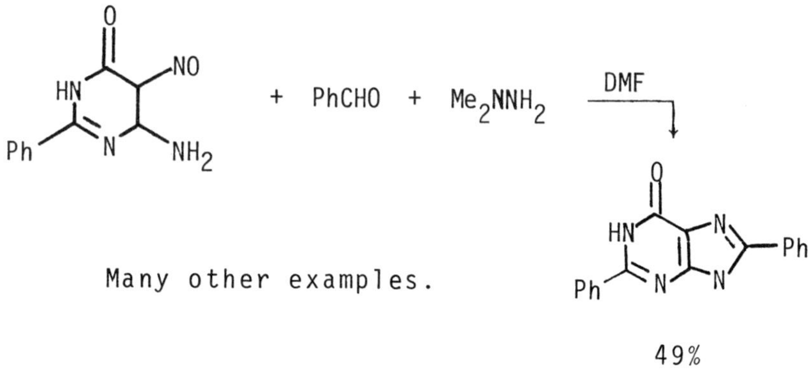

[Pyrimidine with Ph, HN, C=O, NO, NH₂ substituents] + PhCHO + Me₂NNH₂ →(DMF) [purine product with Ph groups] 49%

Many other examples.

VI.A.15-6 G. Zvilichovsky and J. Feingers, J. Chem. Soc. Perkin I, 1507 (1976).

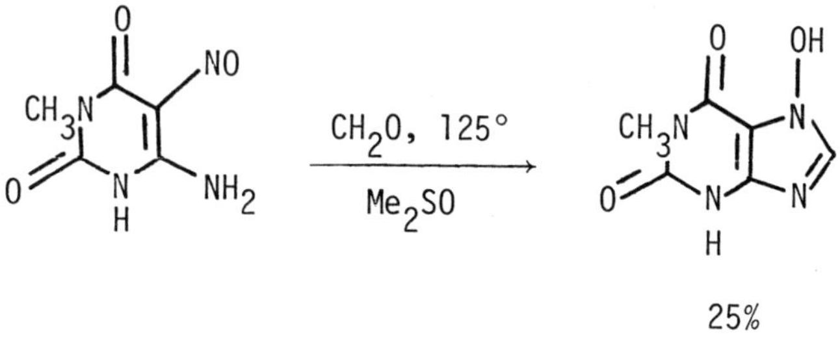

[pyrimidine with CH₃N, C=O, NO, NH₂] →(CH₂O, 125°, Me₂SO) [purine N-OH product] 25%

VI.A.15-7 K. Yamauchi, T. Tanabe, and M. Kinoshita, J. Org. Chem., 41, 3691 (1976).

Use of trimethyl phosphate for the direct N-alkylation of cytosine (N-1), thiamine (N-1 and N-3), uracil (N-1 and N-3), adenine (N-9 and N-3), and guanine (N-1).

USEFUL SYNTHETIC PREPARATIONS

VI.A.15-8   A. D. Broom, J. L. Shim, and G. L. Anderson, *J. Org. Chem.*, **41**, 1095 (1976).

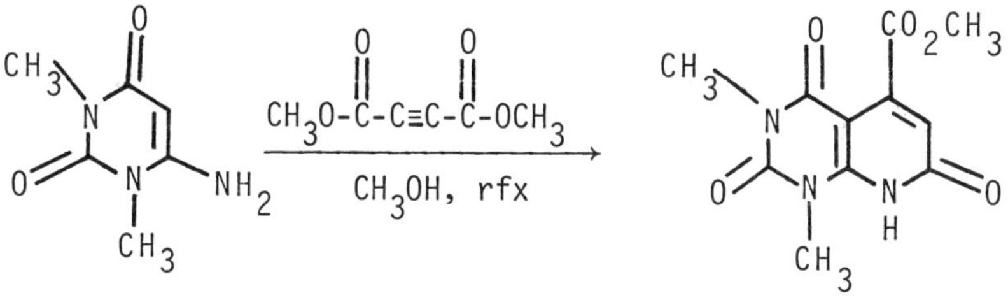

[Reaction: 1,3-dimethyl-6-aminouracil + $CH_3O-C(=O)-C\equiv C-C(=O)-OCH_3$, CH_3OH, rfx → pyrido[2,3-d]pyrimidine product, 64%]

VI.A.15-9   T. Fujii et al., *Chem. Pharm. Bull.*, **24**, 655 (1976).

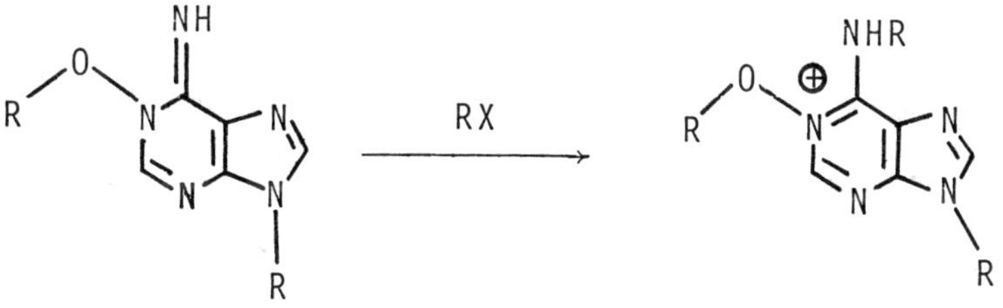

[Reaction: N1-alkoxyadenine derivative + RX → N1-alkoxy-N6-alkyl adeninium salt]

R = Me, Et, Bz   40-80%

VI.A.15-10   J. Herscovici, M. Bessodes, and K. Antonakis, *J. Org. Chem.*, **41**, 3827 (1976).

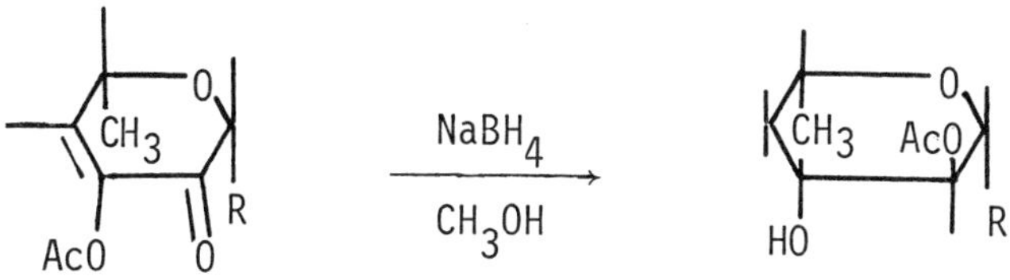

[Reaction: keto sugar with AcO, CH_3, R → $NaBH_4$, CH_3OH → hydroxy sugar with AcO, CH_3, R]

VI.A.15-11 K. Senga, H. Kanazawa, and S. Nishigaki, J. Chem. Soc., Chem. Commun., 155 (1976).

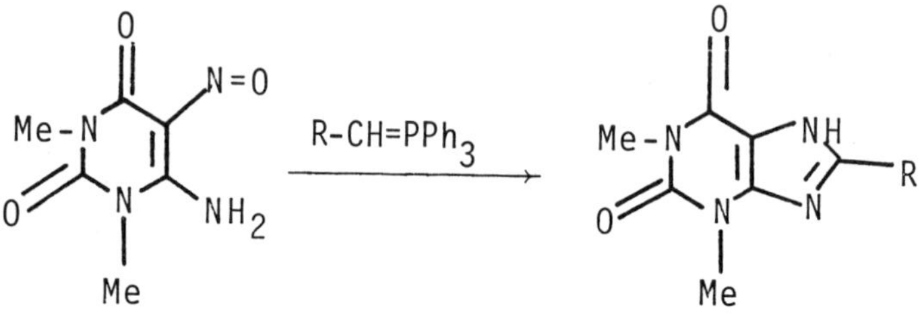

48-75%

VI.A.15-12 J. H. van Boom, P.M.J. Burgers, and P. H. van Deursen, Tetrahedron Lett., 869 (1976).

Use of I for the formation of internucleotide linkages.

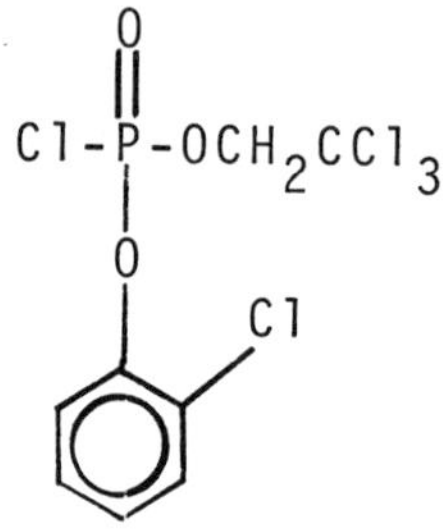

VI.A.15-13 I. Ohtsuka, S. Morioka, and M. Ikehara, Chem. Phar. Bull., 24, 560 (1976).

Synthesis of deoxyribooligonucleotide blocks by an extraction method.

USEFUL SYNTHETIC PREPARATIONS

VI.A.15-14   J. Stawinski, T. Hozumi, and S. A. Narang, Can. J. Chem., 54, 670 (1976).

Use of various arylsulfonyltetrazoles to form the phosphotriester bond in polynucleotide synthesis.

R = H, Me, i-Pr

VI.A.15-15   K. Furusawa, M. Sekine, and T. Hata, J. Chem. Soc. Perkin I, 1711 (1976).

Use of nucleoside phosphoric di-n-butylphosphinothioic mixed anhydrides as precursors to nucleoside di- and tri-phosphates and nucleotide coenzymes. Stable to H_2O, but react with phosphoric acid, pyrophosphoric acid, or nucleotides in the presence of Ag^+.

VI.A.15-16   M. J. Gait and R. C. Sheppard, J. Am. Chem. Soc., 98, 8516 (1976).

Use of a polydimethylacrylamide resin as a solid-phase support for oligonucleotide synthesis.

VI.A.16-1 S. Baba, D. E. VanHorn and E. Negishi, Tetrahedron Lett., 1927 (1976).

$$R-C\equiv CH \xrightarrow[\text{2. n-BuLi}]{\text{1. i-Bu}_2\text{AlH}} \left[\begin{array}{c} R \\ H \end{array} C=C \begin{array}{c} H \\ Al(i\text{-Bu})_2\ n\text{-Bu} \end{array} \right]^{\ominus} Li^{\oplus}$$

R = n-butyl, n-pentyl

$$\xrightarrow{R'X} \begin{array}{c} R \\ H \end{array} C=C \begin{array}{c} H \\ R' \end{array}$$

41-73% (GLC)

R' = alkyl, propargyl, benzyl

VI.A.16-2 J. Hooz and R. Mortimer, Tetrahedron Lett., 805 (1976).

$$R-C\equiv CH + R'_3B \xrightarrow[\text{2. Bu}_3\text{SnCl}]{\text{1. n-BuLi}} \begin{array}{c} R \\ H \end{array} C=C \begin{array}{c} R' \\ H \end{array}$$

R = Ph, n-butyl, n-pentyl

45-71% isolated

R' = Et, Bu, Hex, -(CH$_2$)$_4$-[dioxolane]

USEFUL SYNTHETIC PREPARATIONS

VI.A.16-3 C. W. Spangler and T. W. Hartford, *Synthesis*, 108 (1976).

[cyclohexenol with R,R substituents] $\xrightarrow[\text{HMPT, 50°}]{[(PhO)_3PCH_3]^{\oplus} I^{\ominus}}$ [cyclohexadiene with R,R]

R = H, alkyl   58-78%

[3-hydroxy-1,5-hexadiene] $\xrightarrow{\text{same conditions}}$ [1,3,5-hexatriene]

56%

VI.A.16-4 J. Rémion and A. Krief, *Tetrahedron Lett.*, 3743 (1976).

$$R^2-\underset{\underset{RSe}{|}}{\overset{\overset{R^1}{|}}{C}}-\underset{\underset{OH}{|}}{\overset{\overset{R^3}{|}}{C}}-R^4 \xrightarrow{SOCl_2/Et_3N} \underset{R^2}{\overset{R^1}{\diagdown}}C=C\underset{R^4}{\overset{R^3}{\diagup}}$$

R's = H, 1° alkyl, Ph   50-80%

VI.A.16-5 S. M. Makin, *Pure and Appl. Chem.*, **47**, 173 (1976).

Review: "The Enol Ether Synthesis of Polyenes"

VI.A.16-6   R. C. Larock, J. Org. Chem., **41**, 2241 (1976).

$$R-C\equiv CH \longrightarrow \underset{H}{\overset{R}{>}}C=C\underset{HgCl}{\overset{H}{<}} \xrightarrow[HMPA]{PdCl_2, LiCl} \text{(diene product)}$$

R = Ph, n-Bu   >90% (GLC)

VI.A.17-1   H. R. Kricheldorf, M. Fehrle, and J. Kaschig, Angew. Chem. Int. Ed., **15**, 305 (1976).

Use of the following compounds as condensation agents in polypeptide synthesis:

[triazole-P(OEt)$_2$]   [triazole-P(OCH$_2$CH$_2$O)]   [triazole-P(O-catechol)]

VI.A.17-2   R. Appel, G. Bäumer, and W. Strüver, Chem. Ber., **109**, 801 (1976).

Use of PPh_3/CCl_4 as a condensation agent in peptide synthesis.

VI.A.17-3   A. C. Veronese et al., Gazz. Chim. Ital., **106**, 179 (1976).

Use of hydroximino esters (from acetylacetone, benzoylacetophenone, dimethyl malonate, etc.) as coupling agents in peptide synthesis. Amino acids such as serine, threonine, and tyrosine do not require protection of the OH group.

USEFUL SYNTHETIC PREPARATIONS

VI.A.17-4   K. Horiki, Tetrahedron Lett., 4106 (1976).

Use of a polymeric triphenylphosphine/2,2'-dipyridyl disulfide resin as a reagent for anti-Merrifield solid-phase peptide synthesis. In this case, the coupling reagent is bound to the resin (and can be recycled), and the peptide is in solution.

VI.A.17-5   D. Yamashiro, J. Blake, and C. H. Li, Tetrahedron Lett., 1469 (1976).

Use of trifluoroethanol to increase the reactivity of the α-amino group in solid-phase peptide synthesis.

VI.A.17-6   A. G. Jackson et al., Tetrahedron Lett., 3627 (1976).

Activation of carboxylic acids as diphenylphosphinic mixed anhydrides: Application to peptide chemistry.

VI.A.17-7   Yu. A. Davidovich and V. Gut, Coll. Czech. Chem. Commun., 41, 1805 (1976).

Use of polymeric o-nitrobenzenesulfenylaminoacyl N-hydroxysuccinimide esters as active esters in the synthesis of model protected peptides.

VI.A.17-8   S. Matsuura, C.-H. Niu, and J. S. Cohen, J. Chem. Soc., Chem. Commun., 451 (1976).

Use of HF in pyridine to remove completed peptides from the Merrifield resin. Also cleaves BOC protecting groups.

VI.A.17-9   A. E. Jackson and R.A.W. Johnstone, Synthesis, 685 (1976).

Removal of benzyloxycarbonyl groups from peptides by catalytic transfer hydrogenation.

VI.A.17-10   S. A. Kahn and K. M. Sivanandaiah, Synthesis, 614 (1976).

The use of a toluene medium and 1-hydroxybenzotriazole to accelerate coupling rates in solid-phase peptide synthesis.

VI.A.17-11   M. Akiyama, K. Shimizu, and M. Narita, Tetrahedron Lett., 1015 (1976).

Use of N-hydroxysuccinimide ester polymers as polymer reagents in peptide synthesis.

VI.A.17-12   R. C. Orlowski, R. Walter, and D. Winkler, J. Org. Chem., 41, 3701 (1976).

Synthesis and use of p-methoxybenzhydrylamine resin in solid-phase peptide synthesis.

VI.A.17-13 J. K. Chang, M. Shimizu, and S.-S. Wang, J. Org. Chem., 41, 3255, 3258 (1976).

Use of hydroxymethyl resin in a fully automated solid-phase peptide synthesis; also use of an α-methylphenacyl anchoring bond, which may be cleaved by irradiation at 350 nm.

VI.A.17-14 J. T. Sparrow, J. Org. Chem., 41, 1350 (1976).

Use of a long spacer chain between the first amino acid and the polystyrene support to improve yields in solid phase peptide synthesis.

VI.A.17-15 A. R. Mitchell et al., J. Am. Chem. Soc., 98, 7357 (1976).

Use of a phenylacetamidomethyl (Pam) bridge between the peptide and the resin in solid-phase peptide synthesis. Provides for greater stability toward acid, minimizing acidolysis during peptide elongation.

VI.A.17-16 J. Blake and C. H. Li, J. Chem. Soc., Chem. Commun., 504 (1976).

Use of a p-carbamoylmethylbenzyl ester linkage between the polystyrene resin and the peptide in solid-phase peptide synthesis. Relatively stable toward TFA and liquid HF.

VI.A.17-17   R. W. Roeske and P. D. Gesellchen.
Tetrahedron Lett., 3369 (1976).

Use of 18-crown-6 to catalyze the esterification of the potassium salts of Boc-amino acids with chloromethyl polystyrene.

VI.A.17-18   Yu. P. Shvachkin et al., J. Gen. Chem. (USSR), 46, 717 (1976).

A method for the recycling of unreacted N-protected amino acid derivatives in solid-phase peptide synthesis. The reagent is allowed to react twice with portions of the polymer-bound amino component, and finally with another portion of the amino component in solution.

VI.A.17-19   H. Eckstein, Liebigs Ann. Chem., 2145 (1976).

"Synthesis of Tryptophan-Containing Peptides"

VI.A.18-1   P. Savignac and P. Coutrot, Synthesis, 197 (1976).

$$(EtO)_2P\text{-}\overset{\ominus}{C}Br_2 \; \overset{\oplus}{Li} \quad \underset{R'}{\overset{R\diagdown C=O}{\xrightarrow{\hspace{1cm}}}} \quad \underset{R'\diagup}{\overset{R\diagdown}{C}}=CBr_2$$

R,R' = H, alkyl, aryl, cyclic   40-70%

VI.A.18-2 G. A. Wheaton and D. J. Burton, Tetrahedron Lett., 895 (1976).

$$Ph_3\overset{\oplus}{P}-\overset{\ominus}{C}R^1R^2 + HCF_2Cl \longrightarrow R^1R^2C=CF_2$$

R^1, R^2 = H, alkyl, $-C_5H_{10}-$, etc.

Maximum initial yield 50%, but reagent is recycled to give yields of 80-100% (VPC).

VI.A.18-3 R. D. Clark and C. H. Heathcock, J. Org. Chem., **41**, 636 (1976).

$$\underset{R^2}{\text{cyclohexanedione}}\overset{R^1}{\underset{O}{\bigcirc}} \xrightarrow[CHCl_3]{(COCl)_2} \underset{R^2}{\text{cyclohexenone}}\overset{R^1}{\underset{Cl}{\bigcirc}}$$

R^1 = H, alkyl   76-91%

R^2 = H, alkyl, dialkyl

VI.A.18-4 K. Tamao, M. Zembayashi, and M. Kumada, Chem. Lett., 1237 (1976).

$$BrCH=CHOEt + RMgBr \xrightarrow{[Ni(dppp)Cl_2]} RCH=CHOEt$$

R = alkyl, Ph, thiophenyl   53-77%

VI.A.18-5   L. Blanco, P. Amice, and J. M. Conia, *Synthesis*, 194, 196 (1976).

$$R^1-\underset{H}{\underset{|}{\overset{R^2}{\overset{|}{C}}}}-\overset{O}{\overset{\|}{C}}-R^3 \xrightarrow[Et_3N]{ClSiMe_3} R^1-\underset{}{\overset{R^2}{\overset{|}{C}}}=\underset{}{\overset{OTMS}{\overset{|}{C}}}-R^3$$

R's = H, alkyl, cyclic

↓ CHBz₃, t-BuOK

$$R^1-\underset{Br}{\underset{|}{\overset{R^2}{\overset{|}{C}}}}=\overset{}{\overset{}{C}}-\overset{O}{\overset{\|}{C}}-R^3 \xleftarrow[\text{or } H^{\oplus}, MeOH]{\text{benzene, } \Delta} \text{cyclopropane with } R^2, OTMS, R^1, R^3, Br, Br$$

16-80% overall

VI.A.18-6   G. Zweifel and R. A. Lynd, *Synthesis*, 816 (1976).

$$R-C\equiv CH \xrightarrow[\begin{array}{l}1.\ (\underline{i}\text{-Bu})_2AlH\\2.\ EtOCH_2Cl\\3.\ H_3O^{\oplus}\end{array}]{} \underset{H}{\overset{R}{C}}=\underset{OEt}{\overset{H}{C}}$$

72-80%

R = n-Bu, t-Bu, cyclohexyl

VI.A.18-7 H. Klein et al., Tetrahedron Lett., 947 (1976).

$$[RCu(Br \text{ or } R)]MgX + H_2C=C=CHOCH_3 \xrightarrow{\quad\quad\quad} R\text{-}CH_2CH=CHOCH_3$$
$$2.\ H_3O^{\oplus}$$

R = alkyl, phenyl

80-90%

(mixture of E and Z)

VI.A.18-8 V. Subramanyam, E. H. Silver and A. H. Soloway, J. Org. Chem., 41, 1272 (1976).

$$Ph_3P=CHPh + H\text{-}\overset{O}{\underset{\|}{C}}\text{-}OR \longrightarrow Ph\text{-}CH=CH\text{-}OR$$

R = Et, s-Bu, cyclohexyl

VI.A.19-1 T. Mukaiyama, T. Masui, and T. Izawa, Chem. Lett., 1177 (1976).

$$\underset{\substack{N \\ OR}}{\overset{O_2N \quad NO_2}{\bigcirc}} + R'SH \xrightarrow[\text{catalyst}]{CH_2Cl_2} R\text{-}S\text{-}R'$$

60-91%

R = 1°,2° alkyl, containing double bonds, halides, esters

catalyst = $BF_3 \cdot Et_2O$ or $TiCl_4$

R'= Ph, Bz, t-Bu

VI.A.19-2   A.J.H. Labuschagne et al., Tetrahedron Lett., 3571 (1976).

$$R\text{-}CH_2\text{-}Br + 2\ Me_2S \xrightarrow{\Delta} R\text{-}CH_2\text{-}S\text{-}Me$$

R = vinyl, Ph, COOR, COPh   63-100%

$$RCH_2SMe + RCH_2Br \xrightarrow{\Delta} R\text{-}CH_2\text{-}S\text{-}CH_2\text{-}R$$

R = vinyl, Ph, COOR, COPh   43-76%

VI.A.19-3   R. S. Glass, Synth. Commun., 6, 47 (1976).

$$RCHO + R'SSiMe_3 \longrightarrow \underset{R\text{-}CHSR'}{\overset{OSiMe_3}{|}} \xrightarrow[AlCl_3]{LiAlH_4} RCH_2SR'$$

R,R' = Ph, alkyl, cycloalkyl

generally 50-80%

VI.A.19-4   F. Pochat and E. Levas, Tetrahedron Lett., 1491 (1976).

$$\underset{R'}{\overset{R}{\diagdown}}C\underset{SEt}{\overset{SEt}{\diagup}} \xrightarrow{I_2,\ Hg(CN)_2} \underset{R'}{\overset{R}{\diagdown}}C\underset{SEt}{\overset{CN}{\diagup}}$$

R,R' = H, alkyl, aryl

Yields generally 40-70%

USEFUL SYNTHETIC PREPARATIONS

VI.A.19-5   I. Kuwajima, T. Murofushi, and E. Nakamura, *Synthesis*, 602 (1976).

$$\underset{R^2}{\overset{R^1}{\diagdown}}C=CH-\underset{\|}{\overset{O}{C}}-R^3 + R^4SH \xrightarrow[\text{THF or acetone}]{R_4N^\oplus F^\ominus} R^1-\underset{SR^4}{\overset{R^2}{\underset{|}{C}}}-CH_2-\overset{O}{\underset{\|}{C}}-R^3$$

R^1, R^2 = H, alkyl, vinyl, cyclic   generally 80-90%

R^3 = H, Me, -COOR, cyclic

R^4 = Ph, Bz, -CH$_2$COOR

VI.A.19-6   I. I. Lapkin, G. G. Abashev, and F. G. Saitkulova, *J. Org. Chem. (USSR)*, 12, 975 (1976).

$$R-\overset{O}{\underset{\|}{C}}-\underset{\underset{R''}{\diagdown}}{\overset{\diagup R'}{C}}-Br \xrightarrow[\text{2. ArSCl}]{\text{1. Zn}} R-\overset{O}{\underset{\|}{C}}-\underset{\underset{SAr}{|}}{\overset{R'}{\underset{|}{C}}}-R''$$

R's = H, 1° alkyl   ~70%

few details given

VI.A.19-7   D. Seebach and M. Teschner, *Chem. Ber.*, 109, 1601 (1976).

$$\underset{R^2}{\overset{R^1}{\diagdown}}C=\underset{H}{\overset{OLi}{\diagup}} \xrightarrow{RSCl} R^1-\underset{R^2}{\overset{O}{\diagdown}}\underset{H}{\overset{SR}{\diagup}}$$

R = Me, Ph, CSNMe$_2$

R^1 = Me, Ph, *i*-Pr, cyclic   ~60-90%
R^2 = Me, Ph, cyclic

VI.A.19-8 W. T. Flowers et al., J. Chem. Soc. Perkin I, 2394 (1976).

$$ArSCN + ROH \xrightarrow{Ph_3P} Ar-S-R$$

Ar = Ph, p-MeOPh, p-NO_2Ph ~70-90%
R = n-alkyl, allyl, Bz

VI.A.19-9 B. B. Snider, N. J. Hrib, and L. Fuzesi, J. Am. Chem. Soc., 98, 7115 (1976).

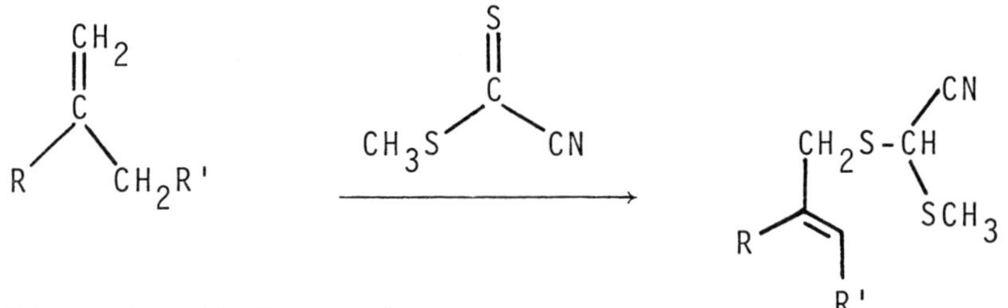

$$\underset{R}{\overset{CH_2}{\underset{|}{C}}}\text{-}CH_2R' \quad + \quad CH_3S\text{-}\underset{\|}{\overset{S}{C}}\text{-}CN \quad \longrightarrow \quad \underset{R\quad\quad R'}{CH_2S\text{-}CH(CN)(SCH_3)}$$

R,R' = H, alkyl, aryl

22-83%

VI.A.19-10 N. Kornblum et al., J. Org. Chem., 41, 1560 (1976).

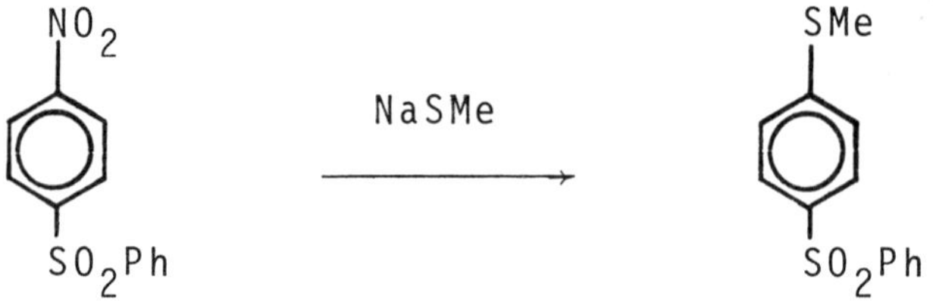

p-O_2N-C_6H_4-SO_2Ph $\xrightarrow{NaSMe}$ p-MeS-C_6H_4-SO_2Ph

61%

USEFUL SYNTHETIC PREPARATIONS

VI.A.19-11   K. Ogura, S. Furukawa, and G. Tsuchihashi, Synthesis, 202 (1976).

$$\underset{R'}{\overset{R}{>}}C=O \;+\; \underset{\text{cyclohexyl}}{\bigcirc}\!\!\!\underset{SCH_3}{\overset{\overset{O}{\|}\ominus}{SCH_2}}\,Li^{\oplus} \longrightarrow \underset{\text{cyclohexyl}}{\bigcirc}\!\!\!\underset{SCH_3}{\overset{\overset{O}{\|}}{SCH_2}}\!-\!\underset{R'}{\overset{OH}{\underset{|}{C}}}\!-\!R$$

1. $H_3O^{\oplus}$
2. $LiAlH_4$

$$\underset{R'}{\overset{R}{>}}\underset{OH}{\overset{CH_2SH}{\underset{|}{C}}}$$

60-90%

VI.A.19-12   P. Vermeer et al., Rec. Trav. Chim. Pays-Bas, 95, 25 (1976).

$$R^1-C\equiv C-S-R^2 \xrightarrow{LiAl(OMe)_3H/Cu_2Br_2} \underset{H}{\overset{R^1}{>}}C=C\underset{H}{\overset{SR^2}{<}}$$

90-95%

R^1 = Me, Et, Ph, vinyl, t-Bu

R^2 = Me, vinyl

$\xrightarrow{LiAlH_4}$

$$\underset{H}{\overset{R^1}{>}}C=C\underset{S-R^2}{\overset{H}{<}}$$

"nearly quantitative"

VI.A.19-13   P. Coutrot et al., Synthesis, 107 (1976).

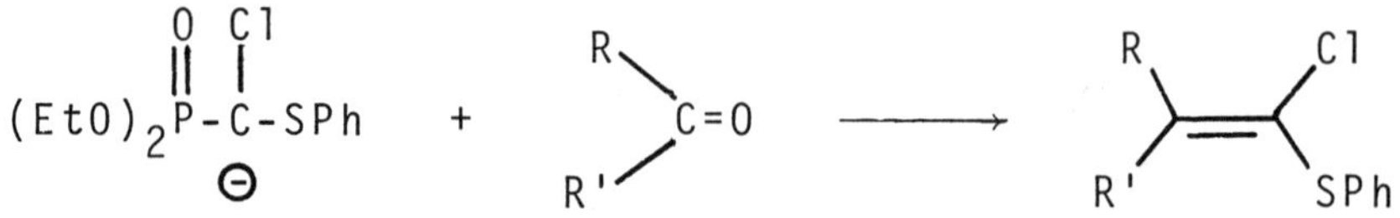

$$(EtO)_2\overset{O}{\underset{\ominus}{P}}-\overset{Cl}{\underset{}{C}}-SPh \;+\; \underset{R'}{\overset{R}{>}}C=O \;\longrightarrow\; \underset{R'}{\overset{R}{>}}C=C\underset{SPh}{\overset{Cl}{<}}$$

R,R' = H, alkyl, aryl, cyclic   46-68%

VI.A.19-14   H. Yoshida, T. Ogata, and S. Inokawa, Synthesis, 552 (1976).

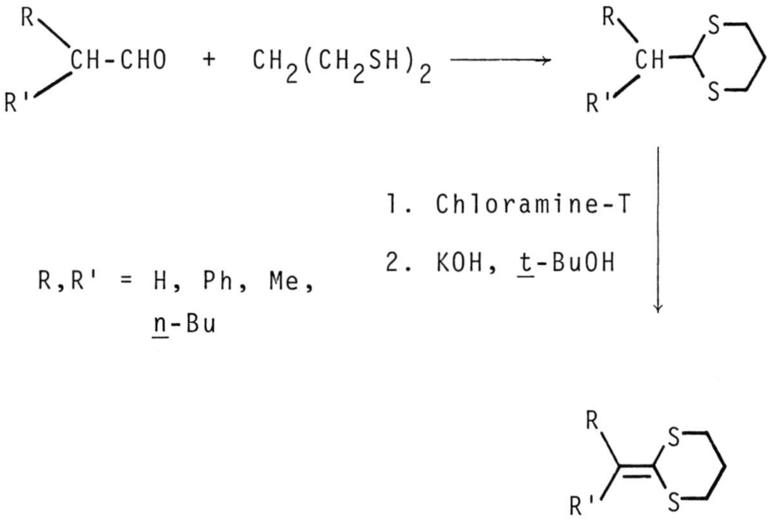

$$\underset{R'}{\overset{R}{>}}CH-CHO \;+\; CH_2(CH_2SH)_2 \;\longrightarrow\; \underset{R'}{\overset{R}{>}}CH-CH\underset{S}{\overset{S}{<}}\bigg]$$

R,R' = H, Ph, Me, n-Bu

1. Chloramine-T
2. KOH, t-BuOH

$$\underset{R'}{\overset{R}{>}}C=C\underset{S}{\overset{S}{<}}\bigg]$$

~60-75%

USEFUL SYNTHETIC PREPARATIONS

VI.A.19-15 H. Yoshida et al., Synthesis, 551 (1976).

$$CH_2(SR)_2 \xrightarrow[\text{Chloramine-T}]{\text{NaN-Ts}} \underset{\underset{NTs}{\overset{\|}{S-R}}}{CH_2}\!\!\!\!-SR \xrightarrow[\text{KOH}]{R'YH} HC\!\!\begin{array}{c}-SR\\-SR\\YR'\end{array}$$

R = alkyl, aryl, cyclic   Y = O, S   ~60-80%

VI.A.19-16 V. I. Cohen, Helv. Chem. Acta, 59, 840 (1976).

$$2\;R\overset{O}{\underset{}{-\!\!\overset{\|}{C}\!\!-}}R' \xrightarrow[\text{pyridine/Et}_3\text{N}]{H_2S} \underset{R'}{\overset{R}{\diagdown}}CH-S-S-CH\underset{R'}{\overset{R}{\diagup}}$$

R,R' = H, alkyl, aryl,   ~60-80%
       cyclic

VI.A.19-17 P. Dubs and R. Stüssi, Helv. Chim. Acta, 59, 1307 (1976).

$$R-S-S-R \xrightarrow{Me_3O^{\oplus}\;BF_4^{\ominus}} R-S-\overset{\oplus}{S}\!\!\begin{array}{c}Me\\R\end{array} \xrightarrow{R'SH} R-S-S-R'$$

R = Me, Et, n-Pr, i-Pr   52-82%

                    R' = alkyl, aryl, hetero-
                         cyclic

VI.A.19-18   J. R. Grunwell and D. L. Foerst, Synth. Commun., **6**, 453 (1976).

$$RCOOH + R'SH \xrightarrow{DCC} R-\overset{O}{\underset{\|}{C}}-SR'$$

R = CH$_3$, Ph   R' = Ar, 1°,2° alkyl   24-74%

VI.A.19-19   F. Souto-Bachiller, G. S. Bates and S. Masamune, J. Chem. Soc., Chem. Commun., 719 (1976).

A summary of the uses of **I** as a condensing agent in the preparation of thiol esters and amides.

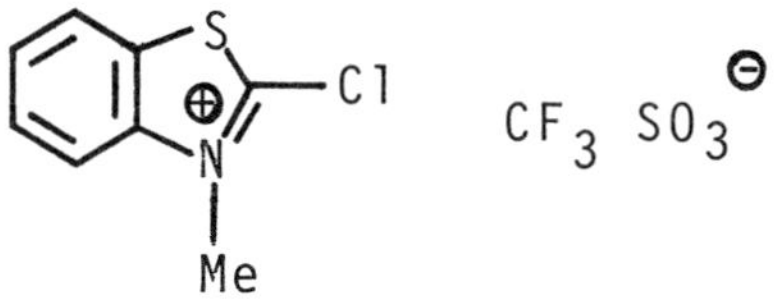

[Structure of benzothiazolium chloride with N-Me and CF$_3$SO$_3^-$ counterion]

I

VI.A.19-20   F. Le Goffic, S. Sicsic, and C. Vincent, Tetrahedron Lett., 2845 (1976).

Use of a membrane-bound pyridine catalyst

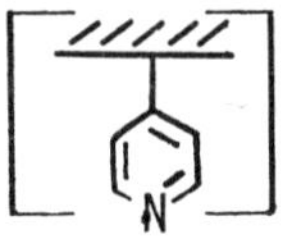

[Diagram of membrane-bound pyridine]

in the synthesis of the thioesters of thiol-amino acids. Most yields ~90%.

VI.A.19-21   W. H. Baarschers, Can. J. Chem., 54, 3056 (1976).

$$Ar-SO_2-OR \xrightarrow{R'Li} Ar-SO_2R'$$

R = Ph, n-alkyl   38-98%

R' = alkyl, aryl, heterocyclic

VI.A.19-22   J. D. Coyle, Chem. Soc. Rev., 4, 523

Review: "Photochemistry of Organic Sulfur Compounds"

VI.B.1-1   H.J.J. Loozen, W.M.M. Robben, and H. M. Buck, Rec. Trav. Chim. Pays-Bas, 95, 248 (1976).

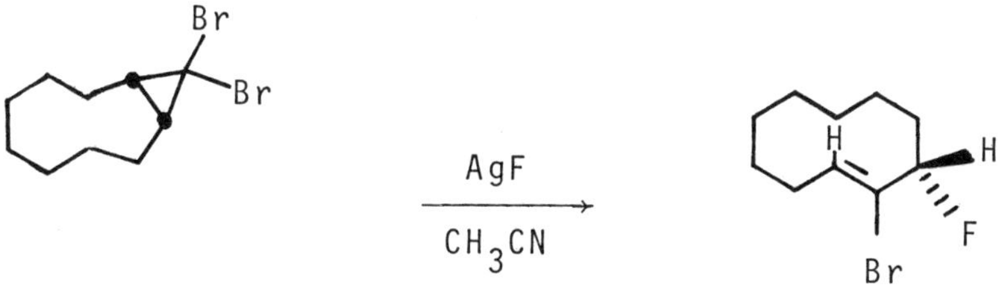

$$\xrightarrow[CH_3CN]{AgF}$$

~30%

VI.B.1-2   H.J.J. Loozen et al., J. Org. Chem., 41, 384 (1976).

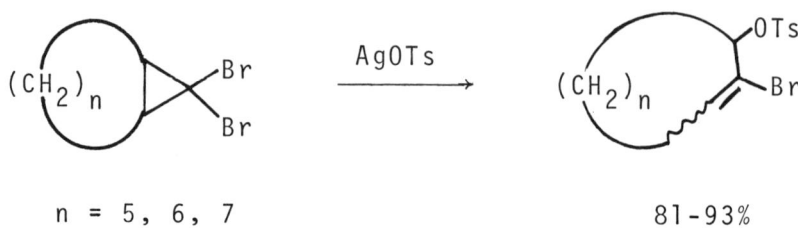

n = 5, 6, 7

81-93%

VI.B.1-3   H. H. Wasserman, M. J. Hearn, B. Haveaux, and M. Thyes, J. Org. Chem., 41, 153 (1976).

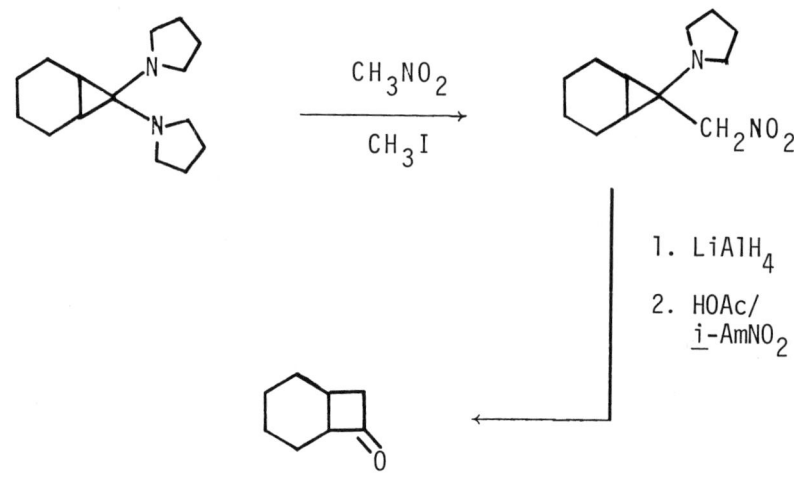

~18% overall

VI.B.1-4 K. Hafner, H. Diehl, and H. U. Süss, Angew. Chem. Int. Ed., 15, 104 (1976).

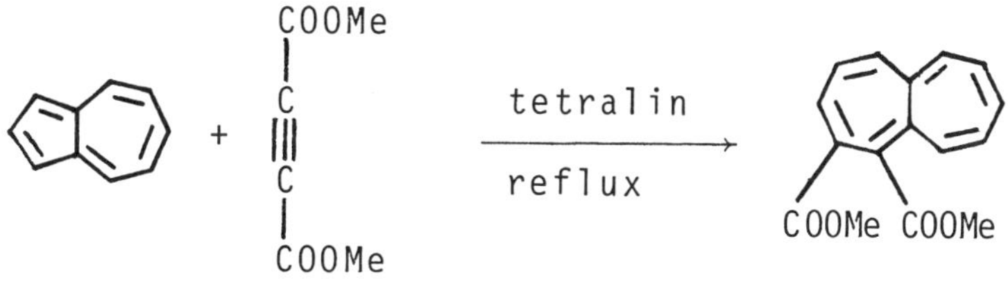
COOMe
C
C
COOMe
tetralin
reflux
COOMe COOMe

15%

VI.B.1-5 M. Sato, J. Tsunetsugu, and S. Ebine, Bull. Chem. Soc. Japan, 49, 2230 (1976).

A study of the following reaction:

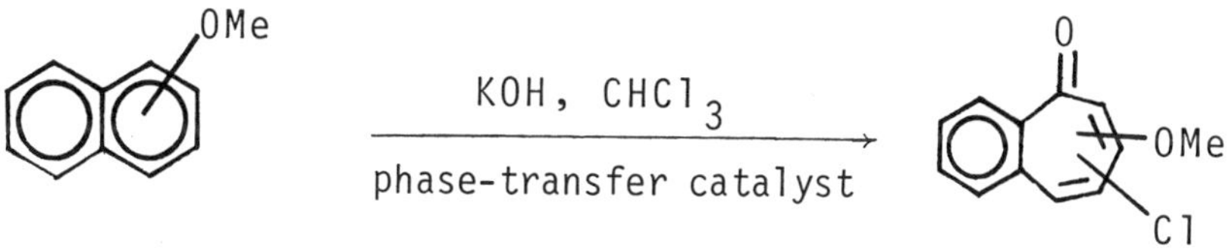
OMe
KOH, CHCl3
phase-transfer catalyst
O
OMe
Cl

Various isomers, widely varying yields.

VI.B.1-6 S. Kwon et al., Synthesis, 249 (1976).

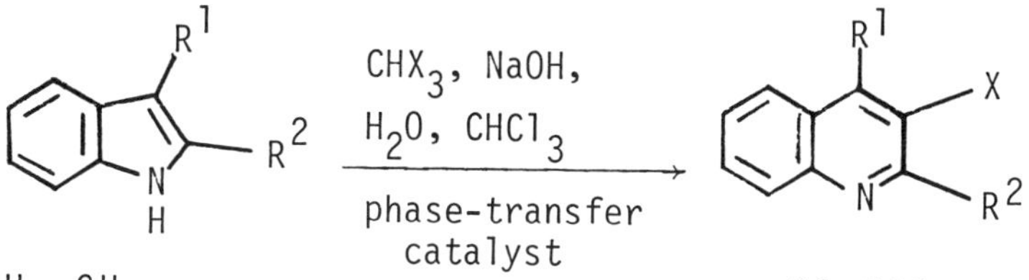
R1
R2
N
H
CHX3, NaOH,
H2O, CHCl3
phase-transfer
catalyst
R1
X
N
R2

R^1 = H, CH_3

R^2 = H, CH_3, Ph

24-68%

VI.B.1-7 M. B. Hasiak, Comptes Rendus (C), 282, 1003 (1976).

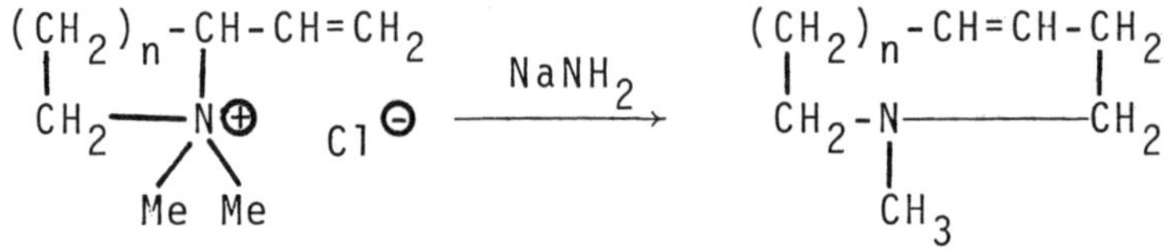

$$(CH_2)_n\text{-}CH\text{-}CH=CH_2 \atop CH_2\text{-}N^{\oplus}(Me)(Me)} \quad Cl^{\ominus} \xrightarrow{NaNH_2} (CH_2)_n\text{-}CH=CH\text{-}CH_2 \atop CH_2\text{-}N(CH_3)\text{-}CH_2$$

94% (n=3)
35% (n=2)

VI.B.1-8 Y. Sato, H. Kojima, and H. Shirai, J. Org. Chem., 41, 195 (1976).

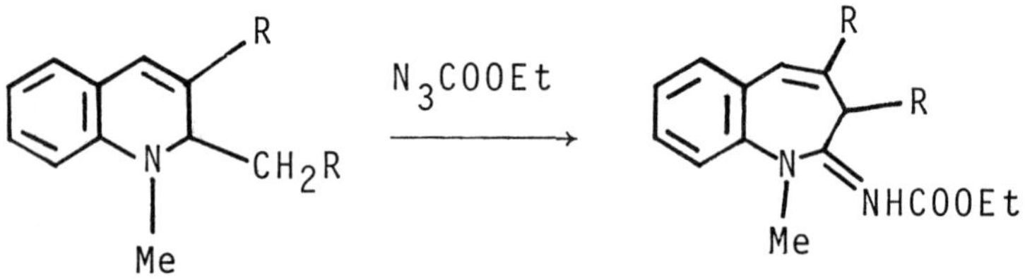

R:	Me	Et	i-Pr
% yield:	67	83	30

VI.B.2-1 G. Ortar and A. Romeo, J. Chem. Soc., Perkin I, 111 (1976).

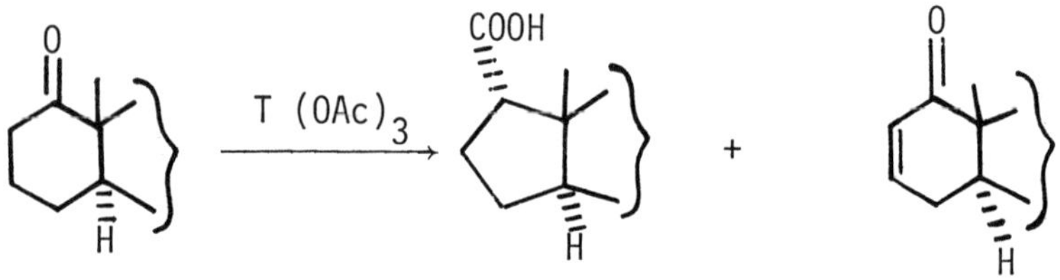

27%   28%

VI.B.2-2 H. Neunhoeffer and V. Boehnisch, Liebigs Ann. Chem., 153 (1976).

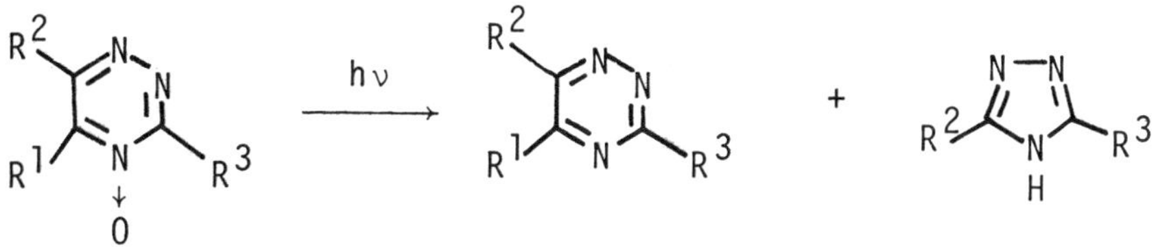

R's = H, Me, Ph   48-82% total yield

VI.C.1-1 S. F. Martin, J. Org. Chem., 41, 3337 (1976).

A novel spiroannelation sequence:

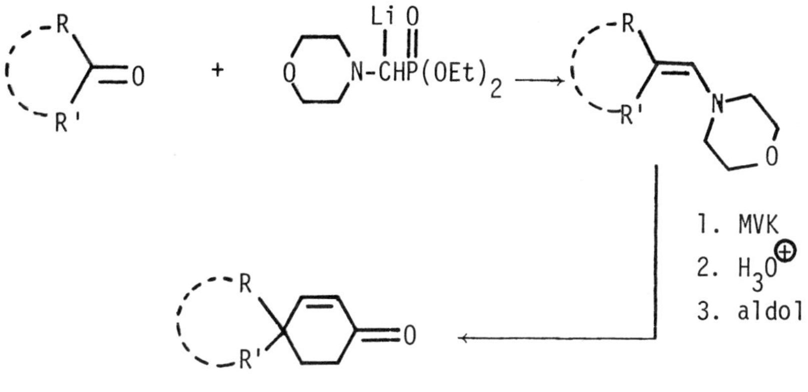

1. MVK
2. $H_3O^{\oplus}$
3. aldol

30-36%

VI.C.1-2 V. V. Kane, Synth. Commun., 6, 237 (1976).

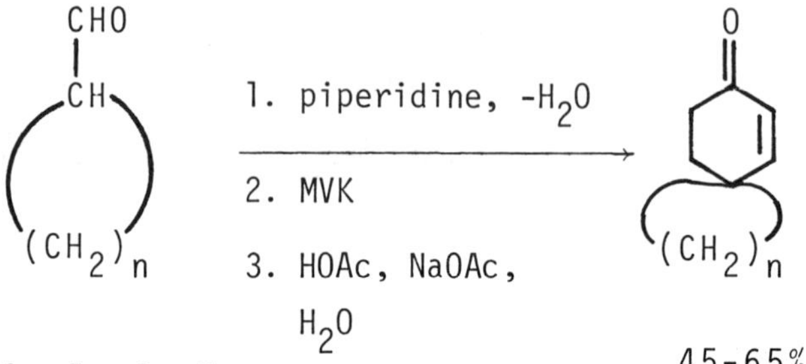

CHO
|
CH
⟨(CH$_2$)$_n$⟩

1. piperidine, -H$_2$O
2. MVK
3. HOAc, NaOAc, H$_2$O

→ cyclohexenone-⟨(CH$_2$)$_n$⟩

n = 4, 5, 6, 7

45-65%

VI.C.1-3 R. R. Wroble and D. S. Watt, J. Org. Chem., 41, 2939 (1976).

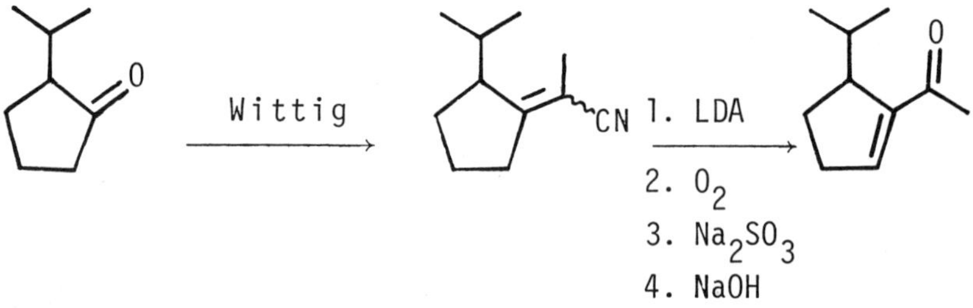

2-isopropylcyclopentanone —Wittig→ 2-isopropyl-1-(1-cyanoethylidene)cyclopentane

1. LDA
2. O$_2$
3. Na$_2$SO$_3$
4. NaOH

→ 5-isopropyl-1-acetylcyclopentene

10 examples, yields generally ~70%

VI.C.1-4 J. E. Baldwin, O. W. Lever, Jr., and
N. R. Tzodikov, J. Org. Chem., 41, 2312 (1976).

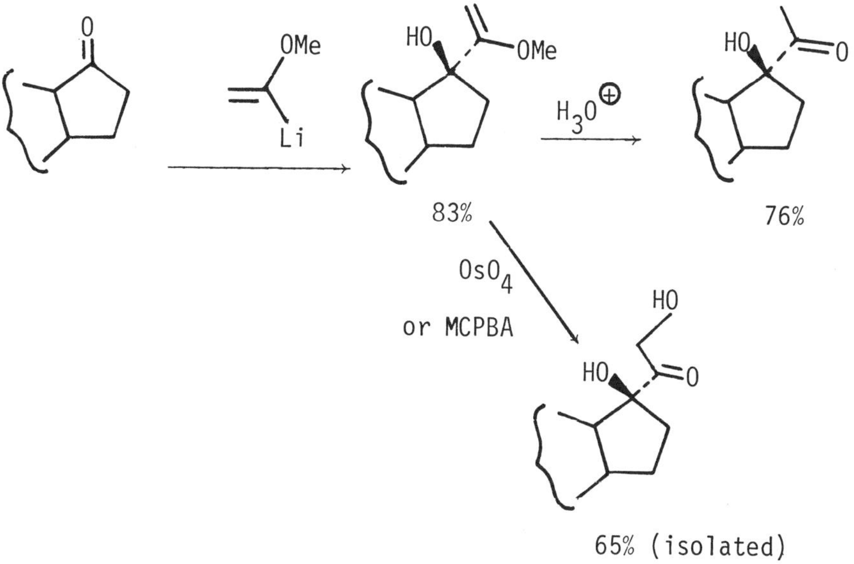

VI.C.1-5 B. M. Trost and D. E. Keeley, J. Am. Chem. Soc., 98, 248 (1976).

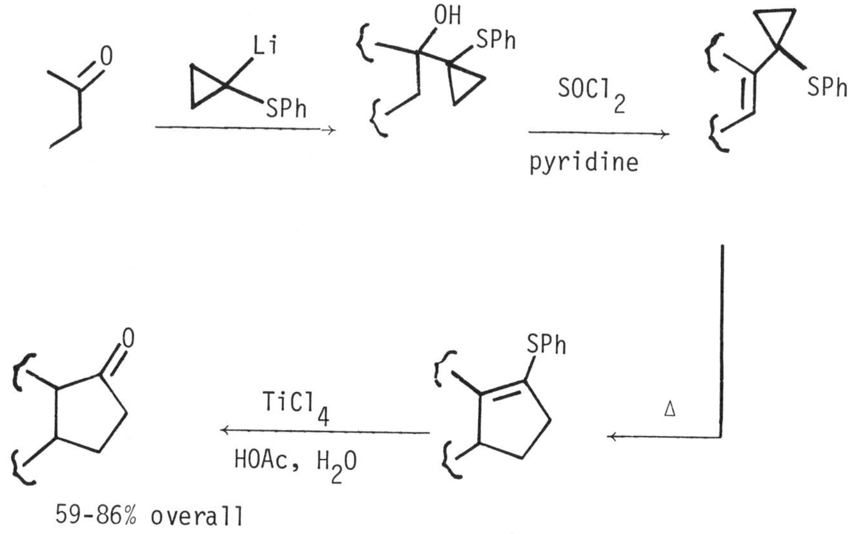

VI.C.1-6   T. Shono, I. Nishiguchi, and M. Nitta, Chem. Lett., 1319 (1976).

An alkylative carbonyl-transposition:

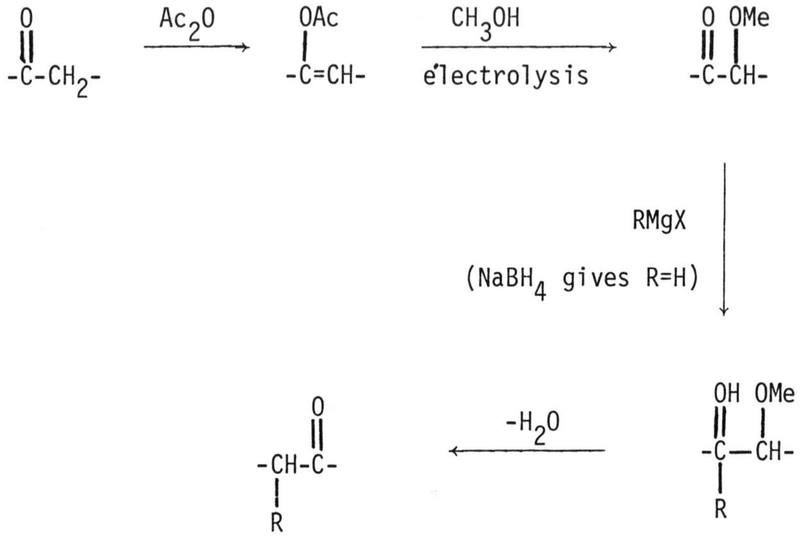

$$-\underset{\underset{O}{\|}}{C}-CH_2- \xrightarrow{Ac_2O} -\underset{\underset{OAc}{|}}{C}=CH- \xrightarrow[\text{electrolysis}]{CH_3OH} -\underset{\underset{O}{\|}}{C}-\underset{\underset{OMe}{|}}{CH}-$$

$$\downarrow \text{RMgX} \quad (NaBH_4 \text{ gives } R=H)$$

$$-\underset{\underset{R}{|}}{CH}-\underset{\underset{O}{\|}}{C}- \xleftarrow{-H_2O} -\underset{\underset{R}{|}}{C}(OH)-\underset{\underset{OMe}{|}}{CH}-$$

VI.C.1-7   G. Stork and G. A. Kraus, J. Am. Chem. Soc., 98, 2351 (1976).

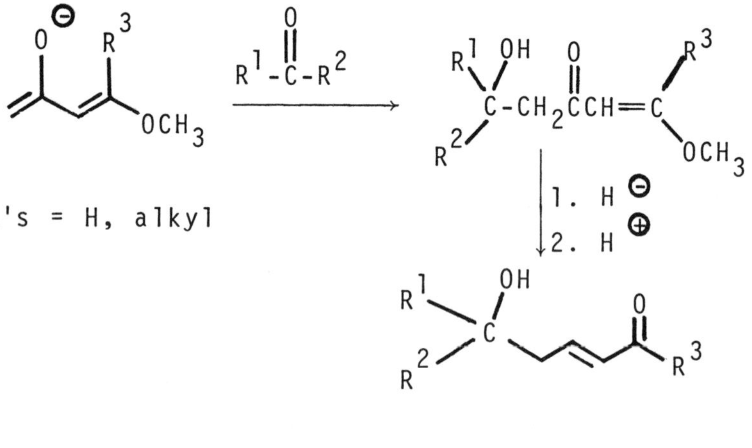

R's = H, alkyl

40-50% overall

VI.C.1-8   B. M. Trost and K. Hiroi, J. Am. Chem. Soc., 98, 4313 (1976).

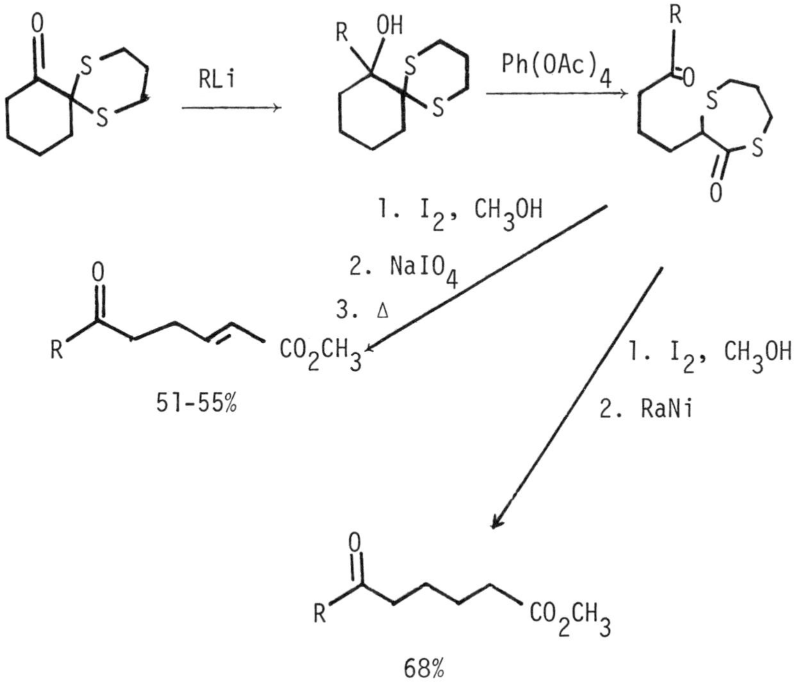

1. I_2, CH_3OH
2. $NaIO_4$
3. Δ

51-55%

1. I_2, CH_3OH
2. RaNi

68%

VI.C.1-9   V. Jäger and H. Grund, Angew. Chem. Int. Ed., 15, 50 (1976).

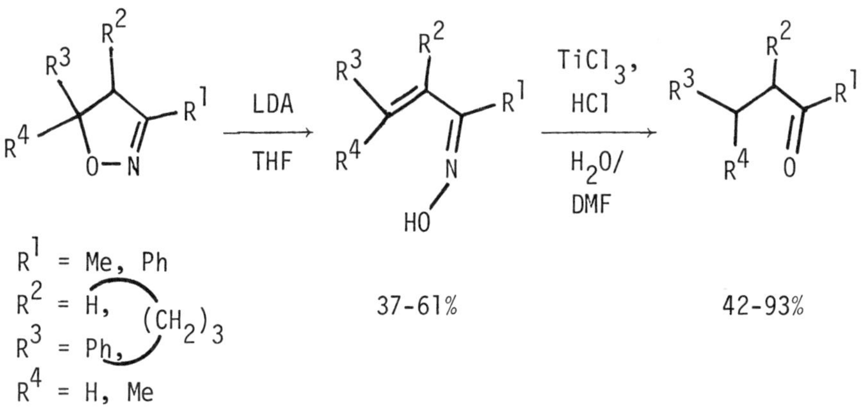

R^1 = Me, Ph
R^2 = H, ⎫
R^3 = Ph, ⎬ $(CH_2)_3$
R^4 = H, Me

37-61%   42-93%

VI.C.1-10 P. Blatcher, J. I. Grayson, and
S. Warren, *J. Chem. Soc., Chem. Commun.*, 547 (1976).

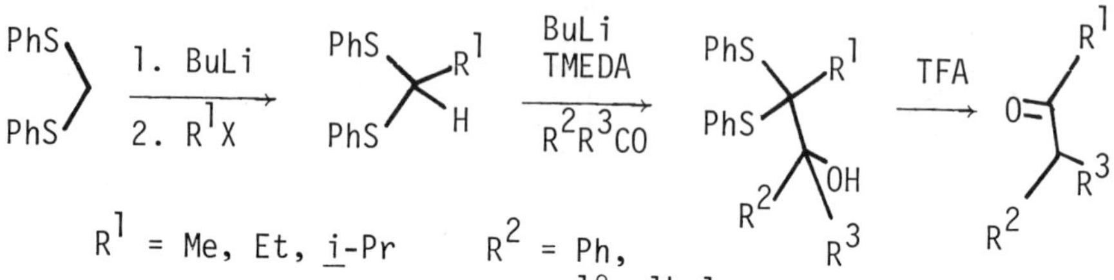

PhS₂CH₂ →[1. BuLi][2. R¹X] PhS–C(R¹)(H)–SPh →[BuLi TMEDA][R²R³CO] PhS–C(R¹)(SPh)–C(R²)(R³)OH →[TFA] O=C(R¹)–C(R²)(R³)

R^1 = Me, Et, i-Pr
R^2 = Ph, 1° alkyl
R^3 = H, -(CH$_2$)$_5$-

Overall yields ~40%

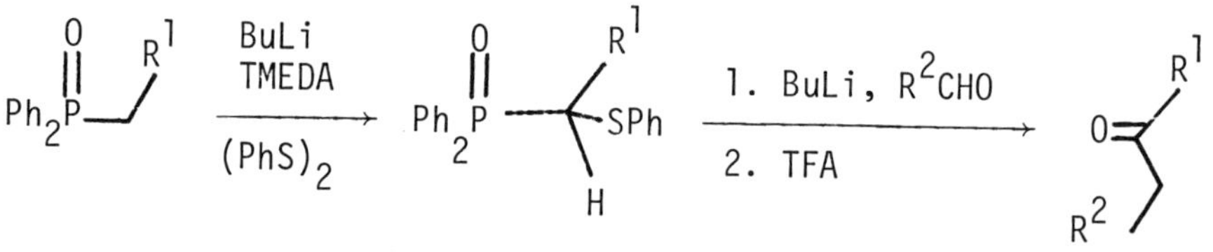

Ph$_2$P(O)–CH$_2$R¹ →[BuLi TMEDA][(PhS)$_2$] Ph$_2$P(O)–C(R¹)(SPh)(H) →[1. BuLi, R²CHO][2. TFA] O=C(R¹)–CH(R²)

R^1 = Et, Bz, i-Pr
R^2 = aryl, hexyl
53-69%

VI.C.2-1 A. Yasuda, H. Yamamoto, and H. Nozaki, *Tetrahedron Lett.*, 2621 (1976).

A 1,3-transposition reaction of allylic alcohols.

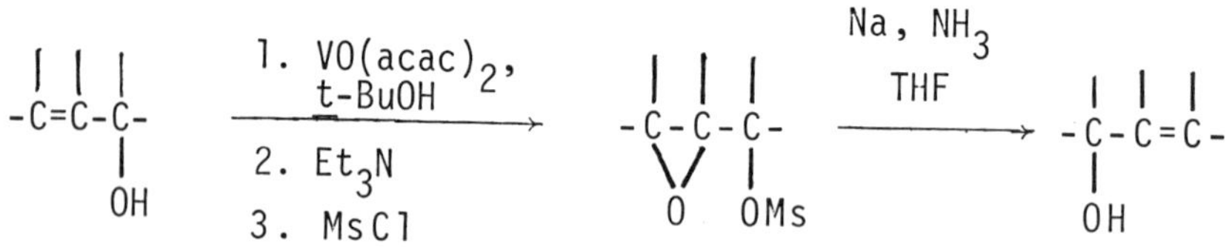

-C=C-C(OH)- →[1. VO(acac)$_2$, t-BuOH][2. Et$_3$N][3. MsCl] -C-C-C- (epoxide, OMs) →[Na, NH$_3$][THF] -C(OH)-C=C-

Yields typically 70%.

VI.C.2-2   S. S. Hall, C.-K. Sha, and F. Jordan,
J. Org. Chem., 41, 1494 (1976).

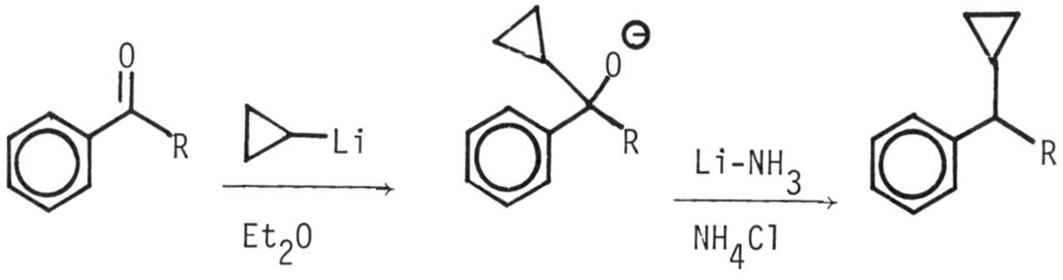

51-98%

VI.C.2-3   B. M. Trost, T. N. Salzmann, and K. Hiroi, J. Am. Chem. Soc., 98, 4887 (1976).

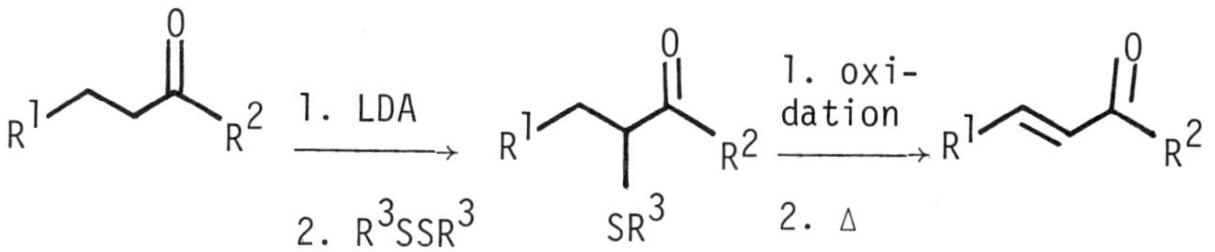

R^2 = alkyl, aryl, alkoxy
R^3 = Me, Ph

overall yields generally 60-80%

VI.C.2-4 L. A. Paquette, W. P. Melega, and J. D. Kramer, Tetrahedron Lett., 4033 (1976).

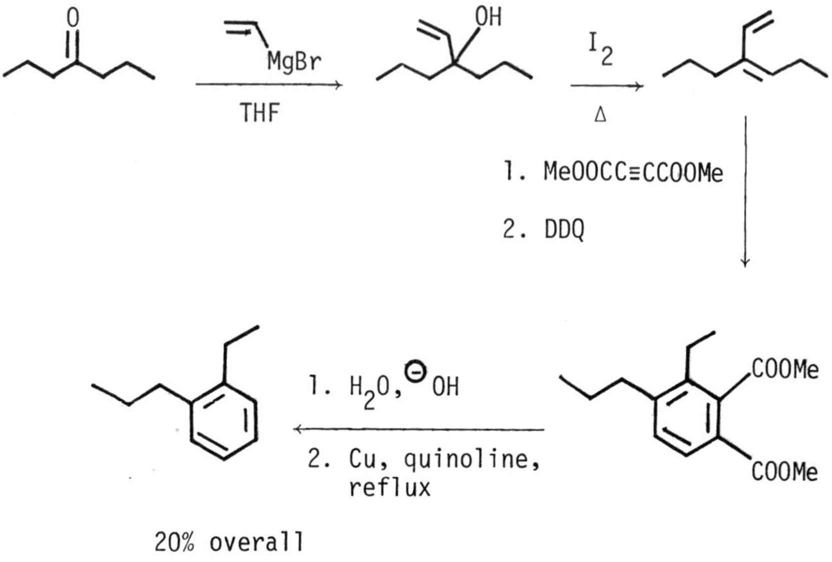
O
MgBr
THF
OH
I_2
Δ
1. MeOOCC≡CCOOMe
2. DDQ
COOMe
COOMe
1. H_2O, ⊖OH
2. Cu, quinoline, reflux
20% overall

VI.C.2-5 J. Rémion, W. Dumont, and A. Krief, Tetrahedron Lett., 1385 (1976).

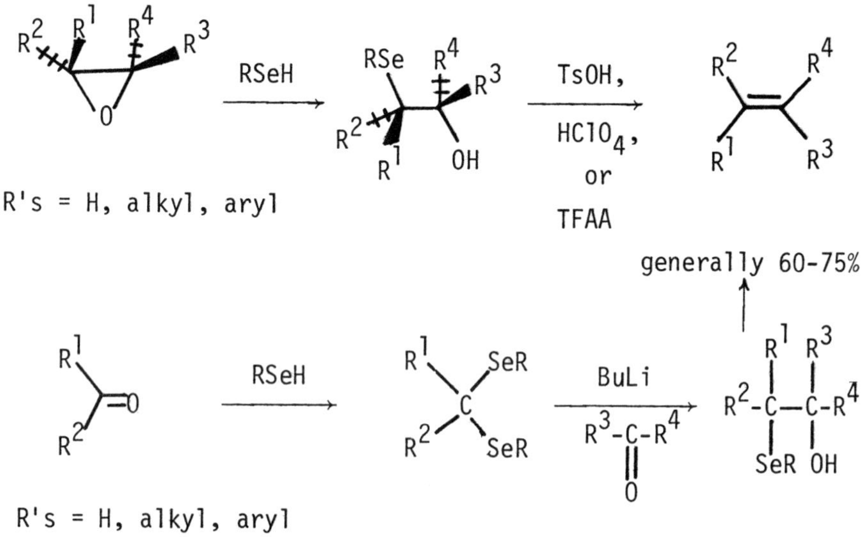
R2
R1
R4
R3
O
RSeH
RSe
OH
TsOH,
HClO4,
or
TFAA
R's = H, alkyl, aryl
generally 60-75%
RSeH
SeR
C
BuLi
R3-C-R4
O
SeR OH
R's = H, alkyl, aryl

VII.1 A. G. Schultz and M. H. Berger, <u>J. Org. Chem.</u>, 41, 585 (1976).

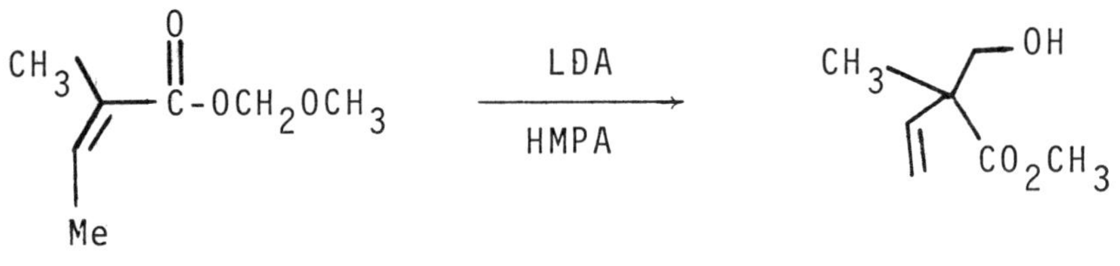

O
CH_3
C-OCH_2OCH_3
Me
LDA
HMPA
CH_3
OH
CO_2CH_3

61%

VII.2 W. Oppolzer, T. Sarkar, and K. K. Mahalanabis, <u>Helv. Chim. Acta</u>, 59, 2012 (1976).

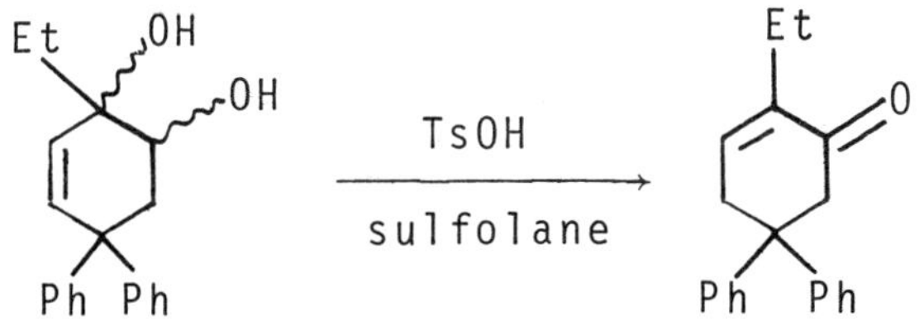

Et
OH
OH
Ph Ph
TsOH
sulfolane
Et
O
Ph Ph

92%

VII.3 C. W. Jefford and A. F. Boschung, <u>Helv. Chim. Acta</u>, 59, 962 (1976).

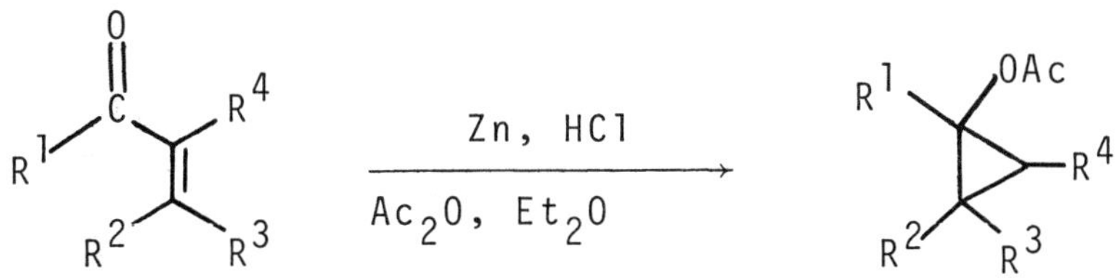

O
C
R^1
R^4
R^2
R^3
Zn, HCl
Ac_2O, Et_2O
R^1
OAc
R^4
R^2
R^3

R's = alkyl, H ~40-50%

VII.4 R.L.N. Harris, J. L. Huppatz, and J. N. Phillips, *Angew. Chem. Int. Ed.*, **15**, 498 (1976).

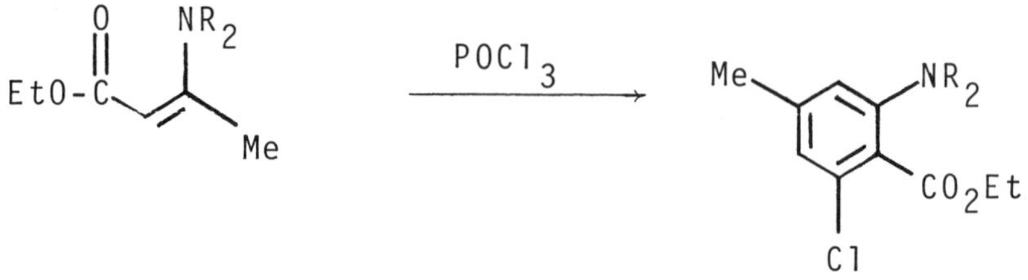

$$EtO-C(=O)-CH=C(Me)(NR_2) \xrightarrow{POCl_3} \text{2-}NR_2\text{-3-}Cl\text{-4-}CO_2Et\text{-6-}Me\text{-benzene}$$

$R_2 = Me_2, Et_2, -(CH_2)_{\overline{4},5}$

78-89%

VII.5 N. Dennis, A. R. Katritzky, and Y. Takeuchi, *Angew. Chem. Int. Ed.*, **15**, 1 (1976).

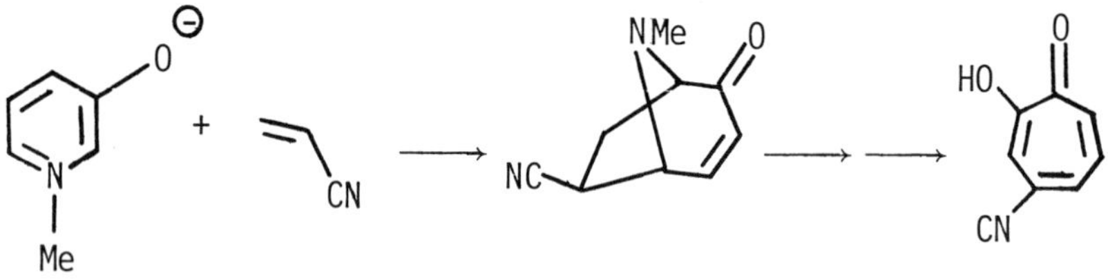

[N-methyl-3-oxidopyridinium] + CH$_2$=CH-CN ⟶ [bicyclic NMe, C=O, CN adduct] ⟶ ⟶ 2-hydroxy-4-cyano-tropone

---

VII.6 B. Ganem, *Tetrahedron Lett.*, 1951 (1976).

Synthesis of hydroxamic acids from oximinoesters:

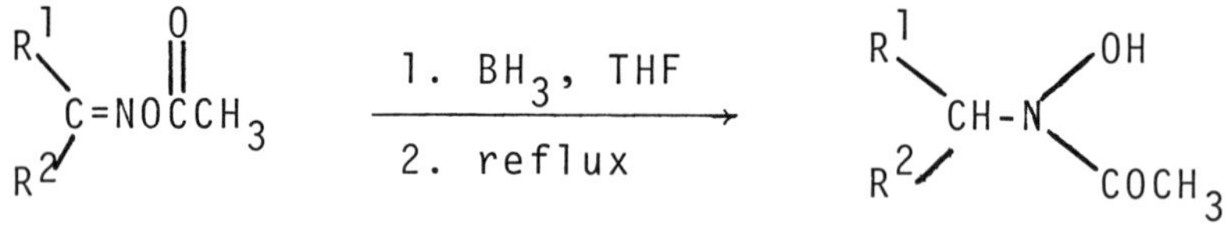

$$R^1R^2C=NOC(O)CH_3 \xrightarrow[\text{2. reflux}]{\text{1. BH}_3\text{, THF}} R^1R^2CH-N(OH)(COCH_3)$$

R^1 = H, alkyl

R^2 = alkyl, aryl

25-60%

VII.7   M. W. Gittos et al., J. Chem. Soc. Perkin I, 141 (1976).

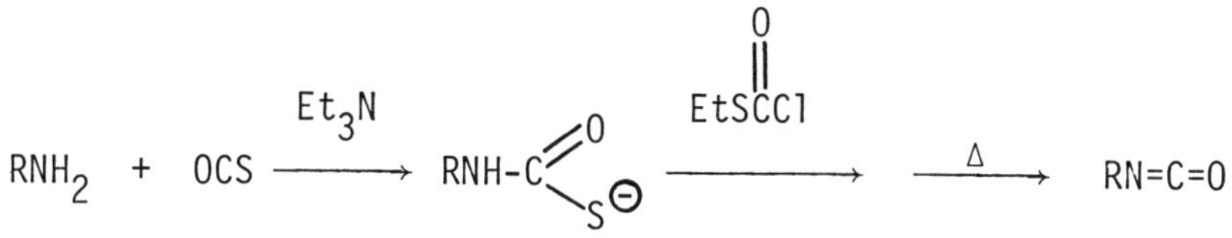

$$RNH_2 + OCS \xrightarrow{Et_3N} RNH-C\begin{smallmatrix}\nearrow O \\ \searrow S^{\ominus}\end{smallmatrix} \xrightarrow{EtS\overset{O}{\overset{\|}{C}}Cl} \xrightarrow{\Delta} RN=C=O$$

R = alkyl, aryl-substituted alkyl   58-96%

VIII.1   J. B. Hendrickson, Fortachritte der Chem. Forschung, 62, 49 (1976).

    Review: "A General Protocol for Systematic Synthesis Design"

VIII.2   M. Bersohn and A. Esack, Chem. Rev., 76, 269 (1976).

    Review: "Computers and Organic Synthesis"

VIII.3   T. Kametani and K. Fukumoto, Accounts Chem. Res., 9, 319 (1976).

    Review: "Total Synthesis of Natural Products by Retero Mass Spectral Synthesis"

VIII.4   G. W. Gokel and H. D. Durst, Synthesis, 168 (1976).

    Review: "Principles and Synthetic Applications in Crown Ether Chemistry"

VIII.5   J. I. Crowley and H. Rapoport, Accounts Chem. Res., 9, 135 (1976).

    Review: "Solid Phase Organic Synthesis: Novelty or Fundamental Concept?"

MISCELLANEOUS REVIEWS

VIII.6 A. Patchornik and M. A. Kraus, Pure and Appl. Chem., 46, 183 (1976).

  Review: "Recent Advances in the Use of Polymers as Chemical Reagents"

VIII.7 B. M. Mikhailov, Russ. Chem. Rev., 45, 557 (1976).

  Review: "Methods of Synthesis and Properties of Allylboranes"

VIII.8 H. C. Brown, Pure and Appl. Chem., 47, 49 (1976).

  Review: "Organoboranes -- The Modern Miracle"

VIII.9 J. Weill-Raynal, Synthesis, 633 (1976).

  Review: "Formation of Carbon-Carbon Bonds by Using Organoboranes"

VIII.10 G. Ohloff, Pure and Appl. Chem., 43, 481 (1975).

  Review: "Singlet Oxygen: A Reagent in Organic Synthesis"

VIII.11   D. H. Kim, <u>J. Het. Chem.</u>, <u>13</u>, 179 (1976).

    Review: "Acetic Anhydride as a Synthetic Reagent"

VIII.12   J. J. Renard and H. I. Bolker, <u>Chem. Rev.</u>, <u>76</u>, 487 (1976).

    Review: "The Chemistry of Chlorine Monoxide (Dichlorine Monoxide)"

VIII.13   J. K. Rasmussen and A. Hassner, <u>Chem. Rev.</u>, <u>76</u>, 389 (1976).

    Review: "Recent Developments in the Synthetic Uses of Chlorosulfonyl Isocyanate"

VIII.14   W. Kirmse, <u>Angew. Chem. Int. Ed.</u>, <u>15</u>, 251 (1976).

    Review: "Nitrogen as a Leaving Group: Aliphatic Diazonium Ions"

VIII.15   T. Mukaiyama, <u>Angew. Chem. Int. Ed.</u>, <u>15</u>, 94 (1976).

    Review: "Oxidation-Reduction Condensation"

MISCELLANEOUS REVIEWS

VIII.16   K. A. Korinek, Chem. and Ind., 931 (1976).

  Review: "Electro-Organic Oxidation and Reduction"

VIII.17   F. Minisci, Fortschritte der Chem. Forschung, 62, 1 (1976).

  Review: "Recent Aspects of Homolytic Aromatic Substitutions"

VIII.18   J. A. Zoltewicz, Fortschritte der Chem. Forschung, 59, 33 (1975).

  Review: "New Directions in Aromatic Nucleophilic Substitution"

VIII.19   R. A. Abramovitch and I. Shinkai, Accounts Chem. Res., 9, 192 (1976).

  Review: "Aromatic Substitution via New Rearrangements of Heteroaromatic N-Oxides"

VIII.20   J. d'Angelo, Tetrahedron, 32, 2979 (1976).

  Review: "Ketone Enolates: Regiospecific Preparation and Synthetic Uses"

VIII.21   R. E. Gawley, Synthesis, 777 (1976).

   Review: "The Robinson Annelation and Related Reactions"

VIII.22   H. Stetter, Angew. Chem. Int. Ed., 15, 639 (1976).

   Review: "Catalyzed Additions of Aldehydes to Activated Double Bonds -- A New Synthetic Approach"

VIII.23   W. S. Johnson, Angew. Chem. Int. Ed., 15, 9 (1976).

   Review: "Biomimetic Polyene Cyclizations"

VIII.24   A. P. Krapcho, Synthesis, 425 (1976).

   Review: "Synthesis of Carbocyclic Spiro Compounds via Rearrangement Routes"

VIII.25   J. F. Liebman and A. Greenberg, Chem. Rev., 76, 311 (1976).

   Review: "A Survey of Strained Organic Molecules"

MISCELLANEOUS REVIEWS

VIII.26   A. Padwa, Angew. Chem. Int. Ed., 15, 123 (1976).

   Review: "Intramolecular 1,3-Dipolar Cycloaddition Reactions"

VIII.27   C. W. Spangler, Chem. Rev., 76, 187 (1976).

   Review: "Thermal [1,j] Sigmatropic Rearrangements"

VIII.28   E. Campaigne and S. W. Schneller, Synthesis, 705 (1976).

   Review: "Cyclization of Ylidenemalonodinitriles"

VIII.29   T. Wagner-Jauregg, Synthesis, 349 (1976).

   Review: "Reactions of Azines and Imines with Dienophiles"

VIII.30   F. J. McQuillin, Chem. and Ind., 941 (1976).

   Review: "Homogeneous Catalysis in the Functionalisation of Alkenes and Alkanes"

VIII.31   I. Ernest, <u>Angew. Chem. Int. Ed.</u>, <u>15</u>, 207 (1976).

 Review: "A Synthesis of Prostaglandins: Strategy and Reality"

VIII.32   N. Cohen, <u>Accounts Chem. Res.</u>, <u>9</u>, 412 (1976).

 Review: "Asymmetric Induction in 19-Norsteroid Total Synthesis"

VIII.33   F. Kienzle, <u>Pure and Appl. Chem.</u>, <u>47</u>, 183 (1976).

 Review: "The Technical Syntheses of Carotenoids"

VIII.34   B.C.L. Weedon, <u>Pure and Appl. Chem.</u>, <u>47</u>, 161 (1976).

 Review: "Synthesis of Carotenoids and Related Polyenes"

VIII.35   H. Pommer and A. Nürrenbach, <u>Pure and Appl. Chem.</u>, <u>43</u>, 527 (1975).

 Review: "Industrial Synthesis of Terpene Compounds"

MISCELLANEOUS REVIEWS

VIII.36   A. J. Floyd, S. F. Dyke, and S. E. Ward, Chem. Rev., 76, 509 (1976).

    Review: "The Synthesis of Phenanthrenes"

VIII.37   N. V. Averina and N. S. Zefirov, Russ. Chem. Rev., 45, 544 (1976).

    Review: "Advances in the Synthesis of Heteroadamantanes"

VIII.38   P. Vollhardt, Fortschritte der Chem. Forschung, 59, 113 (1975).

    Review: "Cyclobutadienoids"

VIII.39   W.-H. Kunau, Angew. Chem. Int. Ed., 15, 61 (1976).

    Review: "Chemistry and Biochemistry of Unsaturated Fatty Acids"

VIII.40   H. G. Aurich and W. Weiss, Fortschritte der Chem. Forschung, 59, 65 (1975).

    Review: "Formation and Reactions of Aminyloxides" (nitroxides)

VIII.41 A. P. Kozikowski and H. F. Wetter, Synthesis, 561 (1976).

>Review: "Transition Metals in Organic Synthesis"

VIII.42 J. Schwartz and J. A. Labinger, Angew. Chem. Int. Ed., 15, 333 (1976).

>Review: "Hydrozirconation: A New Transition Metal Reagent for Organic Synthesis"

VIII.43 R. D. Rieke, Fortschritte der Chem. Forschung, 59, 1 (1975).

>Review: "Use of Activated Metals in Organic and Organometallic Synthesis"

VIII.44 P. Beak, Accounts Chem. Res., 9, 230 (1976).

>Review: "Silver-Assisted Reactions of Chloroformates: A New Route to Reactive Carbocations"

VIII.45 T. J. Marks, Accounts Chem. Res., 9, 223 (1976).

>Review: "Actinide Organometallic Chemistry"

MISCELLANEOUS REVIEWS

VIII.46   K. C. Bishop III, Chem. Rev., 76, 461 (1976).

  Review: "Transition Metal Catalyzed Rearrangements of Small Ring Organic Molecules"

VIII.47   I. P. Beletskaya, G. A. Artamkina, and O. A. Reutov, Russ. Chem. Rev., 45, 330 (1976).

  Review: "The Interaction of Organometallic Derivatives with Inorganic Halides"

VIII.48   P. J. Smith, Chem. and Ind., 1025 (1976).

  Review: "Organic Synthesis via Organotin Intermediates -- Some Recent Developments"

VIII.49   A. McKillop, Pure and Appl. Chem., 43, 463 (1975).

  Review: "Applications of Thallium(III)-nitrate (TTN) to Organic Synthesis"

AUTHOR INDEX

Abashev, G.G.-389
Abdulla, R.F.-290
Abenhaim, D.-54
Abramovitch, R.A.-413
Achiwa, K.-341
Achmatowicz, Jr., O.-137
Adam, W.-195
Adamopoulos, S.-300
Adinolfi, M.-207
Adlington, M.G.-230
Agawa, T.-106,182,250
Agranat, I.-201
Ahmad, A.-368
Ahmad, V.U.-199
Aida, T.-203
Aikawa, Y.-67,335
Akhrem, A.A.-273
Akita, Y.-268
Akiyama, M.-382
Akutagawa, S.-183
Alexandrou, N.E.-300
Alexis, A.-72,174
Ali, M.-342
Al-Jobour, N.H.-256
Alkabets, R.-32
Allen, G.F.-245
Al Neirabeyeh, M.-193
Alsofrom, D.-302
Ames, D.E.-247
Amice, P.-199,386
Anciaux, A.J.-124
Anderson, Jr., A.G.-327

Anderson, G.L.-375
Anderson, R.C.-73
Anderson, W.K.-244
Andrews, D.A.-357
Ansell, J.M.-116
Antonakis, K.-375
Antonowa, A.-368
Aoyama, H.-249
Appel, R.-380
Arackal, T.J.-11
Arai, M.-35
Arase, Y.-33
Arnold, Z.-143
Arnould, J.C.-341
Arpiani, M.P.-196
Artamkina, G.A.-187,419
Artaud, I.-121
Asahara, T.-33
Ashby, E.C.-217
Atabekyan, V.G.-206
Ateunis, M.-86
Auerbach, J.-337
Aumann, R.-160
Aurich, H.G.-417
Averina, N.V.-417
Azzaro, M.-241

Baardman, F.-133
Baarschers, W.H.-395
Baba, A.-182,250
Baba, S.-31,148,164,325,378
Babler, J.H.-97,357

AUTHOR INDEX

Back, T.G.-107
Badet, B.-352
Baeckstrom, P.-125
Baer, T.A.-29
Bair, K.W.-367
Baird, M.S.-119
Baker, Jr., J.D.-238
Baker, R.-182
Bakuzis, P.-356
Baldwin, J.E.-401
Bal'yan, K.V.-176
Ban, Y.-237
Banks, C.M.-321
Bapat, J.B.-367
Barber, G.N.-120,326,346
Bargetzi, J.P.-313
Barnes, M.W.-371
Barreiro, E.-112
Barsky, L.-362
Bartlett, P.D.-205,211
Barton, D.H.R.-107,333
Bartsch, H.-246
Basha, A.-199,233,358
Bass, R.J.-273
Bates, G.S.-394
Baudy, M.-299
Bauer, D.P.-114
Bauer, H.-274
Baukov, Y.I.-77
Baum, J.S.-132
Bäumer, G.-380
Baumes, R.-281

Baumstark, A.L.-105
Beak, P.-418
Beam, C.F.-19,47,48,283
Beaucage, S.L.-320
Bechara, E.J.H.-105
Becher, J.-255
Beck, J.R.-288
Beck, W.-8,39
Becket, G.J.P.-274
Beckhaus, H.-337
Behm, R.-303
Beletskaya, I.P.-187,419
Ben-Ishai, D.-144,343
Bennett, D.A.-19
Benoiton, N.L.-340
Bentley, T.W.-152
Berbalk, H.-340
Bergbreiter, D.E.-63
Berger, M.H.-37,407
Bergstrom, D.E.-373
Bernard, D.-65,169
Bernardon, C.-54
Bernhardt, J.C.-173,361
Bernstein, Z.-144,343
Bersohn, M.-410
Bertelo, C.A.-159,358
Bertisen, M.A.-134
Bertrand, M.-259
Bessodes, M.-375
Best, W.-160
Bestmann, H.J.-4,86,88,106,256
Bettoni, G.-251

Beyerman, H.C.-231
Bhagwatheeswaran, H.-254
Bhanu, S.-27
Bhatt, M.V.-108
Bhattacharjya, A.-107
Binkley, R.W.-188,211
Birch, A.J.-5,178
Bischoff, C.-43,255
Bishop III, K.C.-187,419
Black, D.S.-275
Blair, P.A.-14
Blake, J.-381,383
Blanco, L.-199,386
Blaszczak, L.-29
Blatcher, P.-404
Blatt, H.-279
Blondeau, D.-271
Blum, J.-242
Blyumberg, E.A.-194
Bock, M.G.-47,93
Bodor, N.-193
Boehnisch, V.-302,399
Boelens, H.-62,361
Bohlmann, F.-230
Bohme, H.-347
Boireau, G.-54
Boldrini, G.P.-225
Bolker, H.I.-412
Bond, F.T.-22,59
Bonfiglio, J.N.-267
Bongini, A.-51
Boone, J.R.-217

Borel, A.W.-170
Bos, H.J.T.-108
Boscacci, A.B.-275
Boschung, A.F.-128, 407
Boss, R.-309
Botta, A.-284,296
Bottaro, J.C.-244
Boulton, A.J.-233
Bourdais, J.-246
Bozimo, H.T.-184
Brandänge, S.-310
Bransdma, L.-111,176
Braun, H.-127
Braun, M.-126
Bravo, P.-88
Breitholle, E.G.-237
Brenner, S.-18
Brice, V.T.-166
Bridges, A.J.-131
Brieger, G.-213
Britt, R.W.-362
Britten-Kelly, M.R.-107
Broekhof, N.L.J.M.-305,314
Broom, A.D.-375
Broos, R.-86
Brouwer, A.C.-108
Brown, H.C.-151,154,185,215,
    231,238,239,329,330,411
Brownbridge, P.-96
Browne, L.J.-135
Brunelle, D.J.-264
Brzechffa, L.-248

AUTHOR INDEX

Buck, H.M.-395
Buddrus, J.-120,198
Bullivant, M.J.-125
Bunnett, J.F.-2
Burgers, P.M.J.-376
Burgot, J.-L.-127
Burgstahler, A.W.-354
Burmistrova, M.S.-77
Burns, J.R.-133
Burton, D.J.-89,385
Butler, R.N.-212

Cabiddu, S.-300
Cadona, L.-245
Cagniant, P.-244
Cahiez, G.-65,169,359
Cainelli, G.-189,331
Calas, R.-77,172,326
Calo, V.-101,235
Cambie, R.C.-200
Campaigne, E.-415
Capuano, L.-79
Caputo, R.-99,234
Cardillo, G.-189,354,359
Caress, E.A.-218
Carlson, R.M.-261
Carman, R.M.-328
Carney, R.L.-29
Carter, R.H.-262
Cassar, L.-161,323
Catsoulacos, P.-352
Cha, D.Y.-206

Chaabouni, R.-342
Chalk, A.J.-146
Chaloner, P.A.-3
Chamberlin, A.R.-22,59
Chan, K.C.-124
Chan, T.H.-25
Chandrasekaran, S.-168
Chang, J.K.-383
Chang, Y.-M.-132
Chao, E.-156
Chapman, K.J.-203
Chasar, D.W.-223
Chatterjie, N.-221
Chauhan, S.M.S.-286
Cheema, A.S.-44
Chen, C.-M.-197
Chen, F.C.M.-340
Chen, R.T.Y.-149,372
Cherkasov, L.N.-176
Chertkov, V.A.-77
Chiang, C.-S.-184
Chiba, T.-104
Chiu, K.-W.-153
Chiusoli, G.P.-37
Chong, A.O.-208
Chong, H.L.-124
Chorev, M.-313
Chou, S.-K.-111
Christl, B.-278
Christophe, D.-285
Chum, P.W.-163,229
Chung, L.-L.-106

Chupakhin, O.N.-149,240
Chwang, T.L.-26
Clark, J.H.-307
Clark, R.D.-71,385
Clayton, F.J.-170
Clizbe, L.A.-141
Cockerill, A.F.-293
Coghlan, M.J.-357
Cohen, J.S.-382
Cohen, L.A.-281
Cohen, N.-416
Cohen, S.-125
Cohen, T.-19,102
Cohen, V.I.-393
Cohen, Z.-195
Coleman, R.A.-165
Collin, G.J.-98
Collins, D.-J.-179
Collonges, F.-229
Comasseto, J.V.-89
Commercon, A.-72,174
Compagnon, P.-L.-85
Conia, J.M.-199,366,386
Conley, R.A.-184
Cooke, F.-13
Cookson, R.C.-364
Corbel, B.-21,363
Corey, E.J.-16,47,93,168,195
    264,305,317,318
Cornelisse, J.-148
Cossey, A.L.-267
Cossy, J.-341

Coudert, G.-270
Coulton, S.-340
Coutrot, P.-91,384,392
Coutts, I.G.C.-70
Couture, A.-80,125
Coyle, S.-314
Crabbé, P.-112,237
Crandall, J.K.-229
Creary, X.-2
Croce, P.D.-245
Crowley, J.I.-410
Crumrine, A.I.-26
Cruz, A.-237
Cueto, O.-195
Cunico, R.F.-123, 170
Cusack, N.J.-224
Cuvigny, T.-6,16,17,48,327,368
Czarny, M.R.-203,212,367

Dagani, D.-123
Daley, J.D.-353
Daly, P.J.-294
Damasevitz, G.A.-31
d'Angelo, J.-1,413
Danheiser, R.L.-168
Danieli, B.-210
Danishefsky, S.-42,162,258
Dauphin, G.-268
Davidovich, Yu.A.-381
Davidson, A.H.-92
Davis, S.E.-47
Debal, A.-327

AUTHOR INDEX

Deberly, A.-54
DeBuyck, L.-85
Decker, E.-340
Decouzon, M.-241
Degani, I.-30,355
de Groot, Ae.-36
Dehmlow, E.V.-118
Dejonghe, J.-P.-64
DeKimpe, N.-85
Delbecq, F.-110
Deleris, G.-172,326
Delia, T.J.-304
DeLue, N.R.-330
de Mayo, P.-80,125
Denis, J.N.-52
Dennis, N.-408
Deol, B.S.-218
Derguini-Boumechal, F.-171
Dervan, P.B.-100,234
Descoins, C.-94
Descotes, G.-225
Devasia, G.M.-280
Dhawau, K.L.-15
DiBiase, S.A.-34,370
Diehl, H.-397
Dinizo, S.E.-92
Distler, W.-106
Doleschall, G.-355
Dollat, J.-M.-112
Dolphin, D.-317
Dombrovskii, A.V.-294
Dominic, W.J.-133

Domschke, G.-259
Dorofeenko, G.N.-30
Doutheau, A.-122
Dowd, P.-100,236
Doyle, M.P.-232,331
Drabowicz, J.-222
Dubs, P.-296,393
Dumont, W.-52,53,95,406
Dundulis, E.A.-9
Dunn, L.C.-132
Dunoguès, J.-77,172,326
Durst, H.D.-410
Durst, T.-21,334
Duus, F.-278
Dyall, L.K.-203
Dyke, S.F.-417
Dynak, J.-162,258

Eberle, M.K.-248
Ebine, S.-397
Eckert, H.-310,315
Eckstein, H.-384
Edwards, M.-210
Eguchi, S.-122
Ehmann, W.J.-183
Ehrig, V.-195
Eichinger, K.-340
Eijsinga, H.-176
Eisch, J.J.-24,31,62,109
Eistert, B.-11
Elam, E.U.-286
Eldawy, M.A.-243

Elkik, E.-70,348
Ellis, G.P.-274
El-Naggar, G.M.-243
Enders, D.-16,41,47,93,318,339
Endo, M.-173
Engler, D.A.-102
Entmayr, P.-96
Entwistle, D.W.-320
Ermilova, E.V.-22
Ernest, I.-416
Esack, A.-410
Espinosa, J.-130
Eto, H.-12
Evans, D.A.-40
Evans, G.-269,321

Fairhurst, J.-349
Falck, J.R.-102
Falk, L.C.-361
Fallis, A.G.-237
Fatiadi, A.J.-212
Favorskaya, I.A.-22
Favre, B.-184
Fehrle, M.-380
Feingers, J.-374
Feiring, A.E.-143
Fell, B.-160
Fenzl, W.-155
Ferdinand, G.-109
Ferrey, M.-297
Ferris, J.P.-280
Fetizon, M.-190

Ficini, J.-262,348
Filosa, M.P.-135
Fischer, E.-303
Fischer, R.H.-287
Fizet, C.-366
Fleming, I.-76,96,131
Fleming, M.P.-105,235
Fletcher, H.-307
Flowers, W.T.-390
Floyd, A.J.-417
Foa, M.-161,323
Fochi, R.-30,355
Foerst, D.L.-394
Fogel, E.R.-102
Fomina, T.N.-176
Foote, R.S.-19,48
Ford, M.E.-184,277
Ford, W.T.-83
Formica, G.-215
Fortunato, J.M.-2,226
Foster, C.H.-286
Foucaud, A.-297
Fountain, K.R.-351
Fowler, K.W.-147
Francis, T.-352
Frandsen, E.G.-255
Frangin, Y.-175
Fraser, R.R.-15,272
Fraser-Reid, B.-73
Freche, P.-282
Fréchet, J.M.J.-306
Freedman, H.H.-188,202

AUTHOR INDEX

Freerksen, R.W.-92
Freter, K.-248
Fridman, A.L.-107,332
Friedrich, P.-365
Fritz, H.-366
Fry, A.J.-106
Fry, J.L.-230
Fryer, R.I.-269
Fu, T.-H.-213
Fuchigami, T.-302
Fuchs, P.L.-83,319
Fujii, S.-167
Fujii, T.-375
Fujimoto, Y.-231,232
Fujino, M.-313
Fujioka, A.-165
Fujisawa, T.-213,215,222,224
Fujita, E.-309,320
Fukumoto, K.-410
Fullerton, D.S.-197
Funk, R.L.-180
Furukawa, S.-391
Furusawa, K.-377
Fuzesi, L.-390
Fyles, T.M.-306

Gabitov, F.A.-107
Gait, M.J.-377
Gajewski, R.P.-288
Galle, J.E.-24,62,109
Gallemaers, J.-P.-285
Ganem, B.-2,226,408

Gaoni, Y.-20
Garanti, L.-301
Garcia, B.A.-332
Gardano, H.-161,323
Gasparoni, F.-37
Gaudemar, M.-112,175
Gaur, S.P.-254
Gawley, R.E.-65,414
Geiss, K.-H-55
Genco, N.-177
Germain, C.-54,246
Gesellchen, P.D.-384
Gesson, J.-P.-144
Ghosez, L.-64
Giesecke, H.-145,357
Gilchrist, T.L.-113,346
Girault, Y.-241
Girgis, N.S.-281
Girnth, M.-8,39
Gisler, M.-335
Gittos, M.W.-409
Glass, R.S.-369,388
Glotter, E.-206
Glowinski, R.-147
Glushkova, N.E.-168
Goh, S.H.-124
Gokel, G.W.-34,370,410
Goldschmitt, E.-140
Gomez-Parra, F.-190
Gorbutenko, V.I.-79
Gore, J.-110,122,329
Gorgues, A.-115,282

Gorjan, S.-301
Gorski, R.A.-6
Gotthardt, H.-278
Gould, K.J.-152
Gradeff, P.S.-215
Graefe, J.-83
Gramain, J.-C.-251
Grandi, R.-214
Granoth, I.-32
Gras, J.-L.-305
Grayson, J.I.-404
Greenberg, A.-414
Greene, A.E.-237
Griebsch, U.-18,50
Grieco, P.A.-28,183
Griffen, G.W.-345
Grignon-Dubois, M.-77
Grimm, K.G.-40
Gringore, O.-136
Grosjean, B.-85
Grossberg, H.-302
Grozinger, K.-248,316
Gruber, J.M.-209,353
Grund, H.-104,403
Grundon, M.F.-219
Grunwell, J.R.-394
Grutzner, J.B.-308
Grzejszczak, S.-90,91
Guida, W.C.-240
Gurria, G.M.-97
Gut, V.-381
Gutowski, F.D.-29

Guziec, Jr., F.S.-107

Haas, C.K.-123
Habeeb, J.J.-82
Hafner, K.-397
Hagihara, N.-166
Hagiwada, H.-68
Haines, A.H.-212
Hajek, M.-3
Hall, J.H.-335
Hall, J.L.-166
Hall, S.S.-405
Halpern, B.-311
Hamblin, M.-70
Hammond, J.J.-331
Han, Y.-K.-123
Hancock, W.S.-316
Hanna, J.-42
Hansen, J.F.-282
Hansson, C.-308
Harada, T.-271
Harding, D.R.K.-140,316
Harris, P.G.-316
Harris, R.L.N.-408
Harris, T.M.-37
Harris, T.V.-165
Harrison, C.R.-152,204,205
Hartford, T.W.-94,379
Hartman, G.D.-298
Hartmann, H.-44
Hartzell, S.L.-38,39
Hasegawa, T.-249

AUTHOR INDEX

Hashimoto,-162,163
Hasiak, M.B.-398
Haslouin, J.-136
Hasma, H.-85
Hassner, A.-412
Hata, T.-377
Hatanaka, N.-243
Haugwitz, R.D.-297
Hauptmann, S.-268
Hauser, C.H.-19,48
Haveaux, B.-13,396
Havinga, E.-148
Hayashi, T.-217,227
Hearn, M.J.-13,396
Heathcock, C.H.-71,385
Heber, D.-12
Heber-Brunschweiger, E.-12
Heck, R.F.-146
Hegedus, L.S.-245
Heine, H.-G.-129
Heldeweg, R.F.-161
Hendrickson, J.B.-367,410
Henning, R.-14,46
Henninger, M.-338
Hennion, G.F.-111
Henrici-Olive, G.-187
Hepburn, D.R.-328
Hercouet, A.-90,127
Herma, H.-43,255
Herscovici, J.-375
Hess, F.-248
Hicks, D.R.-73

Higa, T.-279
Hilbrich, R.G.-1
Hindley, N.C.-357
Hirai, H.-224
Hirayama, T.-285
Hiroi, K.-103,403,405
Hirth, M.-137
Hiyama, T.-28
Hjeds, H.-276
Ho, K.-80
Ho, T.-L.-191,192,219,223,233,
    308,315,319,338
Hoberg, H.-18,50
Hocker, J.-145,357
Hodge, P.-204,205
Hoffman, R.W.-198
Hogeveen, H.-161
Hojo, M.-78,204,320
Holland, H.L.-307
Holmes, A.B.-3
Holy, N.L.-56
Hooz, J.-154,378
Hope, A.P.-311
Hori, H.-309
Horiki, K.-381
Hortmann, A.G.-107
Horwell, D.C.-349
Hoshino, M.-80,125
Hosomi, A.-172,173
Houk, K.N.-132
House, H.O.-275
Hoy, R.C.-369

Hozumi, T.-377
Hrib, N.J.-390
Hsu, C.K.-198
Huang, S.L.-189
Huber, G.-127
Hubert, A.J.-124
Hudson, H.R.-328
Huegi, B.S.-23
Huffman, J.W.-316
Hughes, R.-151,156
Huisgen, R.-134
Hullot, P.-6
Huppatz, J.L.-408
Hutchins, R.O.-226,328
Hutchison, G.I.-98
Hyami, J.-12

Iemura, S.-162,163
Iida, K.-227
Ikehara, M.-376
Ikekawa, N.-231
Ikota, N.-341
Imai, K.-241
Imamura, T.-81
Imanaka, T.-175
Imbeaux-Oudotte, M.-70,348
Inokawa, S.-392
Inoue, T.-35,43
Inturrisi, C.E.-221
Inukai, N.-199
Ireland, R.E.-138
Isaacs, N.S.-253

Ishikawa, H.-239
Ishikawa, N.-146,336
Ito, Y.-167
Itoh, M.-32,72,157,158
Itoh, N.-221,222
Ittah, Y.-71,242
Ives, J.L.-196
Iwakuma, T.-221,222
Iwasaki, S.-109
Izawa, T.-387
Izawa, Y.-283,295,296
Izumi, G.-81

Jackson, A.E.-382
Jackson, A.G.-381
Jackson, W.R.-179
Jacob III, P.-154,209
Jacquelin, C.-4
Jacquesy, J.-C.-144
Jacquier, R.-281
Jäger, V.-104,403
Jain, P.C.-254
Jain, T.C.-321
Jansen, B.J.M.-36
Jefford, C.W.-128,407
Jennings, P.W.-166
Jesthi, P.K.-57
Johnson, B.F.G.-321
Johnson, C.R.-276
Johnson, W.S.-414
Johnstone, R.A.W.-382
Jonczyk, A.-121

AUTHOR INDEX

Jones, II, G.-126
Jones, L.D.-61,232
Jones, S.R.-338
Jordan, F.-405
Juillerat, M.-313
Julia, M.-20,352
Jung, F.-136
Jung, M.E.-14,130,192,194
Junghaus, K.-227
Junjappa, H.-286
Just, G.-316

Kabalka, G.W.-185,238,239
Kagan, H.B.-240
Kahn, S.A.-382
Kaiser, E.M.-22
Kaiser, L.-278,297
Kaji, A.-12,52
Kakushima, M.-130
Kalvoda, J.-87,208
Kamata, K.-323
Kamel, M.-120
Kametani, T.-308,325, 410
Kaminski, J.J.-193
Kamitori, Y.-79
Kamper, F.-200
Kanazawa, H.-376
Kanazawa, R.-220
Kane, V.V.-400
Kaneda, K.-175
Kaneko, K.-320
Kanfer, S.-147

Kanno, M.-226
Karanewsky, D.S.-138
Kasahara, A.-249,283
Kasal, A.-99
Kaschig, J.-380
Kato, T.-104,287
Katritzky, A.R.-408
Katzenellenbogen, J.A.-26
Kaufmann, H.-208
Kawabata, N.-121
Kawata, K.-65
Kawazoe, T.-82,353
Kawazoe, Y.-241
Kay, I.T.-303
Keana, J.F.W.-289
Keehn, P.M.-367
Keeley, D.E.-401
Keir, W.F.-254
Kelly, R.C.-206
Kempe, T.-99,234
Kenner, G.W.-311
Khalil, H.-279
Khan, E.A.-27
Khan, N.H.-342
Kharitonov, N.P.-168
Khodakov, Y.S.-228
Khirpach, V.A.-273
Kieczykowski, G.R.-17
Kienzle, F.-416
Kikugawa, Y.-213,281
Killough, J.M.-63
Kim, D.H.-412

Kimpenhaus, W.-120
Kinast, G.-41
Kincaid, K.-353
King, A.O.-164,325
King, F.D.-309
King, J.F.-140
Kinoshita, M.-374
Kirchoff, R.-202,359
Kirmse, W.-412
Kirsch, G.-244
Kise, H.-33
Kishi, Y.-138,254
Kisielowski, L.-106
Kiso, Y.-165
Kitagawa, Y.-162,163
Kitahara, T.-42,162,258
Kitamoto, M.-218
Kitatani, K.-28
Klausner, Y.S.-313
Klein, H.-176,387
Kleine, B.-169
Knapp, S.-195,318
Knecht, J.-160
Knittel, P.-74
Knunyants, I.L.-277
Knuppen, R.-197
Kobayashi, S.-116,335,350
Köbrich, G.-96
Kocienski, P.J.-116
Kodama, S.-165
Kodama, T.-68
Kofron, W.G.-15,46,69

Kohara, T.-159
Koizumi, T.-82,353
Kojima, H.-398
Kokel, B.-324
Kokuryo, K.-182
Kokuryo, Y.-78
Kollenz, G.-257
Kollonitsch, J.-330
Kolobov, N.A.-332
Konen, D.A.-38,108
Könneche, A.-303
Konoike, T.-167
Korinek, K.A.-413
Kornblum, N.-372,390
Korte, F.-137
Koshutin, V.I.-207
Kosley, Jr., R.W.-139
Köster, R.-155
Kosugi, H.-68
Kouwenhoven, A.P.-133
Kovar, R.-225
Kozikowski, A.P.-187,418
Kraatz, U.-137
Kramer, J.D.-406
Krapcho, A.P.-9,414
Krapf, H.-137
Kraus, G.A.-36,402
Kraus, M.A.-411
Kretchmer, R.A.-147,294
Kreutzberger, A.-284
Kricheldorf, H.R.-380
Kricks, R.J.-119

AUTHOR INDEX

Krief, A.-52,53,95,214,349, 379,406
Krishnamurthy, S.-215,231,239
Kröhnke, F.-270
Krubsack, A.J.-279
Kruse, C.G.-305,314
Ksander, G.M.-103
Kuhl, P.-83
Kuhlmann, H.-73
Kumada, M.-165,171,217,385
Kumagai, M.-172
Kunau, W.-H.-417
Kunerth, D.C.-354
Kunz, H.-312
Kurata, K.-261
Kurihara, T.-10,265
Kuroda, Y.-118
Kurz, M.E.-149,372
Kuwajima, I.-8,35,53,389
Kwon, S.-270,397

Labinger, J.A.-186,418
Labuschagne, A.J.H.-388
Lakhvich, F.A.-273
Lane, C.F.-185,239
Lane, R.M.-364
Langer, U.-129
Lansbury, P.T.-362
Lapkin, I.I.-389
Larcheveque, M.-6,16,17,48,327
Larock, R.C.-167,173,179,260, 361,380

Larsen, P.K.-276
Latif, N.-281
Lau, C.K.-135
Laub, R.J.-152
Laurenco, C.-91,349
Laurent, A.-342
LaVoie, E.J.-244
LeBorgne, J.F.-16,17,48,259
Lecas-Nawracka, A.-366
Le Coq, A.-115
Le Corre, M.-90,127
Lee, A.O.-24,60
Lee, G.A.-188,202
Lee, L.F.-275
Lefebvre, A.-324
Le Goffic, F.-394
Lehr, F.-14,45,46
Leonard-Coppens, A.M.-95
Leong, A.Y.M.-9,69
L'Eplattenier, F.A.-257
Lerner, L.M.-8
Leroux, Y.-4
Leroy, J.-272
Lespagnol, C.-324
Levas, E.-172,282,388
Lever, Jr., O.W.-401
Levisalles, J.-117,184
Levy, A.B.-157,327
Lewis, J.-321
Leznoff, C.C.-306
Li, C.H.-381,383
Liebman, J.F.-414

Liebscher, J.-44
Lin, J.J.-225
Linckens, A.-365
Lindblom, L.-310
Linstrumelle, G.-27,171
Lipisko, B.A.-34,370
Lippmann, E.-303
Lipshutz, B.H.-253
Liu, H.-J.-210
Liu, K.-T.-184
Lochow, C.F.-181
Loozen, H.J.J.-395,396
Lopez, L.-101
Lotts, K.D.-120,346
Louis, J.-M.-190
Loustalot, M.F.G.-200
Louw, R.-204
Lowe, J.A.-14
Lowe, O.B.-204
Lubinskaya, O.V.-77
Lubosch, W.-19,41,49
Luche, J.-L.-112
Luftmann, H.-200
Lukas, J.H.-133
Luteri, G.F.-83
Lynd, R.A.-30,64,386
Lythgoe, B.-92

Maak, N.-198
Maccioni, A.-300
Macdonald, T.L.-58
Macomber, R.S.-114

Magennis, S.A.-146
Mageswaran, S.-236
Magnus, P.-13
Mahajan, J.R.-266
Mahalanabis, K.K.-407
Mairanovsky, V.G.-322
Majetich, G.-28
Maki, Y.-221
Makin, S.M.-379
Makisumi, Y.-247
Makosza, M.-121,347
Malek, J.-3
Manescalchi, F.-331
Manhas, M.S.-252,253
Manmade, A.-345
Manning, M.J.-61
Marburg, S.-330
Marek, P.J.-170
Marino, J.P.-135
Marinovic, N.-183
Marks, T.J.-418
Marquet, B.-342
Marten, D.-177
Martens, J.-274
Martin, S.F.-260,399
Marwaha, L.K.-100,236
Maryanoff, C.A.-328
Masamune, S.-394
Masilamani, D.-328
Maslov, S.A.-194
Masson, J.-127
Masuda, A.-65

AUTHOR INDEX

Masuda, R.-78,204,320
Masui, T.-387
Masuyama, Y.-5,360
Mathew, J.-69
Mathey, F.-63,84
Matskovskaya, E.S.-30
Matsui, M.-190,351
Matsumoto, H.-176,251,291
Matsumoto, K.-272
Matsumoto, M.-106
Matsuo, S.-78
Matsuura, S.-382
Matteson, D.S.-57
Mayer, R.-64
McCleery, D.G.-219
McCloskey, J.E.-321
McCombs, C.A.-130
McDermott, J.X.-160
McDonnell, L.P.-126
McIntosh, J.M.-279
McKean, D.R.-332
McKee, R.-42
McKillop, A.-184,210,419
McLane, R.C.-56
McManus, S.P.-293
McMinn, D.G.-287
McMurray, J.E.-103,105,235
McOsker, C.C.-232
McQuillin, F.J.-415
Meijer, J.-111,174,176
Melega, W.P.-406
Mellor, J.M.-338

Mel'nichenko, N.V.-79
Mel'nikov, V.F.-207
Melpolder, J.B.-146
Melvin, Jr., L.S.-7
Merten, R.-145,357
Metzner, P.-127
Meyer, N.-23,57
Meyers, A.I.-263,277,293,323
Mezheritskaya, L.V.-30
Midgley, J.M.-34
Midura, W.-91
Mihelich, E.D.-293
Mikhailov, B.M.-185,411
Mikolajczyk, M.-90,91,222
Miller, J.A.-332
Miller, J.M.-307
Miller, L.L.-201
Miller, R.G.-181
Millership, J.S.-34
Millon, J.-27
Mil'vitskaya, E.M.-135
Minachev, K.M.-228
Minami, N.-35
Minami, T.-106
Minato, A.-165
Mincuzzi, A.-101
Minisci, F.-148,413
Miocque, M.-291
Mise, T.-217
Mishriky, N.-281
Mitchell, A.R.-383
Mitchell, M.A.-179

Mitchell, T.N.-169
Mitsudo, T.-182
Mitsunobu, O.-10,265
Miyake, K.-150
Miyashita, M.-68
Miyaura, N.-72,157
Miyazawa, Y.-172
Möller, F.-10
Mong, G.M.-209
Montellano, P.R.O.-198
Moore, D.R.-260
Moore, G.A.-311,340
Moormann, A.E.-97
Moran, T.A.-92
Moreau, J.-L.-112
Mori, M.-237
Morikawa, A.-191
Morioka, S.-376
Morley, J.O.-223
Mortimer, R.-152,378
Mueller, R.H.-138
Muetterties, E.L.-230
Mühlstädt, M.-83
Mukaiyama, T.-1,7,35,42,43,65, 66,67,191,239,264,324,335, 350,387,412
Mukhina, M.V.-206
Munslow, W.D.-304
Münzenmaier, W.-181
Mura, Jr., A.J.-19,102
Murahashi, S.-I.-153
Murofushi, T.-389

Murphy, G.P.-37
Murphy, R.-155
Murray, T.F.-258
Musso, H.-129

Nagahara, K.-289
Nagai, Y.-176
Nagakura, I.-135
Nagamatsu, T.-374
Nagao, Y.-320
Naka, M.-121
Nakajima, I.-165
Nakajima, T.-343
Nakajima, Y.-10,265
Nakamura, E.-8,53,318,389
Nakamura, H.-42,43
Nakamura, Y.-324
Nakata, F.-122
Nakayama, J.-11,356
Nakhshunov, V.S.-228
Nambudiry, M.E.N.-92
Naples, J.O.-265
Narang, S.A.-377
Narasaka, K.-67
Narita, M.-304,382
Nasipuri, D.-216
Negishi, E.-31,148,153,156,164, 186,325,378
Neumann, H.-59
Neunhoeffer, H.-302,399
Nikado, T.-176
Nimgirawath, S.-364

AUTHOR INDEX

Nishigaki, S.-376
Nishiguchi, I.-82,402
Nishimura, O.-313
Nishio, O.-32
Nishiyama, K.-220
Nishizawa, M.-183
Nitta, M.-402
Niu, C.-H.-382
Nivard, R.J.F.-134
Nobbs, M.S.-182
Node, M.-309
Noels, A.F.-124
Nojima, M.-336
Nomura, Y.-243
Norcross, B.E.-116
Norin, T.-99,234
Normant, H.-6,16,17,48,349
Normant, J.F.-64,65,72,169,174
Norton, J.R.-258
Nozaki, H.-28,31,32,95,152,158,
    162,163,404
Nunn, M.J.-332
Nürrenbach, A.-416
Nuyens, L.J.-306
Nyberg, K.-97,347

Obayashi, M.-95,158
Odaira, Y.-149
O'Dell, C.A.-288
Odo, K.-302
O'Doherty, C.M.-165
Oediger, H.-10

Oeser, E.-137
Ogata, M.-251,291
Ogata, T.-392
Ogawa, T.-190,351
Oglivie, K.K.-320
Oguar, K.-365,391
Oguri, T.-344
Ohashi, M.-150
Ohloff, G.-411
Ohno, M.-122
Ohshiro, Y.-182
Ohta, H.-213,215,222,224
Ohtsuka, I.-376
Ohtsuka, Y.-291
Ojima, I.-172,217
Okada, K.-152
Okano, M.-150
Okawara, M.-5,191,360
Okogun, J.I.-364
Okuhara, K.-62,117
O'Kuhn, S.-29
Olah, G.A.-99,105,167,191,192,
    219,223,233,308,315,319,338
Olive, S.-187
Oliver, J.E.-98,333
Olofson, R.A.-120,326,346
Omizu, H.-82
Omote, Y.-249
Omura, K.-188,189
Ong, B.S.-25
Ono, N.-12
Ooms, P.H.J.-134

Oppolzer, W.-25,58,407
Orena, M.-189,354,359
Orfanopoulas, M.-230
Orlowski, R.C.-382
Ortar, G.-196,398
Osaki, M.-7,66
Osborn, J.A.-229
Oshima, K.-208
Oshiro, Y.-250
Osowska, K.-207,242
Ott, W.-257
Overmann, L.E.-141,341
Oyler, A.R.-261

Pabst, W.E.-92
Paddon-Row, M.N.-98
Padwa, A.-415
Paine, III, J.B.-317
Palmisano, G.-210
Pankova, M.-143
Papadopoulos, E.P.-142
Paquette, L.A.-406
Parham, W.E.-61,76,232
Park, C.A.-292
Parker, K.A.-139
Parrilli, M.-205
Passannanti, S.-272
Pasto, D.J.-111
Patchornik, A.-411
Pattenden, G.-125
Patterson, J.M.-273,371
Pauling, H.-357

Pearce, A.-76,96
Pearson, A.J.-177,178
Pearson, D.E.-201
Pearson, D.J.-113,346
Pedersen, E.B.-268
Pelter, A.-151,152,156,238
Percival, A.-131
Perfetti, R.B.-190
Perkins, L.M.-330
Perry, D.H.-210
Pesce, G.-101
Peseke, K.-299
Pète, J.P.-341
Petragnani, N.-89
Petrova, J.-91
Petty, J.D.-22
Pfab, J.-333
Pfeffer, P.E.-38,108
Pfeiffer, G.-274
Phillips, C.J.-56
Phillips, J.N.-408
Piers, E.-135
Pierschbacher, M.-351
Pietraszkiewicz, M.-137
Pillsbury, D.G.-166
Pittman, Jr., C.U.-304
Place, P.-329
Plachky, M.-101,234
Planat, D.-268
Plante, M.S.-153
Plate, A.F.-135
Platonov, V.E.-335

AUTHOR INDEX

Plettenberg, H.-198
Pocar, D.-280
Pochat, F.-172,388
Pochinok, V.Y.-304
Pohmakotr, M.-55
Poisel, H.-345
Poje, A.J.-37
Pollak, A.-330
Pommer, H.-416
Ponaras, A.A.-55,83
Ponomareva-Stepnaya, M.A.-250
Posner, G.H.-97,190
Postovskii, I.Ya.-149,240
Praefcke, K.-274
Prager, R.H.-155
Pratt, T.M.-223
Prestidge, R.L.-316
Priester, W.-26
Promel, R.-285
Prudchenko, A.T.-33
Przhiyalgovskaya, N.M.-33

Raber, D.J.-240
Raghu, S.-177
Rahman, A.-199,222,233,358
Ramage, R.-311,340
Rao, C.G.-108
Rao, Y.S.-265
Rapoport, H.-410
Rasmussen, J.K.-412
Rathke, M.W.-38,39,156
Ravindran, N.-238

Reynolds, P.W.-61
Reese, C.B.-114
Reetz, M.T.-101,234
Reeves, W.P.-1,367
Reichen, W.-116
Reichman, U.-373
Reinach-Hirtzbach, F.de-334
Rembarz, G.-303
Rémion, J.-95,379,406
Remizova, L.A.-22
Remuson, R.-251
Renard, J.J.-412
Rengaraju, S.-108
Rettig, M.F.-128
Reutov, O.A.-187,419
Ribeiro, O.-247
Ricca, A.-88
Richey, Jr., H.G.-56
Ridley, D.D.-218
Ried, W.-278,297
Riedl, P.-137
Riefling, B.-260
Rieke, R.D.-418
Rio, G.-366
Ripoll, J.L.-113
Ritchie, E.-364
Robben, W.M.M.-395
Robert, A.-297,299
Robey, R.L.-184
Robins, M.J.-372
Robinson, D.T.-182
Rodrigues, R.-89

Roeske, R.W.-384
Romeo, A.-196,398
Rosenblum, M.-177
Rosenfeld, J.M.-353
Rouessac, F.-136
Roumestant, M.-L.-329
Rubottom, G.M.-209,353
Rudler, H.-117,184
Ruffner, R.J.-102
Runquist, A.W.-190
Ruppert, J.F.-54,326
Russ, P.L.-218
Ruston, S.-92
Ruth, J.L.-373

Saalfrank, R.W.-4,88
Sachdev, H.S.-21,356
Sachdev, K.-21,356
Sacks, C.E.-319
Saegusa, T.-167
Sagar, A.J.G.-27
Saigo, K.-7,65,66,191,264
Saitkulova, F.G.-389
Saito, I.-211
Sakai, N.-158
Sakakibara, T.-149
Saksena, V.-194
Sakurai, H.-172,173
Salomon, R.G.-133
Salzmann, T.N.-103,405
Samain, D.-94
Sammes, P.G.-253

Sandifer, R.M.-19,47,48,295
Sandmeier, D.-256
Sandri, S.-189,354,359
San Filippo, Jr., J.-314
Sarkar, T.-407
Sasaki, T.-122
Sataty, I.-144,343
Sato, M.-397
Sato, T.-35,42,43,81,264
Sato, Y.-398
Sauer, J.-137
Savignac, P.-63,84,91,384
Savoia, D.-51
Savost'yanova, I.A.-77
Sayed, Y.A.-76,232
Scandroglio, A.-301
Schamp, N.-85
Schank, K.-109
Scharf, H.-D.-243,365
Scheeren, H.W.-134
Scheffold, R.-309
Scheinmann, F.-27
Schiemenz, G.P.-142
Schlessinger, R.H.-17
Schmid, G.-256
Schmidt, M.-51
Schmidt, R.R.-18,50
Schmidt, U.-142,345
Schmitt, A.-11
Schmitt, G.-160
Schmitz, E.-342
Schnekenburger, J.-12

AUTHOR INDEX

Schneller, S.W.-415
Schöfer, H.J.-200
Schöllkopf, U.-49,292,337
Scholz, H.-U.-49
Schrekenberg, M.-73
Schreurs, P.H.M.-176
Schrock, R.R.-93,229
Schuda, P.F.-42
Schug, R.-134
Schultz, A.G.-37,407
Schulz, R.C.-336
Schulze, U.-274
Schuttenberg, H.-336
Schwartz, A.-206
Schwartz, J.-159,186,358,418
Schwartz, M.A.-369
Schwartz, S.J.-157,327
Schwyzer, R.-311
Scott, C.A.-355
Scott, L.T.-265
Secci, M.-300
Seebach, D.-14,19,23,41,45,46, 49,55,57,59,126,389
Seidel, C.-314
Seitz, E.P.-139
Sekine, M.-377
Sekiya, A.-146
Sekiya, M.-346,373
Semigran, M.J.-105
Senga, K.-376
Senō, M.-33
Sevrin, M.-214

Seyden-Penne, J.-121
Seyferth, D.-123
Seymour, E.-306
Sha, C.-K.-405
Shaefer, C.G.-354
Shahak, I.-71,242
Shandala, M.Y.-256
Sharma, A.K.-188
Sharma, T.C.-194
Sharpless, K.B.-208,339
Shashkov, A.S.-77
Shaw, A.-114
Shaw, I.M.-328
Shaw, J.E.-354
Shealy, Y.F.-288
Sheffer, H.-302
Sheppard, R.C.-377
Shevchuk, M.I.-294
Shim, J.L.-375
Shimiza, N.-205
Shimizer, N.-211
Shimizu, I.-360
Shimizu, K.-382
Shimizu, M.-8,383
Shinkai, I.-413
Shin-Ya, S.-336
Shioda, H.-78
Shioiri, T.-344
Shippey, M.A.-100,234
Shirai, H.-398
Shiraishi, S.-33
Shono, T.-82,402

Shpak, S.T.-294
Shull, D.W.-102
Shults, R.H.-111
Shvachkin, Yu.P.-384
Sicsic, S.-394
Siddiqui, A.A.-342
Sidorenko, E.N.-33
Siegfried, B.-331
Silbert, L.S.-38,108
Silveira, Jr., A.-153
Silver, E.H.-87,387
Simpson, G.W.-218
Sinclair, J.A.-151
Singer, S.P.-208
Singh, H.-44
Sinou, D.-225
Sisak, A.-228
Sivanandaiah, K.M.-382
Sket, B.-80
Skold, C.N.-60
Skuballa, W.-230
Sliwa, H.-271
Slobbe, J.-5
Smirnov, V.A.-207
Smit, W.A.-77
Smith, C.A.-308
Smith, D.G.-355
Smith, D.J.H.-355
Smith, F.X.-269
Smith, K.-40,151,156
Smith, P.J.-186,419
Smith, R.L.-345
Snider, B.B.-390

Snowden, R.L.-25,58,76
Soai, K.-67
Soja, P.-202
Soloway, A.H.-87,387
Sonnet, P.E.-98,329,333
Sonoda, A.-153
Sonoda, T.-116
Sonogashira, K.-166
Sopova, A.S.-206
Souto-Bachiller, F.-394
Spangler, C.W.-94,141,379,415
Sparrow, J.T.-383
Speltz, L.M.-194
Spitulnik, M.J.-298
Staffeldt, J.-230
Stawinski, J.-377
Stein, J.-301
Steiner, G.-134
Stemke, J.E.-22,59
Stephenson, L.M.-361
Stetter, H.-66,73,414
Still, W.C.-58
Stork, G.-9,36,69,83,402
Stradi, R.-280
Stransky, W.-86
Straub, H.-181
Streeting, I.T.-303
Streith, J.-366
Strüver, W.-380
Stubenrauch, G.-197
Stuessi, R.-296,393
Suau, R.-80
Subrahmanyam, C.-152

AUTHOR INDEX

Subramanyam, V.-87,387
Suggs, J.W.-317
Sugimoto, K.-213,215,224
Sullivan, D.F.-252
Sulsky, R.B.-17
Sultanbawa, M.U.S.-236
Sumitani, K.-165
Sundberg, J.E.-2
SuryaPrakash, G.K.-99,105,167, 223
Süss, H.U.-397
Suzuki, A.-72,157,158
Suzuki, J.-373
Suzuki, N.-283,295,296,369
Suzuki, S.-43
Suzuki, T.-116
Suzuki, Y.-166
Swaninathan, K.-40
Swanson, S.B.-354
Swenton, J.S.-61
Swern, D.-188,189

Tabata, M.-151
Tabushi, I.-118,343
Tadema, G.-111
Taguchi, T.-32,241
Takada, A.-287
Takada, S.-247
Takagi,K.-289
Takahashi, K.-261
Takahashi, S.-166
Takahashi, Y.-158,371

Takaishi, N.-344
Takayanagi, H.-370
Takegami, Y.-182
Takeuchi, K.-336
Takeuchi, Y.-243,408
Takimoto, S.-350
Takken, H.J.-62,361
Talbiersky, J.-18,50
Tam, W.-128
Tamao, K.-165,171,385
Tammer, T.-79
Tamura, R.-12
Tamura, Y.-241,271,348
Tanabe, T.-374
Tanaka, K.-52,276
Tanaka, S.-150,261
Tanaka, T.-1,324
Taniguchi, H.-116
Tanikaga, R.-52
Tarakanova, A.V.-135
Tarrago, G.-281
Tatsuno, T.-232
Taylor, E.C.-184
Taylor, W.C.-364
Telschow, J.E.-196
Teranishi, S.-175
Terao, Y.-373
Terasawa, M.-175
Terashima, S.-218
Tesch, U.-H.-284
Teschner, M.-389
Teyssie, P.-124

Thies, R.W.-139
Thomas, C.W.-19,48
Thomas, M.T.-237
Thompson, D.W.-165
Thompson, H.W.-23
Thorne, M.P.-352
Thyes, M.-13,396
Ticozzi, C.-88
Tietze, L.-F.-41
Timms, R.N.-179
Toda, F.-226,227
Toda, H.-350
Tokoroyama, T.-220
Tokuda, M.-32,158
Tokura, N.-336
Toshida, Y.-370
Touzin, A.M.-9,69
Traas, P.C.-62,361
Tregubova, N.L.-22
Trimmer, R.W.-280
Troin, Y.-251
Trost, B.M.-7,103,131,351,401, 403,405
Truce, W.E.-170
Truesdale, L.K.-40
Trybulski, E.J.-317
Tsai, M.-162,258
Tsuchihashi, G.-365,391
Tsuji, J.-360,370
Tsujimoto, K.-150
Tsunetsugu, J.-397
Tuck, D.G.-82

Tun-Kyi, A.-311
Turcant, A.-127
Tweedy, H.E.-165
Tzodikov, N.R.-401

Uchida, K.-31,32
Uchida, T.-272
Uchiyama, T.-175
Uda, H.-68
Ueda, T.-287,289
Uemura, S.-150
Ueno, Y.-5,191,360
Ugi, I.-310,315
Uguen, D.-20
Ulrich, P.-305
Umani-Ronchi, A.-51,225
Umans, J.G.-221
Umino, N.-221,222
Ungvary, F.-228
Usui, M.-264
Utimoto, K.-31,32,95,152,158

Vajna de Pava, O.-88
Valenta, Z.-130
van Boom, J.H.-376
van der Gen, A.-305,314
van Deursen, P.H.-376
Van Ende, D.-214,349
Van Horn, D.E.-31,378
Van Rheenen, V.-206
Varma, V.-258
Vecchia, L.D.-24,60

AUTHOR INDEX

Vedejs, E.-102,196
Verhe´, R.-85
Vermeer, P.-111,174,176,391
Veronese, A.C.-380
Vialle, J.-127
Viehe, H.G.-18,132
Vietti, D.E.-282
Villemin, D.-184
Villieras, J.-64,72,174
Vincent, C.-394
Viola, H.-64
Viout, P.-121
Vlasov, V.M.-335
Vlattas, I.-24,60
Vogev, A.-125
Vögtle, F.-140
Vollhardt, K.P.C.-180,417
Volpe, A.A.-119
Vostrowsky, O.-86

Wagner-Jauregg, T.-415
Wakatsuki, Y.-266
Wakselman, C.-272
Walia, A.S.-363
Walia, J.S.-363
Walker, E.R.H.-240
Walser, A.-269
Walter, R.-382
Walton, D.R.M.-309
Wamhoff, H.-271,290
Wang, C.-L.J.-28
Wang, S.-S.-383

Ward, S.E.-417
Ware, W.R.-80
Warkentin, J.-74
Warren, S.-92,96,404
Warrener, R.N.-98
Washburne, S.S.-186
Wasserman, H.H.-13,196,253,396
Watanabe, Y.-339
Waterhouse, A.-111
Waterman, E.L.-245
Waters, W.L.-190
Watkins, B.F.-201
Watt, D.S.-92,370,400
Weber, U.-350
Weedon, B.C.L.-416
Wehling, B.-271,290
Wehrmeister, H.L.-255
Weigel, L.O.-354
Weill-Raynal, J.-185,411
Weinreb, S.M.-337
Weinstein, B.-312
Weinstock, L.M.-298
Weisel, K.-H.-347
Weiss, W.-417
Weissflog, E.-51
Weisshuhn, M.C.-278
Weitz, H.M.-287
Welch, J.-191,308
Wemple, J.-6
Wender, P.A.-135
Wendisch, D.-129
Wentrup, C.-116,347

West, C.T.-232
West, R.-26
Westmijze, H.-174,176
Westphal, G.-298
Wetter, H.F.-187,418
Whalley, W.B.-34
Wheaton, G.A.-89,385
White, D.A.-75
White, J.D.-54,326
White, M.R.-367
Whitesides, G.M.-29,160
Whitten, C.E.-263
Wiaux-Zamar, C.-64
Wickberg, B.-308
Wieland, P.-334
Wiemann, K.-73
Willard, A.K.-138
Willy, W.E.-332
Wilson, J.W.-219
Wilson, M.E.-160
Wilson, S.E.-163,229
Winkler, D.-382
Winkler, J.-29
Wittig, G.-34
Wolber, G.J.-6
Wolfers, H.-137
Wolloch, A.-292
Wolters, E.-243
Wood, H.C.S.-254
Woodward, R.B.-317
Wroble, R.R.-400
Wu, G.-156

Yabuki, Y.-152
Yabushita, Y.-343
Yakobson, G.G.-335
Yamabayashi, T.-283,295
Yamada, K.-306
Yamada, S.-218,341,344
Yamamoto, A.-159
Yamamoto, H.-162,163,404
Yamamoto, K.-360
Yamamoto, M.-244
Yamamoto, T.-159
Yamamoto, Y.-153,216,324
Yamamura, K.-75
Yamashiro, D.-381
Yamashita, M.-365
Yamashita, S.-121
Yamauchi, K.-374
Yamazaki, H.-110,266
Yanami, T.-68
Yang, D.T.C.-238
Yardley, J.P.-307
Yasuda, A.-404
Yatagai, H.-153
Yeh, M.-K.-15,46
Yeung, D.W.K.-74
Yoneda, F.-374
Yoon, N.M.-239
Yoshida, H.-392,393
Yoshida, T.-156
Yoshida, Z.-118,271
Yoshii, E.-82,353
Yoshikoshi, A.-68

AUTHOR INDEX

Young, G.T.-314
Younglai, E.V.-353

Zamore, M.-337
Zander, R.-79
Zatorski, A.-90,91
Zavada, J.-143
Zbiral, E.-292
Zecchi, G.-301
Zefirov, N.S.-417
Zembayashi, M.-165,171,385
Zenchoff, G.-269
Ziegler, E.-257
Ziegler, F.E.-147
Zimmerman, H.-J.-155
Zoltewicz, J.A.-413
Zoretic, P.A.-202
Zubritskii, L.M.-176
Zupan, M.-80,333
Zvilichovsky, G.-374
Zweifel, G.-30,64,386
Zwierzak, A.-207,242

QD
262
A558
1976

OCT 31 1977

RAYMOND H. FOGLER LIBRARY
DATE DUE